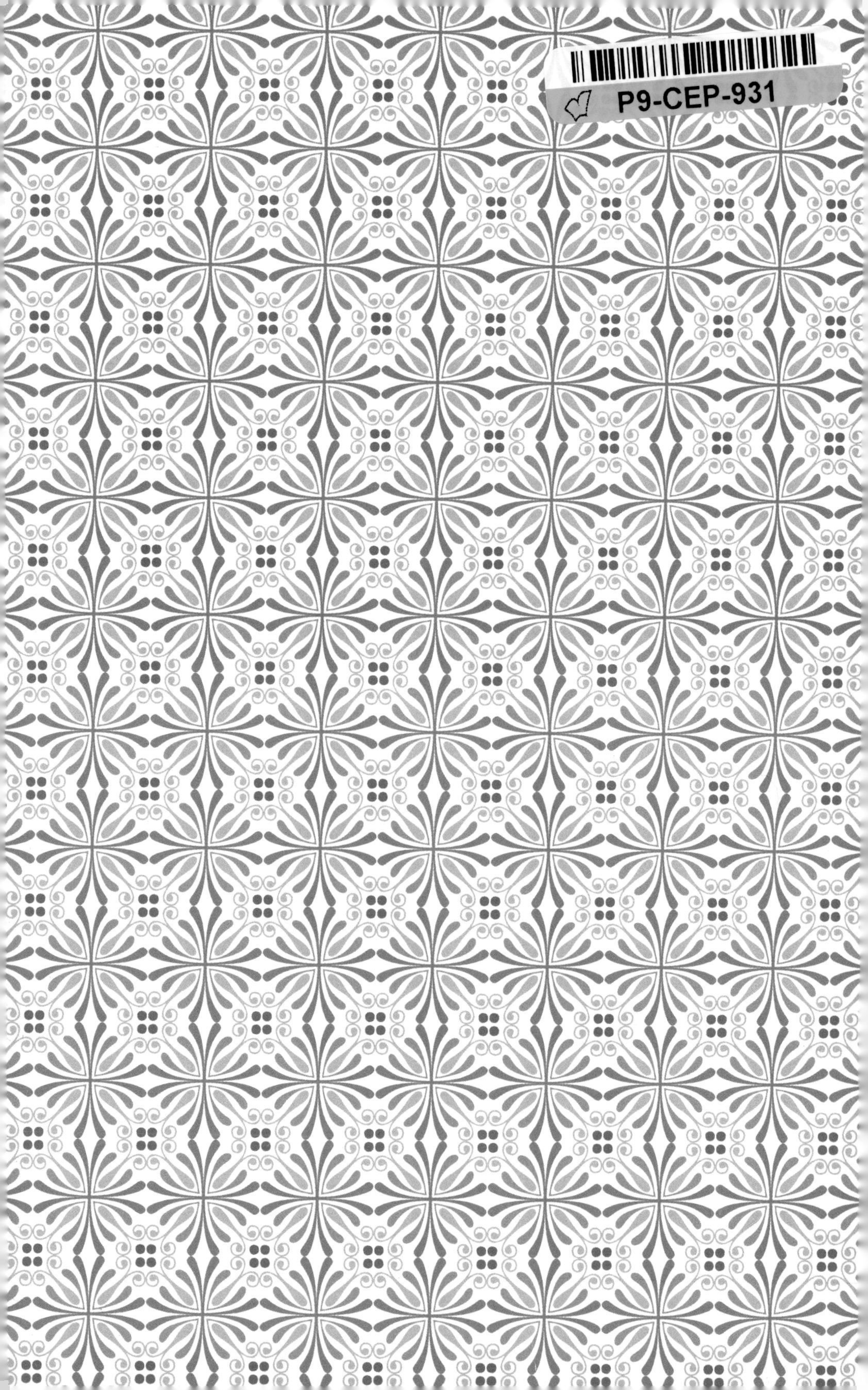
P9-CEP-931

The Book of

Incredible Information

Publications International, Ltd.

Cover image: Shutterstock.com

Interior images: Art Explosion: 29, 44, 52, 126, 147, 170, 174, 214, 261, 272, 291, 294, 378, 382, 395, 408, 411, 426, 443, 449, 473, 510, 517, 533, 540, 578, 589, 590, 625, 627; Clipart.com: 38, 41, 272; Getty: 28, 66; Shutterstock.com: 32

Louis Weber, CEO
Publications International, Ltd.
8140 Lehigh Avenue
Morton Grove, Illinois 60053

ISBN: 978-1-4508-8843-1

Manufactured in China.

8 7 6 5 4 3 2 1

Library of Congress Control Number: 2015930182

Contents

❋ ❋ ❋ ❋

✳ Chapter 1

Extraordinary Explorers

The Real Indiana Jones

Indiana Jones may have roamed the globe, battling Germans and digging up deserts while saving religious relics and drop-dead beauties with aplomb, but in the end he came home to Wisconsin. The real Indiana Jones, that is: Roy Chapman Andrews.

✳ ✳ ✳ ✳

This Story Should Be a Movie!

REAL-LIFE PALEONTOLOGISTS LARGELY agree that Andrews was the inspiration for Dr. Henry "Indiana" Jones Jr. While that has never been confirmed by movie producer George Lucas, the six-foot-tall Andrews wore a holstered revolver and a wide-brimmed hat during his own escapades. He served as a spy in World War I, faced death at least ten times, and was incorrectly reported deceased at least once.

"If you want to argue that he fails to resemble Indiana Jones, throw in the fact that he hated snakes, and think again," says Ann Bausum, author and one of the founders of the Beloit-based Roy Chapman Andrews Society. Her book, *Dragon Bones and Dinosaur Eggs*, published in 2000 by the National Geographic Society, examines Andrews's real-life exploits.

Andrews was an explorer and archaeologist, but also much more. He was a globe-trotting adventurer and one-time director of the American Museum of Natural History in New York.

He never located the Holy Grail or the Lost Ark, as did his fictional counterpart, but he discovered something perhaps just as exciting. He found out how dinosaurs were born.

"Roy Chapman Andrews is best known as the man who discovered fossil dinosaur eggs in Mongolia's Gobi Desert," *Time* magazine reported in 1940. "Before that, no one knew whether dinosaurs laid eggs or bore their young alive. Andrews has done a great deal of other scientific junketing, slaking an insatiable curiosity which he has had ever since he was a Wisconsin boy."

A Boy from Beloit

Born in Beloit in 1884, Andrews trained himself as a naturalist. As a boy, he hunted in woods nearby and learned taxidermy. He attended Beloit College and received a bachelor's degree in English. Later, he would use his education to write more than 20 books. *Discover* magazine named Andrews's *Under a Lucky Star* as one of the 25 greatest science books of all time.

In 1906, Andrews went to New York. He tried to get an interview at the American Museum of Natural History and was almost turned away, but the enterprising Andrews asked, "You have to have somebody to scrub floors, don't you?" He was hired as a janitor. Soon afterward, he joined the museum's collecting staff and earned a master's degree in "mammalogy" from Columbia University. In 1909, he became a collector for the museum and began crossing the globe.

Andrews journeyed to the East Indies, gathering specimens. In about 1920, his travels took him to China on a series of trips. A single Andrews expedition resembled an invading army. His crew traveled in specially built Dodge touring cars with a train of 75 supply-laden camels. For a 1923 expedition, he shipped back two tons of fossils and wound up on the cover of *Time* for his efforts.

In a 1923 article, *Time* reported, "Mr. Andrews owes his position as leader of the Asiatic expedition to a unique combination of scientific authority and practical resourcefulness in big game hunting and open-air life. He is as thoroughly at home in these as the late Theodore Roosevelt."

Just like those of his cinematic counterpart, Andrews's expeditions had to contend with outrageous obstacles like raiders, sandstorms, and packs of wild dogs—not to mention snakes. During one particularly bad night in the Gobi Desert, his expedition killed 47 vipers.

And his reptile problems didn't end when he left the Gobi, either. "Several times he has been on death's brink," *Time* reported. While he was on expedition in Borneo, a boy "yanked him out of range of a huge python which was about to drop on the explorer from a tree."

In 1934, Andrews became director of the American Museum of Natural History, where he once scrubbed the floors. After that, he was mostly forgotten. He was not an embarrassment, exactly, but a reminder of a time when scientists also had to be showmen to gain attention. At the time Andrews entered the field of natural history, there were few special areas of study and scientific academia was still organizing into various formalized disciplines. Meanwhile, explorer-adventurers, often with little education, were going out and getting the actual work done.

Still, Andrews's education and sound scientific practice kept him above the fray. At that time, to be a successful scientist, he had to be a promoter. For example, Andrews's first expedition to Mongolia cost $250,000, and he raised all of it from donations. Thanks to Andrews, science also opened up to corporate sponsorship—a development that made many scientific breakthroughs possible.

In 1942, Andrews left his position at the museum and retired to Colebrook, Connecticut, later moving to Carmel, California. He died in 1960 and, per his request, his cremated remains were buried in Beloit. Today, the memory of Roy Chapman Andrews is alive in his many books, and perhaps in the exploits of Indiana Jones, his fictional twin. Lately, he's getting a little attention again from his former employers, too.

A 2007 IMAX movie, *Dinosaurs Alive!*, features footage of Andrews in the field, and the Roy Chapman Andrews Society offers prestigious annual awards to the top modern explorers, such as oceanographer Robert Ballard, discoverer of the wreck of the R.M.S. *Titanic*, and entomologist-ecologist Mark Moffett. The scientists partner with educators in Andrews's native Beloit to reach out to schoolchildren—and possibly inspire the next globe-trotting explorer.

As Andrews said, "Always there has been an adventure just around the corner—and the world is still full of corners!"

Da Verrazzano and New Angoulême

In 1524, the famed Italian explorer brought Europe its first eyewitness description of New York Harbor and its friendly natives. And he hadn't even been mugged!

❋ ❋ ❋ ❋

GIOVANNI DA VERRAZZANO's New York stopover was part of a lengthy seagoing effort to find the ever-elusive passage to the Pacific Ocean for King François I of France. What land should France claim? Only by knowing the sea passages could François decide.

The expedition originally included four ships and departed Normandy in 1523. Two of the original ships turned back off Brittany with storm damage, however, and one went privateering. Only the multimast carrack *La Dauphine* made it to the Portuguese archipelago of Madeira, where it wintered and took on supplies for the next leg of the voyage across the Atlantic.

Aboard *La Dauphine,* the expedition made its first North American landfall in early spring 1524 off modern-day North Carolina and then followed the coast northeast. Da Verrazzano somehow missed Chesapeake Bay on his way to modern New York Harbor.

But He Didn't Get His "I ™ NY" T-Shirt

The explorer's entry into the harbor indeed passed through the narrows (yes, those with the bridge bearing his name today). Loath to risk *La Dauphine* in tricky exploration of the shoreline, however, da Verrazzano set out across Upper New York Bay in a small boat. He mistook the Hudson for part of a lake and didn't explore further, but he did meet with cordial Lenape Native Americans. His recollections about their dress and agriculture remain valuable today.

Da Verrazzano continued northeast to what later became New England and finally returned to France in July. He named the NYC area *Nouvelle-Angoulême,* which means New Angoulême. (Didn't stick, but nice try.)

Aftermath

Da Verrazzano's quick-and-dirty tour of the modern northeast U.S. coast eventually faded into obscurity, which is unfortunate given its pioneering nature. The explorer glossed over numerous inland waterways (none of which actually led to China, of course, but he failed to confirm this). Unluckiest of all, da Verrazzano had the misfortune to operate shortly after Cortez and Magellan, whose feats upstaged his.

Had he been thorough and written more, we might remember da Verrazzano differently. But as New York's first European visitor, a distant kinsman of some who would one day call the city home, Giovanni da Verrazzano has a secure parking place in New York's history.

- The Abenaki American Indians of Maine were far less friendly than those around New York Harbor. They mooned da Verrazzano.
- His brother Girolamo, brought along as cartographer, jumped to the conclusion that North America was two halves divided by a mythical "Sea of Verrazzano." The error wasn't cleared up for more than a century.
- Two other bridges are named for da Verrazzano: one in Rhode Island's Narragansett Bay and one connecting Assateague Island to the mainland of Maryland.
- The correct spelling of the explorer's name is definitely "da Verrazzano." The I-278 bridge that stands in his honor spells the name wrong, as does the bridge in Maryland. Rhode Island's bridge spells it correctly, though everyone leaves out the "da," which is like referring to the famous NYC mayor as "Fiorello Guardia."
- Some believe da Verrazzano's letter to King François I describing the voyage was a fake and that he never came to America. Most historians accept his travels as genuine.
- "Old" Angoulême is a pleasant but unremarkable region in southwestern France about an hour's drive east of Bordeaux.

Stanley and Livingstone

Whether or not the famous words were really spoken, the two men really did meet in the heart of the "Dark Continent."

* * * *

DR. DAVID LIVINGSTONE went to Africa from Scotland to win Christian converts. Finding little success, Livingstone reinvented himself as an explorer. Although an intrepid seeker of knowledge, he proved a lousy expedition chief. After his 1858 Zambezi Expedition flopped, Livingstone had a difficult time finding donors willing to fund his expeditions. Regardless, he set out in search of the Holy Grail of 1800s African exploration—the Nile River's source.

Livingstone never did find the source of the Nile, though he did discover the source of the Congo River. In 1866, Livingstone took sick in the wilds of south-central Africa and lost touch with civilization for nearly six years. As a publicity stunt, the New York Herald news paper sent journalist Henry Morton Stanley on a Livingstone hunt. In 1871, Stanley found Livingstone near Lake Tanganyika and (supposedly) greeted him with the famous question, "Dr. Livingstone, I presume?" They became colleagues and friends.

Stanley (1841–1904) outlived Livingstone (1813–73) by 30 years. One of Stanley's last wishes was to be buried next to Livingstone in London's Westminster Abbey, but the British government refused permission.

Forty Days on the Rio Grande

Alonso Álvarez de Pineda was the first known European visitor to Texas. In 1519, he and his company explored and mapped the Texas coast, occupying the mouth of the Rio Grande for more than a month while repairing their ships. Despite his map—Texas history's earliest document—surprisingly little is known about Álvarez de Pineda beyond the nature of his grisly death.

* * * *

UNDER ORDERS FROM Francisco de Garay, the Spanish governor of Jamaica, Alonso Álvarez de Pineda embarked on a reconnaissance trip in March 1519. His four ships carried his party of 270 and sailed from Jamaica to explore the gulf coast in the hope of finding a strait leading directly to the Atlantic Ocean. Many historians believe that Álvarez de Pineda's company sailed north along the gulf coast of Florida before attempting to turn east. If that sounds strange, it is due to the erroneous reports of earlier Spanish explorer Juan Ponce De

León. In 1513, De León reported a northern passage that separated Florida from the North American mainland. Therefore, Álvarez de Pineda fully expected to be the first to prove that Florida was an island.

Never coming upon a point at which to turn east, Álvarez de Pineda's company instead headed west along today's Florida panhandle and Alabama coastline. While there is some dispute among historians, it appears that around the time of the feast day of Espíritu Santo, or Pentecost, on June 2, 1519, Álvarez de Pineda became the first European to see the Mississippi River. He named it Río del Espíritu Santo in reference to the religious holiday and managed to sail upstream for several miles before continuing his voyage along the gulf coast.

First into Texas

Álvarez de Pineda mapped the outer islands off the coast of current-day Corpus Christi, Texas, claiming the bay and the land beyond for his king and naming it with the Spanish term meaning "body of Christ." After mapping the entire Texas coastline, Álvarez de Pineda finally made land. His fleet landed slightly south of Boca Chica, at the mouth of the Rio Grande, making his party the first Europeans to set foot in Texas. The Spaniards explored the region for 40 days, the time it took to overhaul their ships and conduct repairs after the extensive voyage. This also means that the Rio Grande was the second place to be visited by Europeans in what is now the United States.

Grisly Demise

Álvarez de Pineda sailed on down the Gulf of Mexico to the mouth of the Panuco River, near where Cortés would later found the town of Tampico, in what is now the Mexican state of Tamaulipas. Here, Álvarez de Pineda's company appears to have suffered a heavy defeat at the hands of the Aztecs. Some reports indicate that Álvarez de Pineda and many of his crew were killed, flayed, and eaten; their skins were then displayed as trophies in the Aztec temples. The Aztecs also burned some of

Álvarez de Pineda's fleet; only one ship, under the command of Diego de Camargo, managed to leave successfully.

Álvarez de Pineda's Legacy

Álvarez de Pineda's map of the entire gulf coast made it to safety with Camargo and was presented to Governor Garay, who went on to take all the credit for the expedition and obtain a grant for the territory it explored. Still, it was Alonso Álvarez de Pineda who discovered that there was no strait linking the Gulf of Mexico to the Atlantic Ocean. While many details of his voyage remain shrouded in mystery, there is little doubt that he was the first European to set foot in Texas.

Cross Country

Preacher Arthur Blessitt made history when he set out on a four-decade-long pilgrimage.

* * * *

The "Minister of the Sunset Strip"

ARTHUR BLESSITT WAS born in Mississippi in 1940, and at age 11 he converted to Christianity. By the time he was a young man in the 1960s, he had moved to California and was preaching the good word on Hollywood's infamous Sunset Strip. On Christmas Day in 1969, Blessitt claims he was called by Jesus to make a pilgrimage on foot to Capitol Hill in Washington, D.C. But he wouldn't make the journey alone: a 12-foot-tall, 45-pound wooden cross would accompany him. Blessitt put the cross on his back and began walking, with a mission to spread the word of Christ throughout the world.

These Shoes Were Made for Walkin'

Not long after he began, Blessitt realized his cross, which dragged on the ground, was losing about an inch of wood every day thanks to rough roads. He added a tricycle wheel to the base of his cross, which helped facilitate the trek.

Blessitt didn't stop his walk once he got to Capitol Hill. He didn't stop at the Atlantic Ocean, either. No, the preacher boarded a plane—he had to break the cross down to get it into a ski bag because it wouldn't fit on the aircraft—and flew overseas.

Over the next decade, Blessitt continued his journey. By 1984, he had walked across 60 countries on six continents. Of course, his quest wasn't easy, and sometimes it was downright dangerous. Still, he garnered quite a following and became an evangelical celebrity of sorts. He met with Yasser Arafat, Muammar al-Gaddafi, former UN Secretary Boutros Boutros-Ghali, and even Pope John Paul II.

Blessitt is credited with witnessing former president George W. Bush's conversion to born-again Christianity, and he proudly admits that over time, his cross-bearing shoulder built up an extra inch of bone. By the end of the pilgrimage, which ended in 2008, Blessitt had covered more than 38,000 miles on all seven continents.

Trek of No Return

Many people attempt to climb Mount Everest—and many never return.

* * * *

IT'S NO SECRET that ascending Everest, the tallest mountain in the world, is a risky feat. It's also no secret that some climbers die trying: In fact, more than 200 people have died in their attempt to summit the 29,035-foot-tall peak. But did you know that many of these dead bodies remain on the mountain, never to receive a proper burial? There are many explanations as to why climbers die on Everest and why their bodies are left behind, yet the issue remains steeped in ethical debate.

The Death Zone

Among the dangers of climbing Mount Everest are avalanches, falling ice, fierce winds, the possibility of falling in a crevasse, severe cold, inadequate equipment, and lack of physical preparation. The majority of deaths, however, are caused by high-altitude sickness. This occurs where the technical climbing begins, far above the final base camp at 26,000 feet. The area is known as the "Death Zone," because the conditions here—specifically the amount of oxygen—cannot sustain human life.

A 2008 study conducted by a group of researchers from Massachusetts General Hospital found that most deaths on Everest were associated with "excessive fatigue, a tendency to fall behind other climbers, and arriving at the summit later in the day." The limited oxygen in the Death Zone contributes to confusion, disorientation, and a loss of physical coordination. Most of the climbers who die do so after reaching the summit, on their way down the mountain.

Frozen in Time

Because of the severe conditions in the Death Zone, most of the dead bodies are left behind. It would be extremely dangerous to attempt to take them off Everest, and there's really nowhere to bury them on the icy upper slopes. In other words, as climbers ascend, along the way they pass the frozen forms of those who have made the trek before them—sometimes decades prior.

Among those frozen in time is George Mallory, who attempted to summit Everest in 1924 but never returned from his climb. When asked why he wanted to climb the tallest mountain on the planet, he famously responded, "Because it's there."

In May 1999, Mallory's body was found below the summit at 27,200 feet, a climbing rope still cinched around his waist. To this day, no one knows if Mallory and his climbing partner Andrew Irvine arrived at the summit before they died. If they had, they would have been the first to do so, preceding Sir Edmund Hillary and Tenzig Norgay's ascent in 1953.

In May 1996, eight climbers lost their lives in a sudden storm on their return from the summit. Two bodies from the expedition were never found. In 1998, Francys Arsentiev collapsed and died on her descent from the summit. For years her body lay close to the trail, in full view of climbers on their way up the mountain. A mountaineer named Ian Woodall, who was unable to help her as she neared her death, returned to Everest in 2007 to bury her in an American flag. Alas, he and his climbing partner were met with harsh weather conditions and only had time for a brief ceremony before dropping her body over the edge of the North Face. Some climbers criticized Woodall for initially abandoning Arsentiev, so perhaps this burial was a gesture of putting the controversy to rest.

Some of the most skilled climbers, the local sherpas, make it a point to avoid corpses. Teams from China have planned a cleanup of the mountain, which has been called "the world's highest garbage dump." In addition to picking up oxygen canisters, camping gear, and other materials discarded on the northern side of Everest, the crew intends to bring back any bodies that can be safely transported.

Roald Amundsen's South Pole Quest

On October 19, 1911, Roald Amundsen's South Pole quest began at the Ross Ice Shelf of Antarctica with 5 assistants, 4 sleds, and 52 dogs. International bragging rights were on the line, as Briton Robert F. Scott was also heading for the Pole.

* * * *

AMUNDSEN HAD ORIGINALLY planned to go for the North Pole but quietly changed plans after hearing of the recent North Pole expeditions by Frederick Cook and Robert Peary.

Only en route did he send Scott a terse heads-up. Scott lost the race by five weeks. Deflated, his party turned for home but never made it. One key reason they failed was that Amundsen's party handled their sled dogs better. Scott's crew did the majority of the work themselves and were less efficient. A 1912 search party found Scott's final, tragic camp with the bodies and diaries of his expedition.

Olav Bjaaland played a key role on Amundsen's team. His savvy reengineering reduced the sleds' weight by 75 percent—an important factor in Amundsen's success. A ski champion in his native Norway (a modern U.S. comparison would be a Super Bowl MVP quarterback), the high-spirited trail-breaker was excited to join the race.

Bjaaland was also the last survivor of the expedition. He died at age 88 in 1961.

The *Italia*

A long-distance dirigible ride to the North Pole ends predictably.

⁂ ⁂ ⁂ ⁂

IN 1928, ITALIAN aviation pioneer Umberto Nobile planned several airship flights to the Arctic. He hoped to base his airship *Italia* at the Norwegian island of Spitzbergen, supported by an Italian vessel. If he could moor an airship at the North Pole and bring back valuable scientific information, he could prove the value of dirigibles while burnishing his reputation.

In the past, he had explored (and feuded) with renowned Norwegian explorer Roald Amundsen. The churlish Norseman wasn't invited this time.

En route from the North Pole back to Spitzbergen on May 25, 1928, *Italia* crashed on the icepack. The impact mashed the cabin, ejecting ten crew members (one fatally). Six others floated helplessly away in *Italia's* remaining portion and were never seen again. The nine remaining crew members and Nobile's dog (happily unhurt) camped on the ice, radioing in vain for help while surviving on emergency supplies and a polar bear they had killed.

When the icepack broke up, setting them adrift on a disintegrating floe, the survivors' situation became even worse. Five days after the crash, Italian naval officers Filipo Zappi and Alberto Mariano, with Swedish scientist Finn Malmgren, set out to seek help. The despondent Swede faded and died; the Italians survived until rescue.

A Russian ship finally detected the party's SOS on June 3. Smelling good propaganda, the Soviets sent icebreakers and aircraft to join the international search effort.

A Swedish pilot landed near Nobile's party on June 23 but could bring out only one person. Nobile was both light and injured, so the Swede chose him and his dog. The aging Soviet icebreaker *Krassin* eventually rescued most of the survivors on July 12.

The last survivor of the *Italia* rescue effort was Giulio Bich, one of an elite *Alpini* (mountain troops) rescue team arranged by Nobile in advance. Bich died in 2003 in Cervinia, Italy, at age 95.

Why Is America Called America?

Weren't you paying attention in your eighth-grade world history class? As you were undoubtedly told, the Americas are named for the Italian explorer Amerigo Vespucci.

⁕ ⁕ ⁕ ⁕

But what did he do that was so great? The only fact about his life that anyone seems to recall is that, well, America is named after him. How did a dude who's otherwise forgotten by history manage to stamp his name on two entire continents?

While he didn't make the lasting impression of his contemporary Christopher Columbus, Vespucci was no slouch. As a young man, he went to work for the Medici family of Florence, Italy. The Medicis were powerbrokers who wielded great influence in politics (they ran the city), religion (some were elected to be bishops and popes), and art (they were the most prominent patrons of the Renaissance, commissioning some of the era's most memorable paintings, frescos, and statues).

Like many of the movers and shakers of that age, the Medici had an interest in exploration, which is where Vespucci came in. Under their patronage, he began fitting out ships in Seville, where he worked on the fleet for Columbus's second voyage. Vespucci evidently caught the exploration bug while hanging around the port—between 1497 and 1504 he made as many as four voyages to the South America coast, serving as a navigator for Spain and later Portugal. On a trek he made for Portugal in 1501, Vespucci realized that he wasn't visiting Asia, as Columbus believed, but a brand-spankin' new continent. This "ah-ha" moment was his chief accomplishment, though he also made an extremely close calculation of Earth's circumference (he was only fifty miles off).

Vespucci's skills as a storyteller are what really put his name on the map. During his explorer days, Vespucci sent a series of letters about his adventures to the Medici family and others. Vespucci livened up ho-hum navigational details with salacious accounts of native life, including bodice-ripping tales of the natives' sexual escapades. Needless to say, the dirty letters were published and proved to be exceedingly popular. These accounts introduced the term "The New World" to the popular lexicon.

German cartographer Martin Waldseemüller was a fan, so he decided to label the new land "America" on a 1507 map. He explained his decision thusly: "I do not see what right any one would have to object to calling this part after Americus, who discovered it and who is a man of intelligence, [and so to name it] *Amerige*, that is, the Land of Americus, or *America:* since both Europa and Asia got their names from women."

But there are those who believe that Vespucci's forename wasn't the true origin of the name. Some historians contend that the term "America" was already in use at the time and that Waldseemüller incorrectly assumed it referred to Vespucci. Some have suggested that European explorers picked up the name Amerrique—"Land of the Wind" in Mayan—from South American natives. Others say it came from a British customs officer named Richard Ameryk, who sponsored John Cabot's voyage to Newfoundland in 1497 and possibly some pre-Columbian explorations of the continent. Yet another theory claims that early Norse explorers called the mysterious new land *Ommerike*, meaning "farthest outland."

In any case, the name ended up on Waldseemüller's map in honor of Vespucci. The map proved to be highly influential; other cartographers began to use "America," and before long it had stuck.

Keep this story in mind the next time you're composing a heart-stoppingly boring email—if you spruce it up a bit, you might get a third of the world named after you.

Route 66: The Texas Miles

For cross-country travelers eager to "get their kicks on Route 66," the Mother Road beckons: More than 150 miles of the original 178 miles of Texas Route 66 remain.

* * * *

When route 66 was officially decommissioned in 1985, Interstate 40 took over the job of funneling motorists from east to west and back again across the desolate landscape of the Texas Panhandle. Now, at 70 miles per hour, one could speed in air-conditioned, anonymous comfort across the staked plains and barely catch a glimpse of the unique people, places, and culture that defined the Llano Estacado.

But it wasn't always that way. Once upon a time in the history of cross-country travel, the ride *was* more important than the destination. Travelers made close contact with the places they passed and, as a result, experienced the journey as a part of their trip.

The Mother Road

During the 1950s and '60s, taking a pilgrimage across America—and the Panhandle of Texas—meant that you took the highway designated with the numerals 66. Because it linked towns small and large, Route 66 was at one time referred to as "America's Main Street," a two-lane corridor that allowed the traveler to read the signs up close, smell the barbecue cooking, and hear the music of people who lived and worked along the route.

During its heyday, Route 66 ran a total of 2,448 miles, with the Texas portion grabbing up 178 of them. "Get Your Kicks on Route 66" was more than just a song written by Bobby Troup and sung by Nat King Cole: It was an anthem for those itching to hit the open road and "see the U.S.A. in their Chevrolet."

Re-creating the Past

This is an experience that can still be had, as more than 150 miles of the Texas section of the original "Mother Road" remain, much of it the old concrete surface. Drivers pulling off almost any exit ramp will be sure to find the old road waiting for exploration. These days, the crumbling two-lane thoroughfare wanders back and forth to the north and south of the new interstate alignment. Between Texola and Amarillo (except at McLean) the old route can be found just south of I-40. Moving west from Amarillo to Glenrio, the vintage road claims real estate north of the superslab.

Although the road is easy to find, it's advisable that those interested in tracing it pick up a good Route 66 map. Over the decades, the path of Route 66 has seen an endless number of changes, including rerouting and new alignments. And all should be advised: The old road that used to connect Jericho to Alanreed and Adrian to Glenrio is gone for good. So don't even bother looking.

Getting Started

To begin a Texas Route 66 adventure, travelers should start their trip west of Texola, Oklahoma, where the I-40 frontage road pulls directly into the town of Shamrock, Texas. Here, at the corner of North Main and East 12th Street, can be found the U-Drop Inn, a former café and Conoco service station, recently restored by the Shamrock Chamber of Commerce. Used as the model for Ramone's Custom House of Body Art in the Disney movie *Cars,* the striking art deco structure is a Mother Road jewel—trimmed with glowing neon.

From town, drivers should continue on 12th to the frontage road, which is the most authentic stretch of Route 66 to be found anywhere. From here, they'll pass through the ghost town of Lela and arrive in McLean, home of the Texas Old Route 66 Association, whose restoration projects include the McLean Phillips 66 filling station, the first in Texas. Offbeat attractions in McLean include the Devil's Rope Museum, where interested parties can learn everything they always wanted to know about barbed wire but were afraid to ask. The 1930s Art Deco Avalon Theater also should not be missed.

The next town up is Alanreed, a forgotten burg whose claim to fame is the oldest cemetery and church along Texas 66. Eldridge cemetery was established in 1888 and includes the graves of Confederate soldiers, freedmen, and other local luminaries. Today, it serves as a sad metaphor for a town killed by progress. Once home to 500 residents, the town is a prime example of what happened to communities after they were bypassed by modern freeways.

Watch Out for the Rain

Next up is Jericho Gap, the infamous spot where Route 66 travelers inevitably got bogged down in the mud. When it rained, the area's rich black soil turned into what some called "black gumbo," rendering the roads impassable. Sandwiched between sections of improved road to the east and the west, it was duly christened "Jericho Gap." Take note: Most of the original Route 66 can't be traveled along this stretch, because it either no longer exists or is located on private property.

On the way to Groom, travelers should keep their eyes peeled for the leaning water tower (which leans because one leg is shorter than the other), a planned tourist attraction that once pulled in customers for the now defunct Britten truck stop and restaurant.

In Groom, it's a good thing that people cannot live by bread alone: One of the most impressive sights here is the tallest cross in the western hemisphere, a 190-foot, 1,250-ton inspiration of Steve Thomas. Further west in Conway, trekkers will cross paths with Bug Farm USA, a low-budget, Volkswagen tribute to the more famous Cadillac Ranch.

On the way into Amarillo (the helium capitol of the world), Route 66 once crossed the Amarillo Air Force Base *and* the airport itself. These days, two separate alignments can be explored: the first along Old Third and Sixth, and the second along Amarillo Boulevard.

Either way, a plethora of vintage businesses will have cars pulling over to the side of the road for passengers to snap pictures of the local scene. By the time they reach the Big Texan Steak Ranch, travelers will be more than ready for the challenge of the "Free 72-ounce steak."

Farther Along the Road

From Amarillo, the north service road goes out of town, where it runs into Stanley Marsh's world-famous Cadillac Ranch, just south of I-40. From there, the old road runs through Bushland and Wildorado and on into Vega (Spanish for meadow), where the crumbling remnants of better days populate the roadsides. Despite the near ghost-town atmosphere, the Vega Motel—an original tourist court built in 1947—is still going strong.

Pressing on to Adrian, travelers come to the halfway point of the old Route 66 highway. Driving west, it's 1,139 miles to Los Angeles; driving east to Chicago, it's the same distance. This midpoint town has also seen better days and is home to about a dozen businesses and only 150 residents. The wonders here are low key, punctuated by the local water tower and the Midpoint Cafe, the oldest continuously operating café found along the Texas stretch of the Will Rogers Highway.

Just west of Adrian, the service road comes to an abrupt dead end, where the slab of Route 66 disappears. Don't worry—that's not the end just yet. It resurfaces about a mile or so before Glenrio, the last settlement in Texas before the road crosses over into New Mexico. Here, buildings such as the "First Inn/Last Inn" motel stand like tombstones—stark reminders of what the area was like before the coming of the interstate and the bypassing of "America's Main Street."

Walking on the Moon

On July 20, 1969, Neil Armstrong accomplished the seemingly impossible by climbing down a ladder and placing footprints on the moon. Millions watched on television as Armstrong, followed by pilot Edwin "Buzz" Aldrin, took that "one giant leap," then explored the surface. Armstrong and Aldrin left behind a plaque reading, "Here men from the planet Earth first set foot upon the moon July 1969 A.D. We came in peace for all mankind."

❋ ❋ ❋ ❋

Apollo 11: Facts and Figures

- **Crew:** Neil A. Armstrong, commander; Michael Collins, command module pilot; Edwin Aldrin, lunar module pilot
- **Launched:** July 16, 1969, from Kennedy Space Center Launch Complex 39A
- **Landed on the moon:** July 20, 1969, at 4:17 P.M.
- **Landing site:** Mare Tranquillitatis (Sea of Tranquility)
- **Time spent on the lunar surface:** 21 hours, 38 minutes, 21 seconds
- **Time spent exploring the lunar surface:** 2 hours, 31 minutes
- **Moon rocks collected:** 21.7 kilograms
- **Departed the lunar surface:** July 21, 1969, at 1:54 P.M.
- **Returned to Earth:** July 24, 1969, at 12:50 P.M.
- **Retrieval ship:** USS *Hornet*

A Not-So-Easy Landing

Despite hundreds of hours of practice on Earth, Armstrong and Aldrin found landing on the moon's surface a little tricky. As they approached the designated site, they realized it was covered with huge boulders, which meant they had to find another location quickly or abort the mission. They had just 30 seconds of fuel left when they spotted an area free of debris and successfully brought down the module. Armstrong was unflappable on the communicator, but his heart rate, which jumped to 156 beats a minute during the final seconds of the landing, clearly illustrated the stress he was feeling.

What Did Armstrong Really Say?

Debate has raged for decades over Neil Armstrong's first words as he stepped out of the lunar lander and onto the moon's surface. Armstrong has long maintained that he said, "That's one small step for a man. One giant leap for mankind." However, forensic linguist John Olsson recently analyzed the original magnetic tape recordings made at Johnson Space Center and has concluded that Armstrong, who understandably was under a lot of pressure at that moment, left out the "a" as he spoke those famous first words. Regardless, Armstrong's inspiring observation will be a part of human history forever.

Exploring Cave of the Mounds

If you're traveling through southern Wisconsin, it's well worth a stop at Cave of the Mounds just off Highway 18/151 in Blue Mounds.

LET'S FACE IT; sometimes "educational" trips are boring for kids and the same can be true for adults. Old buildings often smell, well, old. History and science can be interesting, but sometimes it's overwhelming to hear so many facts and figures all at once. And sometimes everything exciting is behind glass, which certainly takes away from the fun. That's why the Cave of Mounds is so much fun.

You don't have to be a spelunker to enjoy this cave. Science and nature come to life in an underground cave just waiting to be explored. The only caveat here is that it's okay to look, to ooh and ahh, and even to take photos, but absolutely, positively DO NOT TOUCH the cave walls and rock formations!

A cave is just a cave, right?

Wrong! The Cave of the Mounds is actually a National Natural Landmark.

How did it get that designation?

The United States Department of the Interior and the National Park Service gave that honor to Cave of the Mounds in 1988 to recognize it as a site that possesses exceptional value in terms of illustrating our nation's natural heritage. Such landmarks must also help visitors to better understand their environment.

What do those people look for?

They do an extensive study of a potential site, which needs to be one of the best examples of a certain geologic or biotic feature within a given region. Cave of the Mounds fit that perfectly!

What is Cave of the Mounds sometimes called?

The "jewel box" of American caves. The variety, colors, and delicacy of the cave's formations make it unique.

What makes it so colorful?

A mixture of minerals found in the water created the cave's colors. Gray and blue formations are caused by magnesium oxide. Iron oxide produces red and brown.

Does anyone sleep in the Dream River Room?

No, it's just the fanciful name given to one of the areas of the cave. The wide array of beautiful colors found throughout the cave inspired other imaginative names such as Gem Room, Cathedral Room, and Painted Waterfall.

Does the cave go any further than where the tour goes?

No one really knows. Some rooms wind for hundreds of feet,

and some people believe that they may lead to other passages and even other caves.

Is the cave old?

You could say that. The cave itself began to take shape one to two million years ago. Yet, that's nothing compared to the rock it was formed in, which dates back 400 million years.

How did Cave of the Mounds get its name?

The name comes from two large hills, called Blue Mounds, in southern Wisconsin. The cave lies under the East Mound.

How was it discovered?

It was actually an accident! On August 4, 1939, some workers were removing limestone from a quarry in the Blue Mounds area. They set off a blast that suddenly revealed the underground cave.

What did they see?

Something totally unexpected: a limestone cave full of mineral formations, which opened into even more rooms.

What happened next?

So many people came to check out the new cave that it was finally closed to the public in order to protect it.

So did they protect it?

Yes. While it was closed, the cave was made accessible without hurting the cave itself. Wooden walkways and lights were installed. In May 1940, Cave of the Mounds was opened to the public.

So what's so cool about it?

Well, everything. You'll get a tour guide who'll tell you all about the rock formations and the underground stream. And there are amazing stalactites and stalagmites to see up close.

Which is which again?

Stalactites are columns that grow from the ceiling down. (Remember, they hold on "tight" to the ceiling.) Stalagmites grow up from the ground (trying with all their "might" to reach the ceiling).

Speaking of cool, what should visitors wear on the tour?

The cave is the same temperature all day, all year round. It's 50 degrees, so you might need a sweater in the summer, but it feels pretty cozy in the Wisconsin winter months.

Is the cave open year round?

Yes. It's open daily in the spring, summer, and fall, but it's only open weekends in the winter.

Where is it?

Cave of the Mounds is in Blue Mounds, just off Highway 18/151. It's about 20 miles west of Madison.

How many people visit Cave of the Mounds?

More than 59,000 visited the Cave in the first eight weeks it was open! Today, thousands visit every year and millions have come through since it opened in 1940.

What else can you do at the cave?

Visitors can take advantage of a beautiful outdoor setting that offers picnic areas, rock gardens, and walking trails.

Going Underground

Under cover of darkness, thousands of runaway slaves were able to escape to a new life of freedom, thanks to a secret network called the Underground Railroad.

* * * *

TICE DAVIDS WAS determined to make it. He had just plunged into the Ohio River's cold waters and was swimming across, away from his life of slavery in Kentucky and toward help and freedom in Ohio. On the shore, Davids's master kept an eye trained on the slave. The slave owner quickly launched a rowboat into the river, keeping tabs on Davids's head as it bobbed in the water. Both men reached shore, but in a flash, Davids was gone. Unable to find him anywhere, the master remarked that Davids "must of gone off on an underground railroad."

Thus in 1831 a formal name was given to an informal system that already existed to help runaway slaves escape their masters' shackles and attain freedom. The Underground Railroad was a vast network of safe houses throughout the United States that helped slaves escape to free states in the North as well as into Canada. For some of these fugitives, that was just the first stop on a journey that took them in the opposite direction, ending in Mexico or the Caribbean.

Dangerous Cargo

People had been helping slaves escape for many years, but the beginning of the 19th century saw a large increase in slaves finding their way to freedom. Some blacks were able to make the trip safely using clever disguises and outright trickery. A light-skinned slave named Ellen Craft disguised herself as a white man traveling with a black servant who was in reality her husband William. Together, they journeyed openly to freedom in the North. In another highly publicized ruse, Henry "Box" Brown escaped by packing himself in a crate and shipping it to the Philadelphia office of an antislavery campaigner. Not everyone had the means or ingenuity to carry out such grandiose plans. The Fugitive Slave Law of 1793 allowed slave owners to recover their escaped slaves—considered their personal property—in any state, so secrecy was essential. Sympathetic whites, Native Americans, and free blacks opened their homes to the escapees, providing supplies and directions to the next safe house. The Fugitive Slave Law of 1850 included the penalty of a six-month prison sentence and $1,000 fine for anyone convicted of aiding a fugitive slave, making the work of those on the Underground Railroad even more dangerous.

All Aboard

In an effort to maintain secrecy, the route was fraught with railroad euphemisms. Safe houses were called stations, escaped slaves were parcels or passengers, owners of safe houses were stationmasters, and the people traveling with slaves to lead the way were conductors. The Promised Land was Canada, while even the Underground Railroad itself had its own code names: Freedom Train or Gospel Train.

One of the most famous conductors associated with the Underground Railroad was Harriet Tubman, a fugitive slave from Maryland who returned to the South 19 times to rescue others and guide them north. During her years of experience, Tubman came up with some clever tricks to improve her chances of success. She always began her northward journey on a Saturday night, because plantation owners would have to wait until Monday's newspaper to post a runaway slave notice. She carried medicine to put a crying baby to sleep. If she encountered possible slave hunters, she'd turn south for a while, knowing they would expect escaped slaves to head north.

Rolling in to the Station

Another famous figure in the Underground Railroad was Levi Coffin, a Quaker from Indiana who helped more than 3,000 slaves find their freedom. A secret bedroom on the third floor of Coffin's large brick home could securely hide runaway slaves for weeks if they needed rest to complete their northward journey. Coffin's efforts earned him the nickname "President of the Underground Railroad."

Other famous conductors and stationmasters along the railroad included newspaper editor and prominent abolitionist William Lloyd Garrison and feminist Susan B. Anthony. William Still, a black civil rights activist, helped as many as 60 slaves a month reach safety. He carefully recorded personal information and biographies for each fugitive he harbored, information that he later published in his book *The Underground Rail Road Records*. During one interview with a fugitive slave, Still discovered the escapee was his own brother, Peter Still, from whom he had been separated in childhood.

Much Remains Hidden

Historians may never fully know all of the secrets of the Underground Railroad. In 2002, archaeologists in Lancaster, Pennsylvania, discovered a secret hiding place for runaway slaves on the property of Thaddeus Stevens, a Radical Republican who served five terms in the House of Representative, becoming a significant power broker and one the era's most important politicians.

Although the journey was incredibly dangerous, it was well worth it to the many who escaped slavery and oppression. Through the combined efforts of these brave souls, an estimated 75,000 blacks escaped to a new life of freedom.

Robert Manry and *Tinkerbelle*

Manry boarded the tiny craft Tinkerbelle and sailed over the Atlantic in what was then the smallest boat to ever cross those vast seas. He made it, but alas, the story does not end well.

* * * *

ROBERT MANRY WAS born in Landour, India, in the Himalayan Mountains, in June 1918. As a child, he savored his first sailing experience alongside his father. His fire for a nautical adventure was fueled by a German sailor who spoke to his high school class and told of his own exciting adventures. Between a seemingly idyllic childhood and a comfortable adulthood that took him to Cleveland, Ohio, as a copy editor for the Plain Dealer, Manry studied for a semester in China; served in Austria, France, and Germany during World War II; and earned a political science degree from Antioch College. Not until he had a family of his own, complete with a son and daughter, did he again tap into his inner sailor.

The Pull of the Sea

In 1958, at age 40, Manry purchased his prized possession, a used 13.5-foot sailboat named *Tinkerbelle* that was nearly as old as he was. Manry devoted his energies to *Tinkerbelle* in an effort to restore its seaworthiness. The vessel earned a place on every family vacation, and it eventually became the craft that transported him on his famous voyage. That ultimate trip was not, however, a goal he had set upon when purchasing the tiny boat. Those plans unfolded after a seemingly innocuous invitation from a friend to join him on his own voyage in a boat nearly twice the size of *Tinkerbelle*. The friend backed out, but the adventure bug had bitten Manry. He told only his family of his plans before deciding to sail the Atlantic alone in 1965.

Packed for almost any possibility, Manry depended on Dexedrine to keep himself awake and aware. Days on the seas, of course, took their toll. At times, he was overcome with hallucinations that forced him to lose consciousness. Vicious nightmares of harm to his family ravaged him during those times; dramatic weight loss and dangerous storms affected him during his waking hours. But Manry and the small but mighty *Tinkerbelle* continued on their way.

At the end of his 78-day journey, Manry was flabbergasted by the fanfare that awaited him. Back at home, his voyage had topped such national news stories as the growing war in Vietnam. Awaiting him at his destination of Falmouth, England, was a huge celebration that even included his family, who were flown over by the Plain Dealer.

Consequences

Alas, the story does not end with the happiness of a goal accomplished. The years that followed were not filled with the excitement and joy of fame; instead, a sickening series of events, at least partially caused by the *Tinkerbelle* voyage, devastated the Manry family. Victims of an onslaught of media attention that surrounded the now-famous clan, the Manry children spiraled downward. The entire Manry family took to bickering in unease over the constant glare of the limelight. In a freak automobile accident, Manry's wife was killed not quite four years after the voyage; Manry himself died of a heart attack only five-and-a-half years after his adventure. The Manry kids, still teenagers and still suffering the inadvertent consequences of the famous voyage, were left parentless. Douglas Manry eventually was homeless for a time, sleeping in the very park named after his famous father. But he and his sister Robin forever remembered the seemingly proud summer that their dad made history for the toll it took on their once-tranquil household.

The Pizarro Brothers

In the 16th century, Francisco Pizarro and his three half-brothers ruled Peru.

⁕ ⁕ ⁕ ⁕

THE PIZARRO BROTHERS were born in Trujillo, a vibrant commercial city in western Spain. Their father was an army captain. Only Hernando was his legitimate son; the other brothers were products of various affairs.

In 1502, Francisco sailed to the West Indies and established himself as a successful explorer. From 1519 to 1523, Francisco lived in the new town of Panama and served as the mayor. He went on to lead an expedition down the west coast of South America in 1523, and it was on this journey that he heard rumors of a wealthy native civilization in Peru. Wanting to see this civilization for himself, Francisco sent to Panama for reinforcements. When his request was denied, Francisco traveled to Spain to appeal to King Charles V. He won the monarch over, and Francisco, his brothers, and 180 others set sail for Peru in January 1531.

When they arrived in Peru, they were greeted by the Incas. Though he at first presented himself in friendship, Francisco soon kidnapped and executed the Incan ruler, Atahualpa. The Pizarro brothers assumed control of the region, and they proved to be particularly brutal overlords, using torture and execution to solidify power, amass vast personal wealth, and put down frequent rebellions.

Francisco, Gonzalo, and Juan all died violently in Peru. When Hernando, the remaining Pizarro brother, returned to Spain, he did not receive a warm welcome. Many Spaniards were angry at him over his role in the execution of Diego de Almagro, a Spanish soldier who was one of the leaders of the conquest of Peru. Hernando was imprisoned for 20 years.

While he was in prison, Hernando married his niece (his brother Francisco's daughter). Through this marriage, Hernando accumulated great wealth, which allowed him to live comfortably until his death in 1578 at around 103 years of age.

Sacajawea's Story

There aren't many tour guides as famous as Sacajawea, but in truth, she wasn't a guide at all—she had no idea where she was going, and she didn't even speak English!

* * * *

Hooking Up with Lewis and Clark

MERIWETHER LEWIS (A soldier) and William Clark (a naturalist) were recruited by President Thomas Jefferson to explore the upper reaches of the Missouri River. Their job was to find the most direct route to the Pacific Ocean—the legendary Northwest Passage. Setting out in 1803, they worked their way up the Missouri River and then stopped for the winter to build a fort near a trading post in present-day North Dakota. This is where they met the pregnant Shoshone teenager known as Sacajawea.

Actually, they met her through her husband, Toussaint Charbonneau. He was a French fur trader who lived with the Shoshone (he is said to have purchased Sacajawea from members of another group who had captured her, so it may be inaccurate to call her his "wife"). Although Sacajawea is credited with guiding Lewis and Clark's expedition to the Pacific, the only reason she (and her newborn baby) went along at all was that her husband had been hired as a translator.

Pop Culture Icon

The myth of Sacajawea as the Native American princess who pointed the way to the Pacific was created and perpetuated by the many books and movies that romanticized her story. For example, the 1955 movie *The Far Horizons*, which starred Donna Reed in "yellow-face" makeup, introduced the fictional plotline of a romance between Sacajawea and William Clark. Over time, she has evolved to serve as a symbol of friendly relations between the U.S. government and Native Americans. In 2000, she was given the U.S. Mint's ultimate honor when it released the Sacajawea Golden Dollar. At the same time, though, the Mint's website incorrectly states that she "guided the adventurers from the Northern Great Plains to the Pacific Ocean and back."

The Real Sacajawea

The only facts known about Sacajawea come from the journals of Lewis and Clark's expedition team. According to these, we know that she did not translate for the group—with the exception of a few occasions when they encountered other Shoshone. But because she did not speak English, she served as more of a go-between for her husband, the explorers, and members of other tribes they encountered in their travels. Concerning her knowledge of a route to the Pacific, Lewis and Clark knew far more about the land than she did. Only when they reached the area occupied by her own people was she able to point out a few landmarks, but they were not of any great help.

This isn't to say that she did not make important contributions to the journey's success. Journals note that Sacajawea was a great help to the team when she took it upon herself to rescue essential medicines and supplies that had been washed into a river. Her knowledge of edible roots and plants was invaluable when game and other sources of food were hard to come by. Most important, Sacajawea served as a sort of human peace symbol. Her presence reassured the various Native American groups who encountered Lewis and Clark that the explorers' intentions were peaceful. No Native American woman, especially one with a baby on her back, would have been part of a war party.

There are two very different accounts of Sacajawea's death. Although some historical documents say she died in South Dakota in 1812, Shoshone oral tradition claims she lived until 1884 and died in Wyoming. Regardless of differing interpretaions of her life and death, Sacajawea will always be a heroine of American history.

Portuguese Explorers

During Portugal's golden age of maritime discovery, Portuguese sailors brought glory and riches back to the mother country.

❋ ❋ ❋ ❋

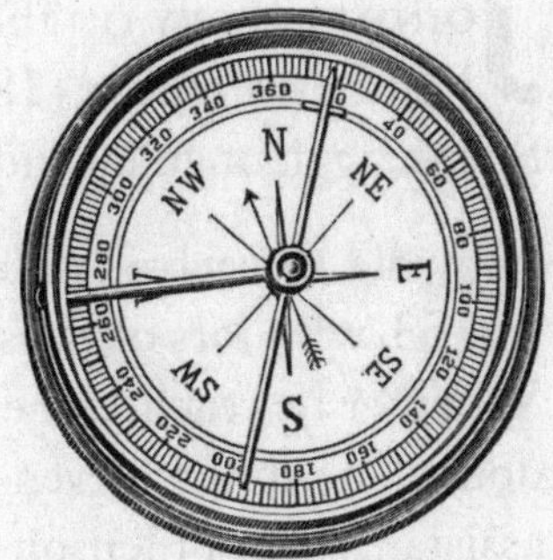

- A Portuguese nobleman who sponsored numerous voyages to West Africa, Henry the Navigator (1394–1460) is noted for fostering exploration, commerce, and the development of new ship designs and navigational techniques.
- Bartolomeu Dias (ca. 1450–1500) was the first European to sail around Africa's Cape of Good Hope, laying the way for Vasco da Gama's voyage to India.
- The first European to sail to Brazil, Pedro Álvares Cabral (ca. 1467–1520) claimed the huge landmass for Portugal, greatly increasing the nation's wealth and influence in the New World.
- Ferdinand Magellan (ca. 1480–1521) sailed from Spain around the southern tip of South America and on to the Philippines, where he died. His expedition continued westward back to Spain and was the first to circumnavigate the globe.
- Vasco da Gama (ca. 1469–1524) was the first to sail directly from Europe to India via the Cape of Good Hope at Africa's southern tip. Leaving Lisbon in 1497, he arrived in India the following year and returned home in 1499. The last of the Portuguese explorers, he died in 1524.

Peary's Journey

Robert Edwin Peary (1856–1920) led an expedition toward the North Pole in 1909. Did he make it?

* * * *

JOINING PEARY ON the final leg were African American Matthew Henson (1867–1955) and four Inuit: Uutaaq, Sigluk, Iggiannguaq, and Ukkujaaq.

Why did Peary bring Henson and the Inuit? He had to. Peary was 53, with lots of missing toes, and he mostly rode a dogsled. Whether he reached the Pole or not (and there's reasonable doubt), he couldn't even have attempted the trip without loyal assistance from Henson and the Inuit.

Henson was the ideal polar explorer: self-educated, skilled in dog mushing, fluent in Inuktitut. At one point Henson slipped off an ice floe into the Arctic water, whereupon Uutaaq quickly hauled him out and helped him into dry clothes, saving his life.

Anyone who guesses that Peary got all the credit while the others were ignored wins a slab of *muktuk* (whale blubber). Not until old age did Henson gain some of the recognition he deserved. Though Henson died in 1955, his remains were not interred in Arlington National Cemetery until 1988. Uutaaq passed away a few years after Henson, making Uutaaq the last survivor of the six who at least got near the Pole.

Byrd's First Antarctic Expedition

U.S. Navy Rear Admiral Richard E. Byrd's 1928 Antarctic Expedition broke a rather long U.S. Antarctic exploration dry spell (88 years, since the last trip had been in 1840). Byrd was the first to integrate aerial photography, snowmobiles, and advanced radio communication into Antarctic study.

* * * *

WITH TWO SHIPS, three airplanes, 83 assistants, lots of dogs, and a large heap of radio gear, Byrd set up a camp on the Ross Ice Shelf called Little America.

The expedition spent all of 1929 and January 1930 in Antarctica, making a flight over the South Pole and exploring Antarctic topography while recording a full year on the icy continent. Its tremendous success made Byrd famous.

Norman Vaughan, Byrd's chief dog musher, dropped out of Harvard and became the first American to drive sled dogs in the Antarctic. Byrd gave his MVP lasting credit by naming a 10,302-foot Antarctic mountain and a nearby glacier after him. (A few days before turning 89, Vaughan became Mount Vaughan's first ascender!)

At the 1932 Winter Olympics at Lake Placid, New York, Vaughan competed in dog mushing as a demonstration sport. Serving in the army in World War II, he became a colonel in charge of Greenland dogsled training and rescue. In one famed exploit, he rescued 26 downed airmen, then returned alone to recover the Norden bombsight, a top-secret device that was able to pinpoint targets using infrared radiation.

Learning to Fly

Back in the 18th century, all you needed to reach the sky was a silk bag, hot air, and a lot of guts.

* * * *

The Heat Is On

AS THE NAME implies, hot-air balloons rely on heat: As heat is produced and ultimately rises, it gets caught within the balloon. Since the heated air is less dense, it causes the balloon to rise. Winds push it along, but the balloon operator can also control the device manually by increasing or decreasing the heat to raise and lower the balloon. A seemingly simple concept—but one that wasn't discovered until the late 1700s.

When Sheep Fly...

The first balloon considered fit for flight was invented by French papermakers Joseph-Michel and Jacques-Étienne Montgolfier. In 1782, the brothers discovered that a silk bag would float to the ceiling of their home when filled with hot air. On April 25, 1783, they successfully launched a hot-air-filled silk balloon. Later that year, their balloon carried a sheep, duck, and rooster into the air. The balloon landed safely, and none of the animals were the worse for the experience.

All animals aside, October 15 of the same year marked another landmark for hot-air-balloon flight in France. The "Aerostat Reveillion" balloon carried scientist Jean François Pilâtre de Rozier 250 feet into the air, though it remained tethered to the ground by a rope. It floated for around 15 minutes and safely landed in a nearby clearing.

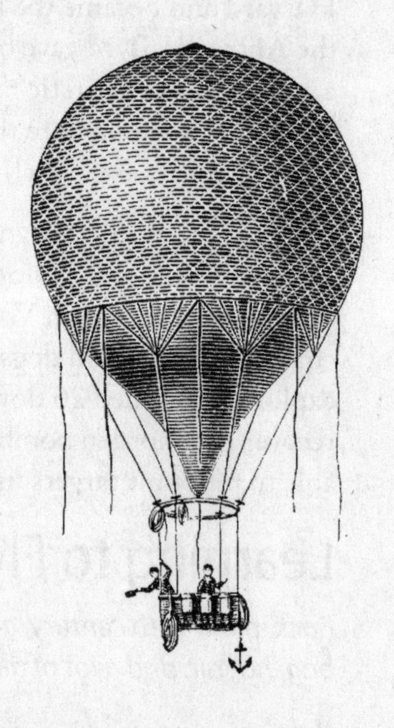

But the true prize came on November 21, 1783, in the Bois de Boulogne in Paris: A 70-foot silk and paper balloon made by the Montgolfier brothers was launched without a tether, carrying its first human passengers: de Rozier and Francois Laurent, the Marquis d'Arlandes. The balloon rose to around 500 feet and flew a distance of 5 1/2 miles, remaining aloft for 25 minutes before the straw used to stoke the hot-air pit set fire to the balloon. Although legend tells of the lofty gentlemen handing bottles of champagne to startled farmers upon landing, the real story is that they landed in a deserted farming area just outside of Paris—with no spectators nearby.

Italian Explorers

Italy contributed its share of explorers to the Age of Discovery.

⁕ ⁕ ⁕ ⁕

- Marco Polo (1254–1324) spent 25 years traveling with relatives through the Middle East and Asia, much of that time serving the Mongol emperor Kublai Khan. His exploits became famous with the publication of *The Travels of Marco Polo (Il milione).*
- Though Italian, John Cabot (ca. 1461–99) sailed to the New World under the British flag. He landed along the eastern coast of Canada and claimed the region for England.
- Undoubtedly the most famous explorer of his era, Christopher Columbus (1451–1506) captained the first voyage to the Americas that led to further exploration and conquest for Spain.
- Amerigo Vespucci (1454–1512) made several voyages along the eastern coast of South America and was among the first to recognize that the New World was in fact a new world, and not Asia.
- Sailing for France, Giovanni da Verrazano (ca. 1485–1528) was the first European to explore New York Harbor.
- Father Eusebio Francisco Kino (1645–1711), missionary and explorer, was the last of the Italian explorers. He explored regions of present-day Mexico, California, and southern Arizona. He introduced wheat, cattle, and sheep to the area and proved that Baja, California, is a peninsula (not an island, as was previously thought).

How Do Cats Always Find Their Way Home?

You can count on two things from your local television news during sweeps week: a story about a household appliance that is a death trap and another about a cat that was lost but somehow trekked thirty miles through a forest, across a river, and over an eight-lane highway to find its way home.

❋ ❋ ❋ ❋

YOU THINK, "NO, I don't think my electric mixer is going to give me cancer, but, oh, that cat..."

What's the deal with felines? How do they always seem to be able to make it home, regardless of how far away home might be? No one knows for sure, but researchers have their theories. One study speculates that cats use the position of the sun as a navigational aid. Another posits that cats have a sort of built-in compass; this is based on magnetic particles that scientists have discovered on the "wrists" of their paws. While these are merely hypotheses, scientists know that cats have an advanced ability to store mental maps of their environments.

Exhibit A is Sooty, one of the felines chronicled on the PBS program *Extraordinary Cats*. Sooty traveled more than a hundred miles in England to return to his original home after his family moved. Sooty's feat, however, was nothing compared to that of Ninja, another cat featured on the program. A year after disappearing following his family's move, Ninja showed up at his old house, 850 miles away in a different state; he went from Utah back to Washington.

But there are limits to what a cat can do—that's why odysseys of felines like Sooty and Ninja are extraordinary. In other words, the odds aren't good that Snowball will reach your loving arms in Boston if you leave her in Pittsburgh.

✳ Chapter 2

Way-Out Professions

Talk to the Expert

Food Sculptor

Q: When did you decide sculpting with food was a good idea?

A: At first I wasn't really sure if it was a good idea, because how serious is food? It's not serious—and I'm a serious sculptor. But I had the opportunity to do Mickey Rooney and Ann Miller in chocolate for the 100th performance of *Sugar Babies* [in 1980]. Then in 1995, my brother saw an ad for a butter sculptor and called me. From butter, I went to cheese. And now I'm even doing things in fruits and vegetables.

Q: What are the easiest and hardest foods to work with?

A: Cheese can be easy to work with, because you just get a block of cheese and carve away—you don't add, you just subtract. But the fact that you can't add does present problems. Butter can be easy, because you can both add to it and subtract from it, but it doesn't have much in the way of support.

Q: What are some of the wildest sculptures you've done?

A: We did a 3,000-pound cheese car. It really was just a skin of cheese on a car—there were maybe three to four inches of cheese on it. The rest was an armature; and all of the cheese had to be screwed onto the car and then caulked.

Q: So, traditional sculpting tools don't always work on food.

A: I use all kinds of different tools. I use woodcarving tools on chocolate. For cheese, I use plaster tools. I use food tools, like a cheese grater. I'm always looking for stuff that I think will work.

Q: Most of your sculptures are done for one-time events. What happens to the food afterward?

A: Right now, a lot of the butter is being turned into biofuel. You can also use it as animal feed. Cheese very often goes to a food bank. And I've had people eat the chocolate sculptures.

Meet the Bug Men

About one in three movies features at least one scene containing an insect, whether it's a single fly landing on a windowsill or a swarm of 30,000 locusts terrorizing an entire town. Stay seated for the end credits and you'll likely spot a credit for an "insect wrangler." Alternately known as a "bug wrangler" or simply "bug man," an insect wrangler is a trained entomologist responsible for not only providing various creepy-crawlies used in a movie or TV show but also for manipulating them onscreen so they swarm, run, or fly on cue.

* * * *

So, how do you train an insect?

YOU DON'T. INSECTS can't be trained—they can only be manipulated. Wranglers have to understand why insects do things and then work out how you can manipulate that behavior to fit the needs of the script. Spiders, for example, refuse to walk on Lemon Pledge furniture wax. If you spray the area you don't want them to go, they will unfailingly avoid it. Similarly, cockroaches will always run from a light source.

To make an insect fly toward a window, wranglers will place a bright light out of shot behind the window. To make an insect fly away and then return to the window, they attach a tiny harness made of very fine silk and control the bug like a puppet.

Is there a casting process for bugs?

It may sound strange, but casting is very important. You need to choose the right insect according to the demands of the shot. In the 1990 movie *Arachnophobia,* in which deadly tropical spiders terrorize a small California town, the insect wranglers deliberately chose to use the New Zealand Avendale spider for the swarm scenes. After testing a variety of other species, they found it was the only spider that would run when it was crowded rather than just attack the other spiders.

What about makeup for insects?

Absolutely! When the insect wrangler on *The Silence of the Lambs* set couldn't obtain specimens of the rare moth needed for the movie, he had to use common moths instead. He anesthetized each moth and painted on the distinctive markings (resembling a human skull) of the death's-head hawk moth onto its body. In *Spider-Man,* the insect wrangler had to paint the tiny blue-and-white Steatoda spider that bites the Peter Parker character. The wrangler used water-based, nontoxic paint, of course, so that it would easily wash off without harming the spider.

Are insects ever harmed during filming?

No. The U.S. Humane Society monitors most movie and television sets. If moviemakers want the "No Animals Were Harmed" end-credit disclaimer, they have to meet the Humane Society's strict guidelines. Wranglers work closely with the actors on how to handle bugs so as not to mistreat them. In the 2005 movie *The Three Burials of Melquiades Estrada,* for example, Tommy Lee Jones's character comes across a dead body covered in ants. In the first shot, the insect wrangler used real ants on a dummy body. But for the shot when Jones sets fire to the corpse, the real ants were replaced with rubber ones so that no ants would be harmed.

Tricks of the Trade

* To create a cockroach death scene, a bug wrangler will administer just the right dose of carbon dioxide. This will make the cockroach run a few feet before flipping onto its back and lying unconscious for several minutes. It will regain consciousness just in time for a second take.
* To get spiders to run up a wall, an insect wrangler will hide out of shot and blow a hair dryer up the wall toward the spider.
* Parts of a floor can be heated or cooled to control the direction of swarming spiders. Electric fields or shivering wires will stop the spiders from swarming too far.

Santa, Is That You?

America's Santa Claus owes a big thank-you to a German artist.

* * * *

THOMAS NAST IS famous in American history as the cartoonist whose drawings brought down Boss Tweed. Less well known is that if it wasn't for Nast, jolly ol' Santa Claus might look very different today.

Ho Ho, Er, *Who* Are You?

Initially Santa's image followed his origins as St. Nicholas, and he was often depicted as a stern, lean, patriarchal figure in flowing religious robes. Around 1300, however, St. Nicholas adopted the flowing white beard of the Northern European god Odin. Years passed, and once across the Atlantic Ocean (and in America), Nicholas began to look more like a gnome. He shrunk in physical size, often smoked a Dutch-style pipe, and dressed in various styles of clothing that made him seem like anything from a secondhand-store fugitive to a character from *1001 Arabian Nights*. One eerie 1837 picture shows him with baleful, beady black eyes and an evil smirk.

Santa Savior

Into this muddled situation stepped Nast. As a cartoonist for the national newspaper *Harper's Weekly*, the Bavarian-born Nast often depicted grim subjects such as war and death. When given the option to draw St. Nicholas, he jumped at the opportunity to do something joyful. His first Santa Claus cartoon appeared in January 1863, and he continued to produce them for more than two decades.

Nast put a twinkle in Santa's eye, increased his stature to full-size and round-bellied, and gave him a jolly temperament. Nast's Santa ran a workshop at the North Pole, wore a red suit trimmed in white, and carried around a list of good and bad children.

Nast surrounded Santa Claus with symbols of Christmas: toys, holly, mistletoe, wishful children, a reindeer-drawn sleigh on a snowy roof, and stockings hung by the fireplace. Nast tied all these previously disparate images together to form a complete picture of Santa and Christmas. Other artists later refined Santa Claus, but it was Thomas Nast who first made Santa into a Christmas story.

Abraham Lincoln Impersonator

Q: May I ask how tall you are?

A: I am 6'3", almost. Mr. Lincoln was closer to 6'4".

Q: Was this something you always wanted to do, or was it more like, "Hey, I'm an actor. I'm tall. This could be a good job for me."

A:I was actually cast as Lincoln in a play back in 1987. Fortunately for me, it turned into a part-time job/avocation; now this is how I earn a living.

Q: You are a former vice president of the National Association of Lincoln Presenters. How many Lincoln impersonators are there?

A: We have about 75, maybe 80 Lincolns, at least two dozen [of Lincoln's wife] Marys, and half a dozen of what we call "teams"—Abraham and Mary. Plus we have some other characters from history, like Frederick Douglass and Ward Hill Lamon, who was Lincoln's bodyguard in Washington, D.C. So the association now has roughly 200 members.

Q: Are there really enough jobs to keep you all busy full-time?

A: There are about six or eight of us that actually do this as a living. The rest do parades and things on Lincoln's birthday.

Q: So do you have to keep the Abe beard going all year, or with enough notice, can you grow it before a gig?

A: Well, I haven't shaved in close to 30 years, so I guess the answer is I keep it going. Plus I have enough events that I couldn't if I wanted to.

Q: What's the toughest thing about being Lincoln?

A: I think it's getting to know the history of the United States up to and around the Civil War. When I'm asked a question, sometimes I have to stop and think, "I can't answer that because that hasn't happened yet."

The Real Sherlock Holmes

The famous fictional sleuth is based on a 19th-century doctor from the University of Edinburgh who had a gift for piecing together tiny details.

* * * *

Paging Dr. Bell

SINCE BURSTING ONTO the scene in 1887, Sherlock Holmes has become quite the celebrity. The members of his fan club, sometimes dubbed the Baker Street Irregulars, number in the hundreds of thousands worldwide. In the 20th century, many readers were so convinced that Holmes was a real person that they sent mail to his address at 221b Baker Street in London. In the 21st century, he has his own page on Facebook.

Holmes is, of course, fictional. But is the detective based on a real person? Author Sir Arthur Conan Doyle claimed that he modeled his famous detective on Dr. Joseph Bell (1837–1911) of the University of Edinburgh. Doyle had been Bell's assistant when Doyle was a medical student at the university from 1877 to 1881.

Like everyone else, Doyle was awed by Bell's ability to deduce all kinds of details regarding the geographical origins, life histories, and professions of his patients by his acute powers of observation. The doctor had what his students called "the look of eagles." Little escaped him. Reportedly, he could tell a working man's trade by the pattern of the calluses on his hands and what countries a sailor had visited by his tattoos.

In 1892 Doyle wrote a letter to his old mentor, saying, "It is most certainly to you that I owe Sherlock Holmes." The resemblance between Bell and Holmes was also noticeable to fellow Scotsman Robert Louis Stevenson. After reading several Sherlock Holmes stories in a popular magazine, Stevenson sent a note to Doyle asking, "Can this be my old friend Joe Bell?"

When queried by journalists about the fictitious doppelganger, Bell modestly replied that "Dr. Conan Doyle has, by his imaginative genius, made a great deal out of very little, and his warm remembrance of one of his old teachers has coloured the picture." Nevertheless, Bell was pleased to write an introduction to the 1892 edition of *A Study in Scarlet*, the tale that had launched Holmes's career as a sleuth—and Doyle's as a writer. By the mid-1890s, Doyle had largely abandoned medicine for the life of a full-time writer.

A Crime-Fighting Legacy

Bell's association with Holmes wasn't his only claim to fame. He was a fellow of the Royal College of Surgeons of Edinburgh, the author of several medical textbooks, and one of the founders of modern forensic pathology. The University of Edinburgh honored his legacy by establishing the Joseph Bell Centre for Forensic Statistics and Legal Reasoning in 2001.

One of the center's first initiatives was to develop a software program that could aid investigations into suspicious deaths. "It takes an overview of all the available evidence," said Jeroen Keppens, one of the program's developers, "and then speculates on what might have happened." Police detectives have praised the software, which is called, fittingly, Sherlock Holmes.

Foley Artist

Q: Why are you called a Foley artist?

A: Adding sound to film was an industry leap that engaged a sense not previously used by the audience: hearing. A man named Jack Foley invented the fine art of background sound effects by using simple objects to create specific noises: Knocking coconut shells together mimics the sound of galloping horses, and snapping a stalk of celery sounds like a bone is being broken.

Q: You must have excellent hearing.

A: Hearing for me is like eyesight for pilots. My special talent is the ability to look at something and hear, in my mind, what sound it would make. I think of my work as audio art.

Q: Describe your work area.

A: It's the biggest collection of junk you've ever seen. My Foley stage has multiple walking surfaces, to produce the sounds of footsteps on anything from sand to snow. I have pie plates, staple guns, boxes of broken glass, crash tubes containing different materials to make particular crashing sounds, bamboo, leather gloves, boxes of cornstarch, you name it.

Q: How, exactly, do you produce some of your sounds?

A: We'll start with kissing sounds: That's me making out with the back of my hand. Someone's walking in the snow? That isn't real snow; I'm just squeezing a pouch of cornstarch. Rusty door hinges? I have an old hinge that I place against different surfaces, pouring on more water or less, depending on the creep factor involved. To create the sound of a crackling fire, I wad up cellophane. The process isn't as technical as you might think.

Sin Eating: A Strange Profession

An ancient custom of "eating" a dying person's sins has made its way into contemporary culture—and brought with it a wave of questions about the unusual approach.

* * * *

To Die Without Sin

IF YOU THINK you've had a rough lunch, wait until you hear about these guys: The so-called sin eaters were a group of people who would perform intricate eating rituals to cleanse dying people of their sins. The idea was to absolve the soon-to-be-deceased of any wrongdoing so they could die peacefully, without guilt or sin.

Much of the history of sin eating is based on folklore, particularly from Wales. Historians, however, have traced mentions of the practice back to early Egyptian and Greek civilizations, and references to sin eating can be found elsewhere in England as recently as the mid-1800s. *Funeral Customs*, a book published in 1926 by Bertram S. Puckle, refers to an English professor who claims he encountered a sin eater as late as 1825.

The sin-eating ritual is believed to have typically taken place either at the bedside of someone who was dying or at the funeral of one who had already passed away. Legend has it that for a small fee, the sin eater would sit on a low stool next to the person. Then, a loaf of bread and a bowl of beer would be passed to them over the body. Sometimes the meal would be placed directly on the body, so that the food would absorb the deceased's sin and guilt. The sin eater would eat and drink, then pronounce the person free from any sins with a speech: "I give easement and rest now to thee, dear [man or woman]. Come not down the lanes or in our meadows. And for thy peace I pawn my own soul. Amen." Family members, it is believed, would then burn both the bowl and platter after the sin eater left.

Odd Job

Sin eating wasn't the most lucrative profession, nor were sin eaters highly regarded among their peers. Many of them were beggars to begin with, and most were considered scapegoats and social outcasts. They would live alone in a remote area of a village and have little or no contact with the community outside of their work. Some accounts say they were treated like lepers and avoided at all costs because of their association with evil spirits and wicked practices. It was also widely believed that sin eaters were doomed to spend eternity in hell because of the many burdens they adopted. The Roman Catholic Church allegedly excommunicated the sin eaters, casting them out of the Church.

Sin Eating Today

Interestingly enough, sin eating lives on—at least in spirit. The idea of the sin eater saw its most recent resurgence with the 2003 movie *The Order*. The film (titled *The Sin Eater* in Australia) starred Heath Ledger and focused largely on a sin eater who was discovered after the death of an excommunicated priest.

In 2007, Hollywood revisited sin eating with *The Last Sin Eater*. That movie, based on the 1998 novel of the same name by Francine Rivers, tells the tale of a young girl whose grandmother has just died. At the funeral, the girl makes eye contact with the sin eater—something suggested as a forbidden act. The girl then spends most of the story trying to figure out how to absolve herself of this act. Sin eating has also made its way into several comic book storylines, most often as the name of villainous characters. These characters have been featured in numerous Spider-Man storylines as well as in some of the Marvel Ghost Rider creations.

While the custom may have died out in modern society, one thing's for certain: The archaic legend of the sin eater is just too delicious to pass up.

Improv Comic

Q: Life is basically one big improvisation—completely unscripted—and yet so many people are uncomfortable with the thought of doing improv. Why is that?

A: It's hard for a lot of people to step out of reality, and that's what improv is. You're using reality, but you're stepping out of it, you're stepping out of yourself, and that's very uncomfortable. Comedy's scary as it is, never mind making it up as you go and hoping the people that paid $22 like it.

Q: Do you think anyone can do improv?

A: I think anyone can do it, but not everyone can do it *well*.

Q: So much of doing improv relies on audience suggestions. What have been some of your most and least favorites?

A: My least favorites are the ones we are able to name before they happen. Like when somebody [famous] dies, we know we're going to get that. When something happens in the news, we know we're going to get that. I've been in more improv "Laundromats" than actual Laundromats. And inevitably, if we're asking for a celebrity who's dead, we'll get Elvis.

Q: Do you always have to take the first suggestion you hear?

A: No, you don't. If I ask for a dead celebrity and someone says Elvis, I'll say "Elvis isn't dead," and everyone giggles enough not to care. And then there's nothing wrong with going, "You know, we get JELL-O all the time here. Let's try to be a little more creative. You're a smarter audience than that." I don't think there's anything wrong with trying to tell your audience you want to give them something you haven't done a thousand times before.

Q: Improv comics often say nothing makes them laugh anymore. Is that true?

A: You really have to catch me off guard to make me laugh. But then again, the people I perform with do that constantly.

Ray Harryhausen: Master of Illusion

Though you may not recognize his name, Ray Harryhausen is revered among fantasy film fans. Through the laborious process known as stop-motion animation, the special effects wizard has brought to life numerous dinosaurs, a giant octopus, a monster from Venus, and everything in between.

⁕ ⁕ ⁕ ⁕

AS A TEENAGER, Harryhausen was enthralled by *King Kong* (1933). He marveled at the giant gorilla's lifelike movements and eventually struck up a friendship with Kong's animator, Willis O'Brien, who taught Harryhausen the tricks of the trade. Harryhausen's first professional job was animating George Pal's Puppetoon shorts for Paramount Pictures. After a stint in the Army Signal Corps, he teamed up with O'Brien to make *Mighty Joe Young* (1949), a big-gorilla flick that starred Terry Moore and Ben Johnson.

From there, Harryhausen was hired by Warner Brothers to animate the giant dinosaur that rampaged through New York City in *The Beast from 20,000 Fathoms* (1953). It was this gig that cemented his reputation as one of the best animators in the business. He soon found himself at Columbia Pictures, where he created the special effects for a variety of innovative science-fiction films, including *It Came from Beneath the Sea* (1955), *Earth vs. the Flying Saucers* (1956), *20 Million Miles to Earth* (1957), *Mysterious Island* (1961), and *First Men in the Moon* (1964).

But it was Harryhausen's lifelong love of mythology that led to some of his most acclaimed films, including three movies featuring Sinbad the Sailor. His crowning achievement, however, was *Jason and the Argonauts* (1963), which featured a remarkable sequence involving seven sword-swinging skeletons. Though it lasts just a few minutes on-screen, the scene took Harryhausen more than four months to animate.

Clash of the Titans (1981), which starred Laurence Olivier and Harry Hamlin, would be Harryhausen's final film. In 1992, the master animator received the Gordon E. Sawyer Award—an honorary Academy Award—in recognition of his remarkable career.

10 Most Deadly Vocations in America

Before you complain about punching the time clock, read this list for some perspective. Maybe the coffee stinks and you don't like your boss, but at least the threat of death or injury isn't perpetually hanging over your head. The order may change from year to year, but these are typically the most dangerous jobs in America.

* * * *

1. Logger
2. Pilot
3. Fisher
4. Iron/Steel Worker
5. Garbage Collector
6. Farmer/Rancher
7. Roofer
8. Electrical Power Installer/Repairer
9. Sales, Delivery, and Other Truck Driver
10. Taxi Driver/Chauffeur

How to Become a Stunt Person

Q: When does a movie need a stunt person?

A: A stunt person stands in for an actor or actress when the scene involves physical danger—anything from fighting or falling from a bicycle to hanging from a skyscraper or driving a car through a burning building. A movie production's insurance company will insist that an actor is never exposed to unnecessary danger. If a lead gets injured, for example, production has to stop and a studio or the investors financing the movie will lose money. It's also quicker and less expensive to employ a skilled stunt double rather than train an actor to perform his own stunts.

Q: Is this a profession for thrill-seekers?

A: Absolutely not! A stunt person needs to be gutsy but not reckless. Performing a stunt involves attention to detail and lots of planning. Those just looking for an adrenaline rush are a danger to themselves and to others on the set.

Q: So what skills are required?

A: You need to be in good physical shape, and it helps to be skilled at sports such as gymnastics or martial arts or even have some military training. There are also workshops that provide basic stunt training. As with acting, aspiring stunt people also need to be good at networking.

Q: How do you break into the business?

A: Landing that first role can be harder than most of the stunts! Serious parties should check out the two largest stunt associations: the United Stuntmen's Association (www.stuntschool.com) and the Stuntmen's Association of Motion Pictures (www.stuntmen.com). You also need to be a member of the Screen Actor's Guild (SAG). You can become a member of this actor's union by working as a movie extra. Then write to stunt coordinators who work in Hollywood regularly. Try to hang out on movie sets, and take any opportunity to introduce yourself to the stunt people.

Q: How much does a stunt person earn?

A: The base rate for SAG members is $500 per day plus extras for repeating a stunt. But remember, employment is often sporadic and the work is dangerous. A desk job might not be as exciting, but it's a lot safer.

Rosie the Riveter

World War II was a time of terror and death for millions, but for many others, it brought undreamed-of opportunities.

* * * *

FOR 18-YEAR-OLD MOTHER Peggy Terry, moving from the Dust Bowl depression of Oklahoma to work at an artillery-shell plant in Kentucky, was, "like a journey to the center of the earth." With her hometown economy still staggered, she remembered, "You did good having bus fare to get across town." Suddenly, she was making $32 a week painting the tips of tracer bullets, trading shifts with her mom and sister to take care of the kids. "Before that, we made nothing." Later, at a munitions plant in Michigan, she pulled in $90 a week, a startling sum for a woman who previously couldn't afford a radio.

The War Brings a New Workforce

In the 1930s, joblessness had averaged 18 percent a year. Wages and output sunk, while attitudes toward women and work remained traditional. In a 1936 Gallup poll, 82 percent of respondents—including 75 percent of women—thought married women shouldn't work if their husbands had jobs.

Almost overnight, the war buried the Great Depression and attitudes changed. Suddenly, with 16 million men in the armed forces and hundreds of new war plants needing triple shifts, there weren't enough workers. Women streamed into the work world, many for the first time.

The number of working women soared from 11,970,000 in 1940 to 18,610,000 in 1945. A quarter of all married women, an unheard-of proportion, were punching time clocks. By the end of the conflict, women made up 36 percent of the civilian labor force. These figures would decline sharply after V-J Day, but remained above prewar levels, setting the stage for the 1970s explosion of workaday women.

In war factories, pay was good. From 1940 to 1944, average yearly wages for all workers increased from $754 to $1,289. A woman in a munitions factory could readily make $50 a week, an excellent salary for the time. Defense work paid women much better than traditional female employment like retail sales, housecleaning, and restaurant work, where the wages averaged about half as much.

The Face of the Working Woman: Rosie the Riveter

This giant occupational shift was helped along by a War Advertising Council campaign. In 1943, the country's newspapers, radio stations, trade journals, and Hollywood movies sang the praises of women in defense jobs. Hundreds of magazines tucked articles with the patriotic theme into their September 1943 issues.

The campaign's "poster child" was Rosie the Riveter—in fact, several Rosies. The first was conjured up by J. Howard Miller, a graphic artist with Westinghouse. Miller created the famous 1942 poster of a woman in work shirt and bandana, determinedly flexing her right bicep, under the screaming slogan, "We Can Do It!" The model for the poster was a real-life Michigan factory worker, Geraldine Doyle.

But the picture was not yet Rosie. The poster became linked to that name the following year, after the hit song "Rosie the Riveter," by Redd Evans and John Jacob Loeb. Its lyrics went:

Rosie the Riveter

Keeps a sharp lookout for sabotage,

Sitting up there on the fuselage . . .

. . . Rosie's got a boyfriend, Charlie.

Charlie, he's a Marine.

Rosie is protecting Charlie,

Working overtime on the riveting machine.

Then, Rosie became a movie star. Actor Walter Pidgeon came across Rose Monroe, an airplane riveter in Michigan's Willow Run assembly plant. She portrayed Rosie the Riveter in a short film that promoted war bonds.

Finally, Rosie was depicted by America's leading commercial artist, Norman Rockwell. On a May 1943 cover of The Saturday Evening Post, he painted model Mary Doyle Keefe as a muscular female worker, the name "Rosie" stenciled on her lunchbox, leaning back confidently toward a giant American flag, a huge rivet gun on her lap, work shoes propped on Hitler's Mein Kampf. This was the toughest Rosie of all.

Due to the advertising blitz, perhaps two million more women signed up for work.

Overcoming Obstacles

Many of the war plants were huge, their social impact tremendous. In Richmond, California, more than 100,000 male and female workers at Ford and Kaiser factories turned out more than 1,500 Liberty and Victory ships, and at a Richmond Ford plant they built some 60,000 tanks and other combat vehicles.

Near sleepy Marietta, Georgia, Bell Aircraft of Buffalo, New York, built a two-million-square-foot factory for constructing P-39s and Bell X-1s. At peak, about 40,000 people worked at the plant, 6,000 of them women. Cobb County's population nearly doubled as workers streamed in from throughout Georgia and other states. Throughout the South and West, such projects set up the postwar Sunbelt boom.

In Alabama, the army built a $40 million factory, the Huntsville Arsenal, which made 27 million units of mustard gas, tear gas, and other chemical warfare agents. Nearby, the $6 million Redstone Arsenal's force of male and female workers churned out 45.2 million grenades, bombs, and artillery shells.

In 1942, some Huntsville officials took a stance against hiring more ladies, as the laboring ability of black women in particular was doubted. An absence of "separate but equal" toilets for whites and blacks was a further concern. At the national level, meantime, a labor expert admonished that, "The employment of millions of untrained workers, including old men, youths, and housewives" will "inevitably result in a material and gradual dilution of labor skills."

Still, by the end of 1942, 40 percent of the workers on Redstone's factory floor were female, which increased to 62 percent by 1945. By 1944, black women made up over 40 percent of the Arsenal's female employees. Workers recall that, even in the land of Jim Crow, blacks and whites often worked together.

Hourly assembly-line pay for Huntsville men producing mustard gas was $5.76, $4.40 for women engaged in the same work. (Entry-level pay was $5.04/$3.06.) Later, the federal War Labor Board enforced the novel principle of equal pay for equal work.

In many plants, time-honored attitudes among men were tested. At Marietta's Bell aircraft plant, 18-year-old Betty Williams turned her childhood interest in flying into a job supervising a crew of 25, most of them men. "I had three strikes against me," recalled the Indiana-born Williams. "I was a Yankee. I was much younger than they—the men would have a wife and three or four kids; there I would be 18." And the view of such fellows, she explained, was, "You're a woman and you should be home doing the pots and pans and cooking cornbread." Though she had been offered the job because there was a dire shortage of managers, Williams persevered. She served as foreman for the top section of plane wings and met her future husband, who worked on the bottom halves.

Under the pressure of wartime production, the concerns became not race or sex, but the dearth of technical and supervisory skills. Factories made changes to keep up. The Huntsville Arsenal founded a Civilian Training Department to let workers "earn while they learned." It arranged with Auburn University to offer tuition-free courses in chemistry, accounting, engineering, drawing, mechanical and electrical maintenance, and industrial management. The University of Alabama gave a class in chemical lab techniques "for women only, who desire to qualify for jobs in defense laboratories." About a hundred black female students from Alabama University worked for the Redstone Arsenal's assembly line. "From all appearances," wrote a local newspaper, "their work and attendance" offer "an example any of us would do well to follow." Through such means, women trailblazed new professions. At the arsenals, women drove trucks and handled drill presses. Female forklift operators, according to work reports of the day, could "handle 124 pound to 150 pound pallets with the ease and efficiency of old timers."

Moreover, the day jobs could be difficult and dangerous. During the war, three women were killed at Huntsville from work mishaps and two were killed at Redstone. Peggy Terry remembered when a thunderstorm knocked out the lights at her munitions plant, and someone knocked a box of detonators onto the factory floor. "Here we were in the pitch dark," Terry explained. "Somebody was screaming, 'Don't move, anybody!' They were afraid you'd step on the detonator. We were down on our hands and knees crawling out of that building in the storm. If we'd stepped on one . . ."

Still, for most "Rosies," the war afforded a desirable means to pay the bills, raise children and, for many, their first opportunity to build new careers. Said Marie Owens of the Huntsville Arsenal, whose husband was in combat overseas, "I am interested in carrying on here while the boys do the fighting over there . . . The harder I work for them here, the sooner they will come home."

Tattoo Artist

Q: What got you interested in the tattoo business?

A: Growing up around tattoo parlors—my mom's heavily tattooed, and my uncle was a tattoo artist.

Q: How do you practice?

A: The best practice is just draw, draw, draw. The apprenticeship generally lasts about a year of you learning all the techniques and skills the tattoo artist has to show you. And when you think you've got a good grasp of everything, you find yourself a client.

Q: So, what happens if you mess up on a client?

A: You don't.

Q: What's your favorite type of tattoo to do?

A: I love sailor-style traditional artwork. Just real bold, bright colors. A lot of swallow birds and anchors and pirate ships.

Q: What's the most common place to get a tattoo?

A: On women, it's generally the lower-back area. On men, it's usually the top of the arm.

Q: Any part of the body you won't tattoo?

A: I don't tattoo people's faces, unless you're in the [tattooing] industry.You've gotta think about your future. I don't tattoo hands or knuckles unless you're a certain age; I don't tattoo necks unless you're a certain age. You can't be 18 and getting your hands or your face tattooed. How are you going to work for the rest of your life?

Q: What are some of the most common tattoo requests?

A: Religious artwork. Or script—oh my gosh, I could do script with my eyes closed.

History of Taxidermy

Want a bear rug, a rat torso, an evil eye, or a dried turkey head? How about a nice end table with the base made from an elephant's foot or a cool thermometer (it really works!) made from a freeze-dried deer foot? Before you decide to buy a mounted, two-headed calf at an auction, take a look at the process that got it there.

* * * *

I'm Stuffed, Thank You

TAXIDERMY, GREEK FOR "the arrangement or movement of the skin," is a general term that describes various procedures for creating lifelike models of animals that will last a long, long time—possibly even forever. The models, or representations of the animals, could include the actual skin, fur, feathers, or scales of the specimen, which is preserved and then mounted.

A quick search on the Internet will certainly reveal the odd, the bizarre, and many sideshow items, some of which are even for sale. The main focus of taxidermy, however, is the integrity of the final product, and getting the best results might require a far greater range of skills and crafts than simply stuffing and mounting. Those in the know consider it a precise art.

Take This Job and Stuff It

By the early 1800s, hunters began bringing in their trophies for preservation. The process usually began with an upholsterer skinning the creature. Then, the hide was removed and left outside to rot and self-bleach. Once this stage was complete, the tanner stuffed the hide with rags and cotton and then sewed the skin closed to replicate the live animal.

These rough methods of taxidermy are a far cry from modern-day processes. Today, professional taxidermists abundantly prefer the term *mounting over stuffing,* and rags and cotton have been replaced by synthetic materials.

They Don't Have to Sing for Their Supper

Dermestid beetles can be used to clean skulls and bones before mounting. As one might guess, these beetles clean by eating anything organic left on the bones, but at least they eliminate the need to use large cooking pots to boil the skulls clean. But wait, before you judge too harshly, it should be pointed out that this same method of turning the beetles loose is used by museums on delicate skulls and bones.

Cryptic as It May Sound

Freeze-drying is the preferred method for preserving pets. It can be done with reptiles, birds, and small mammals, such as cats, large mice, and some dog breeds. However, freeze-drying is expensive and time-consuming—larger specimens require up to six months in the freeze dryer.

But taxidermy can be even more creative than preserving animals. In crypto-taxidermy brand-new animals—dragons, jackalopes, unicorns—can be made. The subjects can also be extinct species, such as dinosaurs, dodo birds, or quaggas.

Some taxidermists have an impish side, as well. One popular practice places animals in human situations, building entire tableaux around them. This is called anthropomorphic taxidermy. Walter Potter has created a number of such works. *The Upper Ten*, or *Squirrels Club*, for example, features a group of squirrels relaxing at an upper-crust men's club. Another well-known work, *Bidibidobidiboo* by M. Cattelan, shows a suicidal squirrel dead at its kitchen table—definitely a conversation-starter in any home or business.

A new trend is the creation of entirely artificial mounts that do not contain any parts of the animal at all: for instance, re-creations of fish taken from pictures by catch-and-release anglers. This technique is called reproduction taxidermy, and although it is not traditional, it is definitely a more environmentally savvy method of preservation.

Michael Jackson: Falling Star

A boy from the Midwest rises to astronomical fame and fortune only to fall from grace years later, as an increasingly eccentric man embroiled in scandal and controversy. Was Michael Jackson a tortured-but-gifted artist or simply a famous-person-turned-freak? Either way, he was a strange yet fascinating man.

* * * *

Boy Wonder

MICHAEL JACKSON WAS born on August 29, 1958, the seventh of nine kids. From the outside, the Jacksons looked like a typical working-class family from Gary, Indiana. But in 1964, Joe Jackson organized his sons into a singing group, and it was clear that the Jackson boys were anything but average.

The group, which Joe named the Jackson 5, was comprised of brothers Jermaine, Jackie, Tito, Marlon, and Michael. With segregation still firmly in place in the early 1960s, the group performed on the "urban music circuit." Venues such as Harlem's famous Apollo Theater and Chicago's Regal Theater gave black performers a place to perform and be seen by industry professionals. Show by show, the Jackson 5 garnered attention with their catchy songs and slick dance moves.

Of all the boys, Michael was clearly the most talented—the consensus among biographers is that the youngest Jackson was a bona fide prodigy. His eye-popping dance moves were matched by his charismatic personality and his innate ability to nail vocals. The more the boys performed, the clearer it became: Michael was in every way a star.

Before long, the Jackson 5 had a deal with Motown Records, and they quickly charted hit after hit, including "I'll Be There," "ABC," and "I Want You Back." American audiences (black and white) embraced the cute singing sensations, and record sales soared. As the group became more successful, Michael received the most attention. He was the unofficial front man of the group, and he was also penning most of their hits. People wanted more Michael, and record producers took note.

Look Ma, No Brothers!

In 1979, Michael teamed up with hit-maker Quincy Jones on a solo album. (The Jackson 5 didn't officially disband until 1984.) *Off the Wall* featured hits such as "Rock With You" and "Don't Stop 'Til You Get Enough" and eventually sold more than seven million copies. The album's disco-inspired sounds were embraced by the boogie-crazed masses, and collaborations with the likes of Paul McCartney and Stevie Wonder gave the album depth. Honors followed, including American Music Awards and a Grammy. All this solidified Jackson as a solo star, but nothing could prepare the world for what came next.

Michael Mania Dominates the World

On December 1, 1982, *Thriller* hit with the force of a tsunami. The album shot to number one on Billboard charts and stayed there for 37 weeks. Never before or since has an album sold as well as *Thriller*—it went platinum 27 times, selling more than 50 million copies worldwide. Even after it relinquished its top spot, it continued to chart for two more years. Total and complete Michael Jackson mania took over the world. He quickly earned the undisputed title "The King of Pop."

One of the reasons for this success was Michael's use of the new cable television channel MTV. The burgeoning station was the perfect medium for visually savvy artists like Michael Jackson, Madonna, Prince, and others to catch the attention of Generation X. Videos for singles such as "Billie Jean" and "Beat It" instantly became classic visuals. But it was the video for "Thriller" that became a "cultural zeitgeist." The video premiered with intense hype on MTV, is still ranked number one on MTV's list of "Greatest Music Videos Ever Made," and was the highest-selling VHS tape ever sold after it was packaged with the documentary *The Making of Michael Jackson's "Thriller."*

At the Grammy Awards in 1984, Michael and *Thriller* were nominated for 12 awards and won a record-breaking 8. The press was touting him as a legend at age 25, and sales of *Thriller* contributed to one of the best years the music industry had ever seen. Fans were wearing Michael's trademark white glove; doing his iconic dance, "the moonwalk"; and paying big bucks to see him perform in concert. No one could dispute it: Michael Jackson was one of the biggest stars the world had ever seen.

Other successful records followed during the 1980s. Michael spearheaded the "We Are the World" charity project and worked with George Lucas and Francis Ford Coppola for the 3-D film *Captain EO*. His next solo album, *Bad*, was released in 1987 and launched another string of hits, including "Man in the Mirror," "The Way You Make Me Feel," and the title track.

What Goes Up...

His accomplishments were astounding, but that doesn't mean Jackson was happy—or that the good times would last. In 1993, two years after his *Dangerous* album was released, Michael was hit with a stunning lawsuit. It was known that the star allowed children to hang out at the home he called Neverland Ranch. The suit alleged that, not only did the kids play in the amusement park and pet the animals at the zoo on Jackson's property, but they also slept over in Michael's own bedroom. The media went crazy over the allegations. Michael was appalled by the accusations and claimed that he was completely innocent of any wrongdoing. The first accuser settled out of court for a reported $20 million. Michael said he did it to avoid a courtroom media circus, but some claimed it was an admission of guilt.

Ten years later, Michael was again charged with child molestation, as well as using intoxicating substances for that purpose and conspiring to hold a boy and his family captive at Neverland Ranch. Two years later, Michael was acquitted of all charges, but further ruin had been done to his reputation. The tremendous amount of negative publicity had a devastating effect on Michael's music, and some say it's what led Michael to become even more reclusive and eccentric.

Leave Me Alone

When someone says "Michael Jackson" these days, it's hard not to immediately think of his appearance. A medium skin-toned African American as a child, Michael's skin is now pale. In 1993, rumors that he bleached his skin were denied in a television interview between Jackson and Oprah Winfrey. He claimed he suffered from a skin disease that caused his skin to lighten. Michael also wrote in his 1988 autobiography, *Moon Walk*, that he had undergone "very little" plastic surgery, claiming only two nose jobs. It was hard to believe, because Michael's entire facial structure (including his lips, nose, and chin) seemed totally different when compared to photos of him as a younger man.

As the years passed, Michael's appearance became even more bizarre. He rarely took off his sunglasses, his complexion paled to a vampirish shade, and his nose became even smaller. Along with the changes in his look, his behavior was truly baffling. A marriage to Lisa Marie Presley seemed to be a publicity stunt to battle rumors that the star was gay or attracted to children. A kiss between the couple at an awards show in 1994 was universally critiqued as being stiff and awkward. The marriage lasted less than two years.

In 1996, Michael married again and fathered two children with his second wife, Debbie Rowe, but the couple divorced within three years. Michael's third child, Prince Michael Jackson II (nicknamed "Blanket"), was born in 2002. While in Berlin that year, Michael made headlines when he presented Blanket to the throngs of press and fans outside his hotel room. However, he didn't just wave with the baby in his arms; he dangled the child over the balcony, causing yet another media frenzy: Was Michael fit to be a parent?

What Happened?

Looking at the events of Michael Jackson's life, it is understandable how a person could become a bit odd over the years. Superstardom before age ten, a demanding father who pushed his kids too hard, a record-setting career unparalleled in world history, and wealth beyond imagination—all of this contributed to the person Michael Jackson became. Jackson died in 2009. Only time will tell how history judges the prodigal son of pop music.

Puppeteer

Q: So you are a "master puppeteer." Is that like having a black belt in puppetry?

A: Well, I do like to think of myself as a ninja sometimes, and one of the skills a puppeteer needs is the ability to disappear. So, yeah, I guess it's kind of like having a black belt.

Q: Puppetry is one of the world's oldest art forms. What do you think continues to be the draw?

A: Why puppetry works specifically for kids is that it's a simple convention, and a convention is just an agreement between the audience and performer. The convention of puppetry is that this piece of fabric or this chunk of wood will represent a particular character, and the audience hopefully enters that agreement. They'll accept that the piece of fabric is Little Red Riding Hood. Obviously kids aren't so stupid that they'll think it's real. It's just their imagination will say, "OK, yeah, for this time, that piece of red fabric is Little Red Riding Hood. Let's play."

Q: How are puppets received differently by kids and adults?

A: There's a lot more doubt in an adult imagination. A lot more acceptance of what's real and a lot more denial of what's impossible. [After an adult show I directed], the comments I got back from the audience were that they were really surprised they had an emotional attachment to the characters, even though they were just puppets. It's interesting to me that it's more natural for them to gain an emotional attachment to an actor that's clearly playing a role just as much as the puppet is playing a role.

Q: So what's your favorite part of your job?

A: Performing for a really good audience. I've done thousands of puppet shows over my career, and it's such a magical moment to think, "They're with me and we're in this together."

Q: Most difficult part?

A: I want to say keeping your arm straight for long periods of time. Yeah, I'm sticking with that.

Best Boy, Dolly Grip, and Other Odd-Sounding Jobs

Making movies requires the skills of dozens of people, and if you've ever read a film's credits, you've probably noticed some strange-sounding job titles. Here's what those crew members really do.

* * * *

Boom Operator

EVER SINCE THE silent era ended, sound has been an important part of the film industry. And if you've got sound, you need to have someone operating the microphones. The *boom* refers to the boom microphone, which is often at the end of a long pole. It can pick up sounds by hanging above the actors, just out of camera range.

Gaffer

The gaffer is the movie's chief lighting technician. This person manages the production's lighting and electrical departments and lights each scene based on direction from the cinematographer.

Best Boy

The best boy is the gaffer's primary assistant in managing the lighting and electrical crews. The best boy can be a female, but the title, which was derived from sailing terminology, remains the same.

Clapper Loader

The clapper loader, also known as the camera loader, handles one of the most recognizable pieces of filmmaking: the black-and-white clapper that marks the beginning of each new take. This person's job is simply to clap the clapper in front of the camera every time a take begins. He or she also loads the film stock into the camera's magazines.

Swing Gang

Don't be frightened—the swing gang isn't a group of dancers that will force you to jump, jive, and wail across the set. This is just a special name for the set decorators—the group that generally works at night to ready the set or soundstage for the upcoming day's filming. They also take down the set when it is no longer needed.

Armoror

If you're going to be scared of any crew member, it should probably be the armoror. This person deals with all the weapons on a movie's set and decides which weapons best suit each scene based on its era and genre, then makes sure everything goes off without a hitch. The armoror is responsible for picking out the right kinds of blanks and teaching the actors how to properly use the weapons.

Key Grip

The key grip is the leader of the grips—lighting and rigging techies who handle equipment that is not attached to the lights (which are the realm of the gaffers). Grips are essentially responsible for the entire cast and crew's safety.

Dolly Grip

A dolly is a platform on wheels that moves the camera during filming to ensure a smooth, fluid shot. The dolly grip moves the dolly around the set as needed.

Focus Puller

The focus puller measures the distance between the subject and the camera lens and uses the marking ring on the camera lens to determine where the lens needs to be set to ensure focus.

Lead Man

The lead man works with the art department to buy, borrow, or scrounge for objects to make the set more believable or atmospheric.

Wrangler

A wrangler deals with animals. He or she is responsible for selecting, training, and taking care of the animals used in a film, which can range from horses and cows to spiders and snakes.

Casino Dealer

Q: Why do some people call you a "croupier"?

A: That's a title more often used among pretentious characters, of whom there are many in Las Vegas. But in my book, "croup" refers to a bad cough, and I don't think I'd enjoy a job as a professional cougher.

Q: No offense, but your job doesn't look much more difficult than coughing. Spin a wheel, wait for a ball to stop, figure out winnings. There has to be more to it than that.

A: If you want to make your car payment, there is. Casino dealing is the ultimate people job. The game itself, whether it's roulette or blackjack, isn't very hard. People are winning or losing money, and they react to that—that's what requires me to have good social skills. People will keep playing a table partly because of the environment the dealer creates.

Q: A fortune flows across your table every workday. How much of it do you get to keep?

A: I won't tell you what I get, but if it weren't for tips, I could do just as well in retail sales. The more players I have and the more winners who come through, the more tips I make—thanks to the tradition of "winners tip."

Q: Is it hard to break into your field?

A: Easy to break in, hard to stay. You can find gambling nearly anywhere in the country, though some training helps you get hired. People don't realize how many jobs there are on a casino staff: cashiers, dealers for all the games, pit clerk, pit boss, hosting, and security—a lot of security. Most casino employees don't deal cards or craps or roulette. Some don't know how to play, and you'd be surprised how many never gamble a nickel.

Q: How's it possible to cheat at roulette?

A: I watch for that very carefully. Some people will try to manipulate the ball. People used to have an accomplice create a distraction, then blow at the wheel through a straw. Casinos ended that by putting clear shields around the wheels. Other people try bumping the table at the right time. I keep my hand on the table at all times while the wheel's going, so if someone bumps the table I'll feel it. The first time I'll use my instincts; maybe it was an accident. The second time is no accident, and we'll ask the person to leave. All casinos work diligently to spot cheaters.

Q: How easy is it for the house to cheat? The most cynical view of the gambling industry is that it's run and rigged by organized crime.

A: The house has a natural statistical advantage, with no need or reason to cheat. In theory, if the house wanted to cheat, it'd use electromagnets. I can think of quite a few ways, but if I were asked to involve myself in actually doing them—or if I knew they were going on—I'd resign without notice. I don't want to go to jail. We get inspected; I think our gambling equipment is subject to much tougher scrutiny than those new voting machines. I'm not sure this will comfort you any, but the casino business is very corporate. We get lectures on how to reduce liability. We even get retirement benefits. It isn't The Sopranos by a long shot.

What It's Like: Crime Scene Decontamination Crew

The next time the big television executives sit around the conference table to decide which new shows will air in the fall, chances are one that will never make the lineup is CSD: Miami (Crime Scene Decontamination). Call us cynical, but it's unlikely that viewers would want to rush home to watch cleanup crews swab up buckets of blood while they eat their dinner. This is a job that's not for the weak of stomach.

* * * *

The Worst of the Worst

DEATH ISN'T THE most pleasant thing to think about, but cleaning up after someone dies is no picnic, either. Particularly if that person died in a violent way. Blood has soaked into the floor, bits of bone and brain matter are everywhere—that mess doesn't go away by itself. That's when Crime Scene Decontamination units (also called CTS—Crime/Trauma Scene–Decon crews) are called in.

CTS decon crews handle the worst of the worst. Hours or even days after a violent crime has occurred or a methamphetamine lab has been busted, CTS decon crews are contracted to come in and clean up. Sorry crime-show fans: You won't find a slightly-grizzled-but-still-handsome detective tiptoeing through this scene in an Armani suit—this is dirty work that would make most people sick to their stomachs.

The Task at Hand

By the time the decontamination crew is called to the trauma scene, the police, fire department, and crime scene investigators have all done their jobs and left (*these* are the people you see on TV). A decontamination lead will be appraised of the situation and alerted to the risks involved. Some are worse than others, such as decomposing bodies. Bodies that are discovered hours, days, or even months after death are considered to be in a state of decomposition or "decomp." During decomp, the body swells, the skin liquefies, and maggots move in to feed off the body. But this all pales in comparison to the smell as ammonia gas from the body fills the room. And if this doesn't make you contemplate a new line of work, consider that all of the creepy-crawlies that made the body their home must be rounded up and destroyed to eliminate any biohazard threat.

Dressing for Success

After putting on disposable head-to-toe biohazard suits, cleaners will don filtered respirators, rubber gloves, and booties before entering the crime scene. They'll also bring tools of their trade. There are no microscopes, centrifuges, or fingerprinting kits; instead, there are 55-gallon plastic containers, mops, sponges, enzyme solvents, putty knives, and shovels.

Once the contaminated material has been cleaned off of the room's surfaces and the site has been returned to normal, the cleaners must dispose of the hazardous waste material. Since federal regulations deem even the smallest drop of blood or bodily fluid to be biohazardous waste, the cleaners must transport all matter in specially designed containers to medical waste incinerators—all at an additional cost to the client.

The Big Business of Violence

To date, there are more than 300 companies that provide professional decontamination of crime scenes. Each one of these companies typically handles up to 400 cases a year and is on call around the clock. Decontamination companies are contracted through police departments, insurance companies, or by the deceased's relatives. And there's big money to be made: Companies charge between $1,000 and $6,000 to clean up a crime scene and return it to its original condition. In addition to cleaning houses, decon crews often clean car interiors and busted meth labs. In the case of the meth labs, virtually all of the carpeting, furniture, cabinets, and fixtures—basically, anything that can absorb dangerous chemicals such as methanol, ammonia, benzene, and hydrochloric acid—must be removed and destroyed.

The TV executives will probably never glamorize decontaminating crews. However, for those families in need of a CTS decon team, they provide a valuable service in helping them get their lives back to normal. It is important, sensitive work and in most cases, worth every penny the cleaners charge.

A Few Good Men (and Women)

A crime scene decontamination company is hardly going to have a booth at the local job fair. So, who on earth would get into this kind of work? Some decon workers find their way into the field from medical backgrounds (say, as an emergency room nurse) or the construction industry. The latter, in particular, proves helpful when dealing with jobs that require walls torn into or taken down. This is often the case in clean-ups that take place in meth labs.

And then there's the paycheck. Depending on the company and its location, decontamination cleaners can make a competitive living. In metropolitan cities where caseloads are high, cleaners can earn as much as $100,000 a year—and that's without a college education. Of course, few actually see that much money; the average career of a decontamination cleaner is less than eight months.

One of the requirements for becoming a decontamination cleaner is a strong stomach and the ability to detach themselves from their environment but still have empathy for family members and friends of the deceased. If the thought of picking up dog poo grosses you out, this is not the job for you. It's also not for you if you are easily depressed or internalize human suffering. Decontamination cleaners must be able to handle the worst-case scenario as "just part of the job." Before crew members are sent on their first job, cleaners must go through training that includes performing heavy physical labor while wearing a biohazard suit, watching videos of prior crime scenes, or cleaning up a room strewn with animal remains—all without tossing their cookies.

Yeoman

Q: Is there such thing as a "yeo"?

A: Yeah, it's how you get people's attention. Or how you show that you're paying attention. If you're an enlisted person in the Navy, though, and an officer calls for you, I wouldn't answer him or her with "yeo." Not good for your career. Some yeomen are called "yeo" for short, the way people say "Chief" for "Chief Petty Officer." In that case it's okay.

Q: Seriously, though, what's your title?

A: My rate is Petty Officer, Third Class, and my rating is Yeoman, so that makes me a Yeoman Third Class—usually just "Yeoman Third." I'm a personnel clerk, secretary, and typist, among other things.

Q: People who have read about Robin Hood might expect you to show up with a bow and arrow. How come we use the same word for old archers as for Navy clerks?

A: The term has evolved over a thousand years in ways that kind of amaze me. It used to refer to a sturdy peasant farmer, and then it seemed to gradually rise in social class. It also came to signify someone who was stout-hearted and capable—it was always a compliment, except perhaps to a lord or some other muckety-muck. Most English farmers were also good archers, and it came to refer to that skill as well. Then it meant an attendant on higher authority, such as a Yeoman of the Guard. I think that's probably where it crossed over from a guy with a bow or pitchfork to a clerk. The Royal Navy picked it up as the name for its clerks, and, eventually, so did the U.S. Navy.

Q: What does it take to be a good yeoman?

A: At the top of the list is literacy. Besides journalist, yeoman is perhaps the main job in the Navy that values a liberal arts degree, because you spend a lot of time fixing officers' and petty officers' grammatical issues. And they will thank you for it, take my word, because it makes them look better to their bosses. At the same time, I think they always regard yeomen as intellectual show-offs on some level. Can't be helped.

Q: You have a political science degree. Is that why you joined the Navy, to be a yeoman?

A: I felt it would combine travel with managerial experience with core clerical and administrative skills. Plus, it beat moving back in with my parents until I got my student loans paid off. You'd be surprised how many college graduates become enlisted men and women in all the services in a tight job market.

Q: Is it true that a yeoman is the person you least want to annoy in the Navy?

A: I think I'd be more careful about annoying a SEAL or the boatswain than a yeoman. SEALs can kill you silently with an old fingernail clipping, and the boatswain can make any sailor's life miserable. But I understand why people think that. An occupational hazard of working directly with officers is dealing with big egos. Yeoman can be officious, even with each other; some are flat-out snobs. And there are some who would unscrupulously "misplace" someone's records long enough to screw them out of a promotion or some other benefit. I wouldn't, but some would.

Q: So, after the Navy, you're trained for clerical work?

A: Yes. I'm prepared to do anything in a company's human resources, secretarial, or administrative departments, and I'm also trained in office management. I don't want to limit myself, though. What I'm really interested in getting into is editing or publishing. My yeoman rating has given me ample experience and information to write something really interesting!

Combat Medics

Among all the brave men who fought the war, medics and corpsmen stand out for special notice. When a wounded soldier cried out for help, he knew that he would be attended by someone willing to risk his own life to save others.

* * * *

On-the-Job Training

MERE SECONDS ARE significant after being wounded, and the U.S. Army instructed all soldiers in the rudimentary care of wounds they might receive. "First aid is self-aid," stated one slogan. All soldiers were issued an individual first-aid kit. On being wounded, they were to sprinkle sulfa powder on the wound, take sulfa tablets, apply a Carlisle bandage, and seek help. But for serious injuries or wounds that were incapacitating, soldiers depended on the care of the medical corps. Trained doctors and surgeons were always in high demand. Civilian physicians and senior medical students could enter the army as commissioned officers and go to work at trauma units and hospitals relatively far from the front lines. But with too few doctors to go around, the bulk of immediate frontline care was the responsibility of the combat medic.

Medics or aidmen were soldiers themselves, having gone through much of the same basic training as their brothers in arms before being given additional training in basic wound care. They were not experienced health care professionals, but they were called on to make immediate life-or-death decisions without a minute's warning. They carried aid kits with more resources than the individual soldier packets, including penicillin and pain-numbing morphine shots—the syringes of which would be pinned to the clothing of the victim after being administered, as a warning to the next medic to avoid overdosing. They also faced a wide range of situations beyond the obvious bullet and burn wounds. Soldiers thrown around in explosions suffered concussions and fractures that protruded from the skin. Gastrointestinal problems and vitamin deficiencies were widespread. Medics in the Pacific had to treat men suffering from malaria and more unusual tropical diseases such as beriberi, which caused some men to lose all feeling in their legs.

Medics' duties didn't end there. They had to treat battle fatigue, which was not always viewed sympathetically by a man's fellow soldiers, forcing the medic to play the role of psychologist as well as that of medical doctor. They provided aid to civilians caught up in the war zone, including women and infants. Medics in Europe at the end of the war were some of the first to encounter the horrors of the concentration camps, and tried to treat the victims as best they could. They also treated enemy prisoners of war. The latter group was entitled to care under the articles of war, but rendering aid to them sometimes caused hard feelings among the soldiers who had just been in a firefight with the prisoner.

Medic!

It was in firefights that medics earned their reputation as heroes. When one of his buddies was hit, the medic sprang into action, putting aside his personal fears to perform his duty. Medics were always in the thickest of the fight, since that was where their services were most needed. They crawled out into enemy fire, or even an artillery barrage, to attempt to save a life. A medic's safety depended in part on the theater of war in which he served—the Germans generally respected the symbol of the Red Cross the corpsmen wore and avoided firing on them. One medic even reported he was able to wave at German troops to get them to delay fire while he treated an injured man. The Pacific was a different story, and the medics there felt as if they drew extra Japanese bullets. As a result, many dyed their armbands green to make themselves less visible.

Regardless of geography, medics regularly came under attack, whether intentionally or through the fog of war, and continued their efforts despite the danger. Aidman Thomas Kelly refused to leave his injured buddies when his platoon was driven back by the Germans, crawling on ten separate trips through 300 yards of enemy machine-gun fire, dragging wounded soldiers to safety. Corpsman Robert Bush stayed at a wounded marine's side as the Japanese overran their position. He killed six of the enemy while managing to hold a life-saving plasma transfusion bag high in one hand, continuing to fight even after he lost his own eye. Navy pharmacist's mate John Harlan Willis performed a similar feat on Iwo Jima, where he picked up and hurled back at the enemy no fewer than eight live grenades that had been thrown at him and his patient, only to lose his life when a ninth one exploded in his hand. All three were awarded the Congressional Medal of Honor for their efforts, as were other medics in the war. Their fellow soldiers knew the medic in their platoon might save their lives one day, and their repeated acts of heroism were a frequent source of awe. As a result, they were singled out for special treatment, their buddies sometimes volunteering to dig a foxhole for them or relieving them of other mundane tasks.

"Thank God for the Medics"

Such devotion was made even more difficult by the fact that medics rarely had the satisfaction of seeing a full recovery. Their first priority was saving lives. After providing immediate aid, their patients were evacuated to areas farther behind the lines, given more care, sent farther back, and so forth. The caregivers had little time to do more than ensure that their patient lived long enough to hand them off to the next caregiver before moving on to the next injury.

Despite the hardships it faced, the medical service performed admirably during World War II compared to earlier conflicts. Injured men who received treatment quickly had a very good chance of survival, with mortality rates around 4 percent—half that of World War I. Although one medic claimed that he thought most of the good he did was psychological, making men feel better just by his presence, many soldiers have another opinion. Always ready to put their own life at risk to save others, the combat medic earned the respect and admiration of his buddies and established his reputation as "the soldier's best friend."

Musicologist

Q: What's a musicologist?

A: Musicologists study the history, theory, and performance of audio art. Nearly all of us teach.

Q: Why would someone become a musicologist, as opposed to a musician?

A: Many of us are also musicians, but it takes a tremendous investment of time and energy to become a professional musician. Talent and creativity are only the beginning.

Q: Is musicology also sociology?

A: Not in the formal academic sense, because sociology relies heavily on numbers and their analysis. Musicology touches upon sociology in the more general sense of placing a finger on the pulse of humanity. I would also consider it allied with history and psychology, in that one can gain great insight into both through the study of music.

Q: How would you describe your teaching approach?

A: I ask my students to speculate: To whom does a particular style of music speak? I want them to look outside themselves, to see what the music says to others, and in so doing, to open their minds to its multiple messages. I want them to be deep listeners. Anyone who studies musicology with me and still listens to music the same way after finals week should by all rights have failed my course.

Q: Can you name someone famous you would embrace as a fellow musicologist?

A: There are many. Here's one who may surprise you: "Weird Al" Yankovic. Most people know Al as a stage performer and parody artist, but his parodies are snapshots of the pop culture of the day. There is no genre he can't touch on if he puts his mind to it.

Psychic Investigators

When the corpse just can't be found, the murderer remains unknown, and the weapon has been stashed in some secret corner, criminal investigations hit a stalemate and law enforcement agencies may tap their secret weapons—individuals who find things through some unconventional methods.

* * * *

"Reading" the Ripper: Robert James Lees

WHEN THE PSYCHOTIC murderer known as Jack the Ripper terrorized London in the 1880s, the detectives of Scotland Yard consulted a psychic named Robert James Lees who said he had glimpsed the killer's face in several visions. Lees also claimed he had correctly forecasted at least three of the well-publicized murders of women. The Ripper wrote a sarcastic note to detectives stating that they would still never catch him. Indeed, the killer proved right in this prediction.

Feeling Their Vibes: Florence Sternfels

As a psychometrist—a psychic who gathers impressions by handling material objects—Florence Sternfels was successful enough to charge a dollar for readings in Edgewater, New Jersey, in the early 20th century. Born in 1891, Sternfels believed that her gift was a natural ability rather than a supernatural one, so she never billed police for her help in solving crimes. Some of her best "hits" included preventing a man from blowing up an army base with dynamite, finding two missing boys alive in Philadelphia, and leading police to the body of a murdered young woman. She worked with police as far away as Europe to solve tough cases but lived quietly in New Jersey until her death in 1965.

The Dutch Grocer's Gift: Gerard Croiset

Born in the Netherlands in 1909, Gerard Croiset nurtured a growing psychic ability from age six. In 1935, he joined a Spiritualist group, began to hone his talents, and within two years had set up shop as a psychic. After a touring lecturer discovered his abilities in 1945, Croiset began assisting law enforcement agencies around the world, traveling as far as Japan and Australia. He specialized in finding missing children but also helped authorities locate lost papers and artifacts. At the same time, Croiset ran a popular clinic for psychic healing that treated both humans and animals. His son, Gerard Jr., was also a professional psychic and parapsychologist.

Accidental Psychic: Peter Hurkos

As one of the most famous psychic detectives of the 20th century, Peter Hurkos did his best work by picking up vibes from victims' clothing. Born in the Netherlands in 1911, Hurkos lived an ordinary life as a house painter until a fall required him to undergo brain surgery at age 30. The operation seemed to trigger his latent psychic powers, and he was almost immediately able to mentally retrieve information about people and "read" the history of objects by handling them.

Hurkos assisted in the Boston Strangler investigation in the early 1960s, and in 1969, he was brought in to help solve the grisly murders executed by Charles Manson. He gave police many accurate details including the name Charlie, a description of Manson, and that the murders were ritual slayings.

The TV Screen Mind of Dorothy Allison

New Jersey housewife Dorothy Allison became a clairvoyant crime solver when she dreamed about a missing local boy as if seeing it on television. In her dream, the boy was stuck in some kind of pipe. When she called police, she described the child's clothing, including the fact that he was wearing his shoes on the wrong feet. When she underwent hypnosis to learn more details, she added that the boy's surroundings involved a fenced school and a factory. She was proven correct on all accounts when the boy's body was found about two months after he went missing, floating close to a pipe in a pond near a school and a factory with his little shoes still tied onto the wrong feet.

Allison considered her gift a blessing and never asked for pay. One of her more famous cases was that of missing heiress Patty Hearst in 1974. Although Allison was unable to find her, every prediction she made about the young woman came true, including the fact that she had dyed her hair red.

Like a Bolt Out of the Blue: John Catchings

While at a Texas barbeque on an overcast July 4, 1969, a bolt of lightning hit 22-year-old John Catchings. He survived but said the electric blast opened him to his life's calling as a psychic. He then followed in the footsteps of his mother, Bertie, who earned her living giving "readings."

Catchings often helped police solve puzzling cases but became famous after helping them find a missing, 32-year-old Houston nurse named Gail Lorke. She vanished in late October 1982, after her husband, Steven, claimed she had stayed home from work because she was sick. Because Catchings worked by holding objects that belonged to victims, Lorke's sister, who was suspicious of Steven, went to Catchings with a photo of Gail and her belt. Allegedly, Catchings saw that Lorke had indeed been murdered by her husband and left under a heap of refuse that included parts of an old, wooden fence. He also gave police several other key details. Detectives were able to use the information to get Steven Lorke to confess his crime.

Among many other successes, Catchings also helped police find the body of Mike Dickens in 1980 after telling them the young man would be found buried in a creek bed near a shoe and other rubbish, including old tires and boards. Police discovered the body there just as Catchings had described it.

Fame from Fortunes: Irene Hughes

In 2008, famed investigative psychic Irene Hughes claimed a career tally of more than 2,000 police cases on her website. Born around 1920 in rural Tennessee, Hughes shocked her church at age four when she shouted out that the minister would soon leave them. She was right and kept on making predictions, advised by a Japanese "spirit guide" named Kaygee. After World War II, Hughes moved to Chicago to take a job as a newspaper reporter. She financed her trip by betting on a few horse races using her abilities. She gained fame in 1967 when she correctly prophesied Chicago's terrible blizzard and that the Cardinals would win the World Series. By 1968, she was advising Howard Hughes and correctly predicted his death in 1976.

Hughes's predictions included the death of North Vietnamese premiere Ho Chi Minh in 1969 (although she was off by a week), the circumstances of Ted Kennedy's Chappaquiddick fiasco, and that Jacqueline Kennedy would marry someone with the characteristics of her eventual second husband, Aristotle Onassis. Hughes operated out of a fancy office on Chicago's Michigan Avenue and commanded as much as $500 an hour from her clients. She hosted radio and TV shows, wrote three books, and in the 1980s and '90s, wrote a much-read column of New Year's predictions for the *National Enquirer*.

Pet Psychic

Q: Why do people call for your services?

A: About 90 percent of the cases I work with are "lost pet" cases. The other work I do is with pets that have behavioral problems, like cats who are not using their litter boxes.

Q: You say humans need to create images to communicate with animals. Can you describe that process?

A: What you want to do is send a picture to the animal. The easiest way to do that is to either describe it out loud or write it down. If you're trying to send a cat a message that you're going to be leaving for several days, but there's going to be a pet sitter coming in, you'd want to say something like, "I'll be leaving through the door, which is brown, and I'll be driving in my car, which is blue, but then somebody will come in." So basically you're using as much description as possible.

Q: Are images also how you perceive what's going on with animals?

A: Right, it's mostly pictures. Sometimes it's the sense of fear, or it's raw emotions. Sometimes it's feelings—if an animal is having pain in the chest, I'll get an empathetic chest pain.

Q: Which animals have the most to say, and which ones are the toughest to "crack"?

A: In my work with lost animals, I won't work with ferrets, snakes, turtles, hamsters, or spiders. Those are animals that often go into holes, and they don't pay attention to what's around them. Figuring out what they're seeing is impossible. Other than that, each species has certain things that are more important to them. Horses talk a lot about the emotions that they have or the feelings that they're getting from their human companions. Dogs will communicate about how they see their role in the house or what they can do to get our attention.

Odd Jobs

Our language abounds in curious words denoting the occupations people have now or have had in historical times. In fact, many occupational terms are so strange that it's a real job figuring them all out. Here are some unusual ones.

⁕ ⁕ ⁕ ⁕

⁕ *agonistarch*—a trainer of participants in games of combat

⁕ *amanuensis*—a secretary or stenographer

⁕ *boniface*—an innkeeper

⁕ *dragoman*—an interpreter or guide for travelers, particularly where Arabic, Turkish, or Persian is spoken

⁕ *famulus*—an assistant to a scholar or magician

⁕ *funambulist*—a tightrope walker

⁕ *hippotomist*—an expert dissector of horses

⁕ *histrio*—an actor

⁕ *indagatrix*—a female investigator

⁕ *moirologist*—a hired mourner

⁕ *plongeur*—a dishwasher in a hotel or restaurant

⁕ *pollinctor*—someone who prepares a body for cremation or embalming

⁕ *rhyparographer*—a painter of unpleasant or sordid subjects

⁕ *tragematopolist*—a seller of sweets

⁕ *venireman*—a juror

⁕ *visagiste*—a makeup artist

⁕ *yegg*—a burglar or safecracker

⁂ Chapter 3

Supernatural and Superweird

The Windy City's Famous Phantom: Resurrection Mary

Most big cities have their share of ghost stories. In the case of Chicago, there is one legend that stands out among the rest. It's the story of a beautiful female phantom, a hitchhiking ghost that nearly everyone in the Windy City has heard of. Her name is "Resurrection Mary" and she is Chicago's most famous ghost.

⁂ ⁂ ⁂ ⁂

The Girl by the Side of the Road

THE STORY OF Resurrection Mary begins in the mid-1930s, when drivers began reporting a ghostly young woman on the road near the gates of Resurrection Cemetery, located on Archer Avenue in Chicago's southwestern suburbs. Some drivers claimed that she was looking for a ride, but others reported that she actually attempted to jump onto the running boards of their automobiles as they drove past.

A short time later, the reports took another, more mysterious turn. The unusual incidents moved away from the cemetery and began to center around the Oh Henry Ballroom, located a few miles south of the graveyard on Archer Avenue. Many claimed to see the young woman on the road near the ballroom and sometimes inside the dancehall itself. Young men claimed that they met the girl at a dance, spent the evening with her, then offered her a ride home at closing time. Her vague directions always led them north along Archer Avenue until they reached the gates of Resurrection Cemetery—where the girl would inexplicably vanish from the car!

Some drivers even claimed to accidentally run over the girl outside the cemetery. When they went to her aid, her body was always gone. Others said that their automobiles passed right through the young woman before she disappeared through the cemetery gates.

Police and local newspapers began hearing similar stories from frightened and frazzled drivers who had encountered the mysterious young woman. These first-hand accounts created the legend of "Resurrection Mary," as she came to be known.

Will the Real Resurrection Mary Please Stand Up?

One version of the story says that Resurrection Mary was a young woman who died on Archer Avenue in the early 1930s. On a cold winter's night, Mary spent the evening dancing at the Oh Henry Ballroom, but after an argument with her boyfriend, she decided to walk home. She was killed when a passing car slid on the ice and struck her.

According to the story, Mary was buried in Resurrection Cemetery, and since that time, she has been spotted along Archer Avenue. Many believe that she may be returning to her eternal resting place after one last dance.

This legend has been told countless times over the years and there may actually be some elements of truth to it—although, there may be more than one "Resurrection Mary" haunting Archer Avenue.

One of the prime candidates for Mary's real-life identity was a young Polish girl named Mary Bregovy. Mary loved to dance, especially at the Oh Henry Ballroom, and was killed one night in March 1934 after spending the evening at the ballroom and then downtown at some of the late-night clubs. She was killed along Wacker Drive in Chicago when the car that she was riding in collided with an elevated train support. Her parents buried her in Resurrection Cemetery, and then, a short time later, a cemetery caretaker spotted her ghost walking through the graveyard. Stranger still, passing motorists on Archer Avenue soon began telling stories of her apparition trying to hitch rides as they passed the cemetery's front gates. For this reason, many believe that the ghost stories of Mary Bregovy may have given birth to the legend of Resurrection Mary.

However, she may not be the only one. As encounters with Mary have been passed along over the years, many descriptions of the phantom have varied. Mary Bregovy had bobbed, light-brown hair, but some reports describe Resurrection Mary as having long blonde hair. Who could this ghost be?

It's possible that this may be a young woman named Mary Miskowski, who was killed along Archer Avenue in October 1930. According to sources, she also loved to dance at the Oh Henry Ballroom and at some of the local nightspots. Many people who knew her in life believed that she might be the ghostly hitchhiker reported in the southwestern suburbs.

In the end, we may never know Resurrection Mary's true identity. But there's no denying that sightings of her have been backed up with credible eyewitness accounts. In these real, first-person reports, witnesses give specific places, dates, and times for their encounters with Mary—encounters that remain unexplained to this day. Besides that, Mary is one of the few ghosts to ever leave physical evidence behind!

The Gates of Resurrection Cemetery

On August 10, 1976, around 10:30 P.M., a man driving past Resurrection Cemetery noticed a young girl wearing a white dress standing inside the cemetery gates. She was holding on to the bars of the gate, looking out toward the road. Thinking that she was locked in the cemetery, the man stopped at a nearby police station and alerted an officer to the young woman's predicament. An officer responded to the call, but when he arrived at the cemetery, the girl was gone. He called out with his loudspeaker and looked for her with his spotlight, but nobody was there. However, when he walked up to the gates for a closer inspection, he saw something very unusual. It looked as though someone had pulled two of the green-colored bronze bars with such intensity that handprints were seared into the metal. The bars were blackened and burned at precisely the spot where a small woman's hands would have been.

When word got out about the handprints, people from all over the area came to see them. Cemetery officials denied that anything supernatural had occurred, and they later claimed that the marks were created when a truck accidentally backed into the gates and a workman had tried to heat them up and bend them back. It was a convenient explanation but one that failed to explain the indentations that appeared to be left by small fingers and were plainly visible in the metal.

Cemetery officials were disturbed by this new publicity, so, in an attempt to dispel the crowds of curiosity-seekers, they tried to remove the marks with a blowtorch. However, this made them even more noticeable, so they cut out the bars with plans to straighten or replace them.

But removing the bars only made things worse as people wondered what the cemetery had to hide. Local officials were so embarrassed that the bars were put back into place, straightened, and then left alone so that the burned areas would oxidize and eventually match the other bars. However, the blackened areas of the bars did not oxidize, and the twisted handprints remained obvious until the late 1990s when the bars were finally removed. At great expense, Resurrection Cemetery replaced the entire front gates and the notorious bars were gone for good.

A Broken Spirit Lingers On

Sightings of Resurrection Mary aren't as frequent as in years past, but they do still continue. Even though a good portion of the encounters can be explained by the fact that Mary has become such a part of Chicago lore that nearly everyone has heard of her, some of the sightings seem to be authentic. So whether you believe in her or not, Mary is still seen walking along Archer Avenue, people still claim to pick her up during the cold winter months, and she continues to be the Windy City's most famous ghost.

Ghosts of the *Queen Mary*

Once considered a grand jewel of the ocean, the decks of the Queen Mary *played host to such rich and famous guests as Clark Gable, Charlie Chaplin, Laurel and Hardy, and Elizabeth Taylor. Today, the* Queen Mary *is permanently docked, but she still hosts some mysterious, ghostly passengers!*

* * * *

The *Queen Mary* Goes to War

THE *QUEEN MARY* took her maiden voyage in May 1936, but a change came in 1940 when the British government pressed the ocean liner into military service. She was given a coat of gray paint and was turned into a troop transport vessel. The majestic dining salons became mess halls and the cocktail bars, cabins, and staterooms were filled with bunks. Even the swimming pools were boarded over and crowded with cots for the men. The ship was so useful to the Allies that Hitler offered a $250,000 reward and hero status to the U-Boat commander who could sink her. None of them did.

Although the *Queen Mary* avoided enemy torpedoes during the war, she was unable to avoid tragedy. On October 2, 1942, escorted by the cruiser HMS *Curacoa* and several destroyers, the *Queen Mary* was sailing on the choppy North Atlantic near Ireland. She was carrying about 15,000 American soldiers.

Danger from German vessels was always present, but things were quiet until suddenly, before anyone could act, the *Queen Mary*'s massive bow smashed into the *Curacoa*. There was no way to slow down, no time for warning, and no distress calls to the men onboard. They had only seconds to react before their ship was sliced in two. Within minutes, both sections of the ship plunged below the surface of the icy water, carrying the crew with them. Of the *Curacoa*'s 439-man crew, 338 of them perished on that fateful day. The *Queen Mary* suffered only minor damage and there were no injuries to her crew.

After that, the *Queen Mary* served unscathed for the remainder of the war. Following the surrender of Germany, she was used to carry American troops and GI war brides to the United States and Canada, before returning to England for conversion back to a luxury liner.

Last Days of an Ocean Liner

After the war, the *Queen Mary* and her sister ship, the *Queen Elizabeth*, were the preferred method of transatlantic travel for the rich and famous. But by the 1960s, airplane travel was faster and cheaper, and so, in late 1967, the *Queen Mary* steamed away from England for the last time. Her decks and staterooms were filled with curiosity seekers and wealthy patrons who wanted to be part of the ship's final voyage. She ended her 39-day journey in Long Beach, California, where she was permanently docked as a floating hotel, convention center, museum, and restaurant. She is now listed on the National Register of Historic Places and is open to visitors year-round.

The Haunted *Queen Mary*

The *Queen Mary* has seen much tragedy and death, so it's no surprise that the ship plays host to a number of ghosts. Because of the sheer number of passengers who have walked her decks, accidents were bound to happen. One such mishap occurred on July 10, 1966, when John Pedder, an engine room worker, was crushed to death when an automatic door closed on him.

There have been other reported deaths onboard, as well. For instance, during the war, when the ship was used for troop transport, a brawl broke out in one of the galleys and a cook was allegedly shoved into a hot oven, where he burned to death. There are also reports of a woman drowning in the ship's swimming pool and stories of passengers falling overboard.

Another strange death onboard was that of Senior Second Officer William Stark, whose ghost has often been spotted on deck and in his former quarters. Stark died after drinking lime juice mixed with cleaning solution, which he mistook for gin. He realized his error, and while he joked about it, he called the ship's doctor. Unfortunately, though, Stark soon felt the effects of the poison. As the young officer's condition worsened, he lapsed into a coma and died on September 22, 1949.

Witnesses have also encountered a spectral man in gray overalls who has been seen below deck. He has dark hair and a long beard and is believed to be a mechanic or maintenance worker from the 1930s.

Another friendly spirit, dubbed "Miss Turner," is believed to have been a switchboard operator on the ship. A ghostly woman known as "Mrs. Kilburn" wears a gray uniform with starched white cuffs. She was once in charge of the stewardesses and bellboys, and she's still watching over the comings and goings on the ship. And although it is unknown who the ship's "Lady in White" might be, she haunts the *Queen*'s Salon and is normally seen wearing a white, backless evening gown. Witnesses say she dances alone near the grand piano as if listening to music only she can hear, then vanishes.

Security guards, staff members, and visitors have also reported doors unlocking, opening, and closing on their own, often triggering security alarms. Other unexplained occurrences include phantom voices and footsteps, banging and hammerings sounds, cold spots, inexplicable winds that blow through closed-off areas, and lights that turn on and off.

During a tour of the ship, one guest felt someone tugging on her purse and sweater and stroking her hair. Cold chills crept down her spine when she realized there was no one near her at the time!

In 1967, some 25 years after the tragic accident with the *Curacoa*, a marine engineer working inside the ship heard the terrible sound of two ships colliding. He even heard screams and shredding steel. Did the terrible events of 1942 somehow leave an impression on the atmosphere of this grand old ship? Or worse, is the crew of the *Curacoa* still doomed to relive that fateful October afternoon for eternity?

Echoing the Present

The stories of mysterious encounters and strange events go on and on. It seems almost certain that the events of the past have left an indelible impression on the decks, corridors, and cabins of the *Queen Mary*, creating a haunting that is rivaled by few others in the annals of the supernatural.

Bigfoot: The King of All Monsters

Let's face it—if you had to pick one monster that stands head (and feet) above all others, it would be Bigfoot. Not only is it the stuff of legends, but its likeness has also been used to promote everything from pizza and beef jerky to amusement park rides and monster trucks.

* * * *

Early Sightings

FOLKTALES FROM NATIVE American tribes throughout the Northwest, the area that Bigfoot traditionally calls home, are filled with references to giant, apelike creatures roaming the woods. They described the beast as between seven and ten feet tall and covered in brown or dark hair. (Sasquatch, a common term used for the big-footed beast, is actually an anglicization of a Native American term for a giant supernatural creature.)

Walking on two legs, there was something human about Sasquatch's appearance, although its facial features more closely resembled that of an ape, and it had almost no neck. With looks like that, it's not surprising that Native American folklore often described the creature as cannibalistic, supernatural, and dangerous. Other tales, however, said Sasquatch appeared to be frightened of humans and mostly kept to itself.

It wasn't until the 1900s, when more and more woodlands were being devoured in the name of progress, that sightings of Sasquatch started to increase. It was believed that, though generally docile, the beast did have a mean streak when feeling threatened. In July 1924, Fred Beck and several others were mining in a mountainous area of Washington State. One evening, the group spotted and shot at what appeared to be an apelike creature. After fleeing to their cabin, the group was startled when several more hairy giants began banging on the walls, windows, and doors. For several hours, the creatures pummeled the cabin and threw large rocks at it before disappearing shortly before dawn. After several such encounters in the same general vicinity, the area was renamed Ape Canyon.

My, What Big Feet You Have!

In August 1958, Jerry Crew, a bulldozer operator, showed up for work at a wooded site in Bluff Creek, California. Walking up to his bulldozer, which had been left there overnight, Crew found giant footprints in the dirt. At first, they appeared to be the naked footprints of a man, but with one major difference—these feet were huge! After the tracks appeared on several occasions, Crew took a cast of one of them and brought it to *The Humboldt Times* in Eureka, California. The following day, the newspaper ran a front-page story, complete with photos of the footprint and a name for the creature: Bigfoot. The story and photographs hit the Associated Press, and the name stuck.

Even so, the event is still rife with controversy. Skeptics claim that it was Ray Wallace, not Bigfoot, who made the tracks as a practical joke on his brother Wilbur, who was Crew's supervisor. Apparently the joke backfired when Crew arrived at the site first and saw the prints before Wilbur. However, Ray Wallace never admitted to faking the tracks or having anything to do with perpetrating a hoax.

Video Evidence?

In 1967, in response to numerous Bigfoot sightings in northern California, Roger Patterson rented a 16mm video camera in hopes of filming the elusive creature. Patterson and his friend, Robert Gimlin, spent several days on horseback traveling through the Six Rivers National Forest without coming across as much as a footprint.

Then, on October 20, the pair rounded a bend and noticed something dark and hairy crouched near the water. When the creature stood up on two legs and presented itself in all its hairy, seven-foot glory, that's when Patterson said he knew for sure he was looking at Bigfoot. Unfortunately, Patterson's horse saw the creature, too, and suddenly reared up. Because of this, it took Patterson several precious seconds to get off the horse and remove the video camera from his saddlebag. Once he did that, he ran toward the creature, filming as he went.

As the creature walked away, Patterson continued filming until his tape ran out. He quickly changed his film, and then both men retrieved their frightened horses and attempted to follow Bigfoot further before eventually losing sight of it.

When they arrived back in town, Patterson reviewed the film. Even though it was less than a minute long and extremely shaky in spots, the film appeared to show Bigfoot running away while occasionally looking toward the camera. For most Bigfoot enthusiasts, the Patterson–Gimlin film stands as the Holy Grail of Bigfoot sightings—physical proof captured on video. Skeptics, however, alleged that Patterson and Gimlin faked the entire incident and filmed a man in an expensive monkey suit. Nevertheless, more than 40 years after the event occurred, the

Patterson–Gimlin film is still one of the most talked about pieces of Bigfoot evidence.

Gone Sasquatching

The fact that many people doubt the existence of Bigfoot hasn't stopped thousands of people from heading into the woods to try to find one. Even today, the hairy creature makes brief appearances here and there. Of course, sites like YouTube have given rise to dozens of "authentic" videos of Bigfoot, some of which are quite comical.

Reconstructing the Bermuda Triangle

The mysterious origin of this famed locus of oceanic tragedy suggests that the curse of the Bermuda Triangle may be more maritime concoction than reality.

* * * *

The Triangle's First Victims

It is 2:10 in the afternoon on December 5, 1945. Five Avenger torpedo bombers take flight from the U.S. Naval Air Station in Fort Lauderdale, Florida. The mission is a standard training flight: 13 students and their commander, Lieutenant Charles Taylor, are scheduled to fly a short, triangular path over the sea and then return to base.

But things did not go according to plan on that fateful afternoon. About an hour and a half into the flight, a transmission was received from Lieutenant Taylor. His compasses were not working properly, and he was lost. In those days, pilots didn't have snazzy technology like GPS to keep a constant update of their precise location, so in the absence of a working compass, a pilot had to fly by the seat of his pants. Taylor was an experienced pilot, but he was disoriented. He was accustomed to flying westward from Florida instead of east toward the Bahamas—the direction he was headed when he got lost. From the snatches of radio transmission that were received, it seems Taylor continued to lead his men farther out to sea, thinking that he was headed for land.

Taylor and his men were never heard from again, and their aircraft were never recovered. The most likely scenario is that they ran out of fuel and had to crash-land in the ocean. This tragedy would likely have disappeared into the annals of naval disaster had something very strange not happened next. A patrol plane, meant to search for the missing Avengers, never returned from its search and rescue mission. A merchant ship off of Fort Lauderdale reported seeing a "burst of flames" in the sky soon after the rescue plane took off. The plane was a PBM Mariner; the aircraft were nicknamed "flying gas tanks" for their propensity to explode from a single spark.

Putting the Triangle Together

The tragedy of 1945 received a good deal of media coverage at the time, and some speculated that mysterious, possibly supernatural, forces were responsible for the disappearance of the planes. Rumors of strange magnetic fields, time warps, and even alien abduction began to circulate. It didn't help that Navy reports on the incident, which were requested by Taylor's mother, concluded the incident was due to "causes or reasons unknown."

Stories developed claiming that the Avengers disappeared over a particularly dangerous portion of the ocean. Not surprisingly, other accidents were found to have occurred in nearby stretches of sea. The exact dimensions of the Bermuda Triangle differ greatly depending on the source, but the three points are usually designated as Bermuda, Puerto Rico, and Miami, Florida. The Triangle's size is also in question; reports and studies list it as anywhere from 500,000 to 1.5 million square miles.

The area wasn't christened "the Bermuda Triangle" until the publication of a 1964 fiction story called "The Deadly Bermuda Triangle," written by Vincent Gaddis and published in *Argosy* magazine. A decade later, the Triangle leapt into popular culture with 1974's best-selling book *The Bermuda Triangle*, a sensationalized account of mysterious accidents that had occurred in the area.

More mundane explanations for the Triangle's deadly powers point to the statistical probability of more accidents occurring in an area that sees high traffic. The waters between Florida and the Bahamas are frequented by pleasure boats, which are often crewed by inexperienced tourists. In addition, aircraft and boats are also sometimes victim to short, unexpected storms that dissipate before reaching shore and thus seem mysterious or fantastic. Still, whether or not the Bermuda Triangle is really more dangerous than other patches of sea, the persistent legends of tragedy have certainly prevented many a weary traveler from entering its dreaded perimeter.

The Enigma of the Crystal Skulls

Once upon a time, a legendary set of crystal skulls was scattered across the globe. It was said that finding one of these skulls would bring the lucky person wealth...or death. The story also goes on to say that if all the skulls were located and placed together, they would begin to speak and reveal prophecies, including the end of the world. Could these skulls really exist?

* * * *

The History...Maybe

ADMITTEDLY, THE BACKGROUND of the crystal skulls is a little patchy. According to the legend, either the Aztecs or the Mayas hid 13 crystal human-size skulls around the world (though the number varies story to story). The skulls are said to possess supernatural powers, including the ability to speak as well as to heal, so perhaps they were hidden to prevent them from falling into the wrong hands.

Incredibly, a few skulls exist—you can even see them in respected museums such as the British Museum and the Smithsonian. But there is no documentation to support that any of the skulls were found during excavations, or even how they were found at all. So where did they come from?

Selling Skulls and Seeing Visions

In the late 1800s, Eugene Boban was enjoying a successful career as a world-traveling antiques dealer. Boban is believed to have owned at least three of the crystal skulls, although it is unclear where he acquired them. However, two of these Boban skulls would end up in museums—one in the British Museum and one in Paris' Musee de l'Homme.

But the most intriguing crystal skull is one that Boban did not own. This skull was discovered in 1924 by Anna Le Guillon Mitchell-Hedges, the adopted daughter of famed British adventurer F. A. Mitchell-Hedges. Anna claimed she found the skull in what is now Belize, inside a pyramid. Interestingly, her father wrote several books, but he never once mentions his daughter finding a crystal skull. Professional jealousy or did he regard the skull as a sham? Regardless, Anna claimed that the skull had magical powers and that she once stared into the skull's eye sockets and had a premonition of President John F. Kennedy's assassination.

Putting the Skulls to the Test

Since the legends say that the skulls were handcarved, or a gift from the heavens (or aliens), scientists were eager to determine how they were formed. When the British Museum conducted tests on the two skulls they owned, they found marks that made it clear the skulls were carved using modern rotary tools. Likewise, Paris' Musee de l'Homme also found that their skull was created using modern tools. Both museums also discovered that the specific type of crystal used to form the skulls wasn't the kind available anywhere in the Aztec or Mayan empires.

At first, Anna Mitchell-Hedges was open to having the skull she found tested by the company Hewlett Packard (HP). They found that the skull was indeed crystal—and one solid block of crystal at that, which is incredibly difficult to carve, whether by hand or using modern machinery. Interestingly, Hewlett Packard also found that the quartz crystal is the same kind of crystal used in making computers.

The Legend Continues

Skeptics dismiss the crystal skulls as nothing more than a silly story. And it is entertaining: Even director Steven Spielberg jumped on the bandwagon with his 2008 movie, *Indiana Jones and the Kingdom of the Crystal Skull.* True believers, on the other hand, firmly believe that just because the current skulls may be fakes, it doesn't mean the real skulls aren't still out there waiting to be found. And, say the believers, once all 13 are placed together in a room, the skulls will begin to speak, first to each other and then to anyone else who might be present. But until then, the crystal skulls are keeping their mouths shut.

The Smurl Incident

In the 1970s, the "Amityville Horror" story ignited a firestorm of controversy that's still debated today. The Smurl haunting is another incident that's not as well known but is equally divisive.

* * * *

Spirit Rumblings

IN 1973, JACK and Janet Smurl and their two daughters Dawn and Heather moved into an unremarkable duplex in West Pittston, Pennsylvania. Jack's parents occupied one half of the home and Jack and Janet took the other. Nothing out of the ordinary occurred during the first 18 months that they lived there, but then odd things started to happen: Water pipes leaked repeatedly, even though they had been soldered and resoldered; claw marks were found on the bathtub, sink, and woodwork; an unexplained stain appeared on the carpet; a television burst into flames; and Dawn saw people floating in her bedroom.

In 1977, Jack and Janet welcomed twin daughters Shannon and Carin to the family. By then, the home had become Spook Central: Unplugged radios played, drawers opened and closed with no assistance, toilets flushed on their own, empty porch chairs rocked back and forth, and putrid smells circulated throughout the house.

Unfortunately, by 1985, events at the Smurl home had taken a dangerous turn. The house was always cold, and Jack's parents claimed they often heard angry voices coming from their son's side of the duplex, even though Jack and Janet were not arguing at the time.

In February of that year, Janet was alone in the basement doing laundry when something called her name several times. A few days later, she was alone in the kitchen when the room became frigid; suddenly, a faceless, black, human-shaped form appeared. It then walked through the wall and was witnessed by Jack's mother.

At this point, the situation became even more bizarre. On the night Heather was confirmed into the Catholic faith, Shannon was nearly killed when a large light fixture fell from the ceiling and landed on her. On another night, Janet was violently pulled off the bed as Jack lay next to her, paralyzed and unable to help his wife as a foul odor nearly suffocated him. Periodically, heavy footsteps were heard in the attic, and rapping and scratching sounds came from the walls. Not even the family dog escaped: It was repeatedly picked up and thrown around.

"Who You Gonna Call?"

Unwilling to be terrorized out of their home, in January 1986, the Smurls contacted psychic researchers and demonologists Ed and Lorraine Warren, who confirmed that the home was haunted by four evil spirits, including a powerful demon. The

Warrens theorized that the emotions generated as the older Smurl daughters entered puberty had somehow awoken a dormant demon.

The Warrens tried prayer and playing religious music, but this only angered the demon even more. It spelled out "You filthy bastard. Get out of this house" on a mirror, violently shook drawers, filled the TV set with an eerie white light, and slapped and bit Jack and Janet.

One day, Janet decided to try communicating with the demon on her own. She told it to rap once for "yes" if it was there to harm them; it rapped once. Next, the entity unleashed a new weapon: sexual assault. A red-eyed, green-gummed succubus with an old woman's face and a young woman's body raped Jack. An incubus sexually assaulted Janet, Dawn was nearly raped, and Carin fell seriously ill with a high fever. Pig noises—which supposedly signal a serious demonic infestation—emanated from the walls.

The Smurls could not escape even by leaving their home. The creature followed them on camping trips and even bothered Jack at his job, giving new meaning to the phrase "work is hell." The family appealed to the Catholic Church for help but to no avail. However, a renegade clergyman named Robert F. McKenna did try to help the Smurls by performing an exorcism in the spring of 1986, but even that didn't help.

Going Public

Finally, in August 1986, the family went to the media with their story. The incidents continued, but the publicity drew the attention of Paul Kurtz, chairman of the Buffalo-based Committee for the Scientific Investigation of Claims of the Paranormal (CSICOP). He offered to investigate, but the Smurls turned him down, stating that they wanted to stay with the Warrens and the Church. They were also concerned that CSICOP had already decided that their story was a hoax.

The Smurls did, however, contact a medium, who came to the same conclusion as the Warrens—that there were four spirits in the home: One she couldn't identify, but she said that the others were an old woman named Abigail, a murderer named Patrick, and a very strong demon.

Another exorcism was performed in 1986, and that seemed to do the trick because the incidents stopped. But just before Christmas of that year, the black form appeared again, along with the banging noises, foul odors, and other phenomena.

Surrender

The Smurls finally moved out of the home in 1988; the next owner said that she never experienced any supernatural events while she lived there.

That same year, *The Haunted*, a book based on the Smurl family's experiences, was released. And in 1991, a TV movie with the same title aired.

But the controversy surrounding the alleged haunting was just beginning. In an article written for *The Skeptical Inquirer*, CSICOP's official magazine, Paul Kurtz cited financial gains from the book deal as a reason to doubt that the incidents were authentic. He also said that for years, residents in the area had complained about foul odors coming from a sewer pipe. He cited other natural explanations for some of the incidents and raised questions about Dawn Smurl's accounts of some of the events. He further claimed that the Warrens gave him a number of conflicting reasons for why he couldn't see the video and audio evidence that they said they'd compiled.

And that's where matters stand today, with the true believers in the Smurl family's account on one side and the doubters on the other. Like the Amityville incident, the Smurl haunting is likely to be debated for a long time to come.

The Haunted Destroyer

Shortly after the Japanese attack on Pearl Harbor, five brothers from Waterloo, Iowa—George, Francis, Joseph, Madison, and Albert Sullivan—enlisted in the U.S. Navy and served together aboard the light cruiser USS Juneau. *Sadly, their inspiring story of family patriotism turned tragic when the Juneau was sunk by a Japanese submarine in November 1942, sending all five Sullivan brothers to a watery grave. Their story was immortalized in the movie* The Fighting Sullivans *(1944) and served as an inspiration for Steven Spielberg's* Saving Private Ryan *(1998).*

In 1943, the navy honored the Sullivan brothers by naming a destroyer after them: USS The Sullivans. *It was a proud ship that served valiantly during the remainder of World War II, in the Korean War, and then in various hot spots around the world as part of the 6th Fleet. But after the vessel was decommissioned in 1965, the navy had a difficult time finding people willing to clean and maintain it. The reason? The spirits of the Sullivan brothers were apparently haunting the ship.*

* * * *

Haunted Happenings Begin

THE GHOSTS WERE quiet while the ship was on active duty, but they started making themselves known upon its retirement. Those who worked aboard *The Sullivans* after it was decommissioned reported seeing flying objects and hearing weird sounds and terrifying moans. Fleeting glimpses of young men dressed in World War II-era naval uniforms were also common sights.One of the first acknowledgments that something bizarre was occurring aboard *The Sullivans* came when an electrician's mate refused an order to make a routine check of the ship. It was Friday the 13th he explained, and the last time he had been aboard the ship on that traditionally superstitious day, an unseen hand had reached out from a bulkhead, grabbed him by the ankle, and tripped him.

More Incidents Revealed

After the sailor's story was made public, others came forward with tales of their own frightening encounters aboard the destroyer. Another electrician's mate reported that something had reached out and snatched away the toolbox he had been carrying, and another sailor claimed that five glowing spheres passed him in a darkened hatchway while he stood paralyzed with fear.

In another account, a sailor assigned to work on the vessel said that he felt a chill and a sense of dread the moment he set foot aboard the ship. Within minutes, he was having trouble breathing, and he experienced an odd buzzing in his ears. "I felt like I had stepped into another world, and it wasn't a world where I wanted to be," said the sailor, who, until that day, had never believed in ghosts. "I knew there and then that I was never going back aboard that ship."

Most of the supernatural phenomena reported aboard *The Sullivans* occurred while the destroyer was docked in Philadelphia. For reasons unknown, removing the ship from active service apparently triggered a tremendous amount of activity from the spirits of the five Sullivan brothers. When the ship was relocated, however, ghost sightings and paranormal activity slowed dramatically.

Now a Museum

In 1977, *The Sullivans* was donated to the Buffalo and Erie County Naval & Military Park in Buffalo, New York, where it was turned into a memorial museum that is open for public tours. In 1986, the fabled vessel was declared a National Historic Landmark.

The story of the five Sullivan brothers and their untimely deaths captured the nation's attention and led to immediate policy changes within the U.S. Navy, which worked to ensure that no American family would ever again suffer such a grave loss. The story of the ship's haunting isn't well known outside of the small fraternity of people who worked aboard the vessel and experienced the brothers' spirited antics firsthand. Why the restless spirits of the brothers manifested when they did, did what they did, and then quieted down remains a mystery.

Unsettled Spirits at the Sanatorium

It was designed to save lives at a time when an epidemic was sweeping the nation. Little did its developers know that they were erecting a building in which scores of people would take their last gasping breaths. Is it any wonder that the halls of the Waverly Hills Sanatorium in Louisville, Kentucky, still echo with the footsteps of those who died there?

⁕ ⁕ ⁕ ⁕

Origins

AROUND 1883, THE first building was erected on the site of what is now the Waverly Hills Sanatorium. Major Thomas Hays, the property owner, decided that the local schools were too far away for his daughters to attend, so he built a small schoolhouse on the land and hired teacher Lizzie Harris to instruct the girls. Because of her love of Walter Scott's *Waverley* novels, Harris named the place Waverly School. Taken by the name, Hays decided to call his property Waverly Hill.

In the early 1900s, an outbreak of tuberculosis spread across the United States. In an effort to confine the highly contagious disease, the construction of TB sanatoriums and hospitals was planned. In 1908, the Board of Tuberculosis Hospitals purchased the Hays property, and in July 1910, a small building was opened; it had the capacity to house nearly 50 patients.

They Just Keep Coming...

Without a cure in existence or any way to slow the disease, little could be done for TB patients at the time. Treatment often consisted of nothing more than fresh air and exposure to heat lamps. More and more patients arrived at the sanatorium; therefore, in the 1920s, expansion of the facility began, and in 1926, the building that stands today opened. This massive five-story structure could house nearly 400 patients. But once again, the rooms quickly filled up. The sad truth was that the sanatorium was only kept from overcrowding due to the fact that, without a cure, many of the patients died. Just how many people passed away there is the stuff of urban legends—some estimates go as high as 65,000. In truth, the number is probably closer to 8,000, but that's still a staggering number when one realizes that tuberculosis causes patients to slowly and painfully waste away over the course of weeks or even months.

In the 1940s, treatments for TB were introduced, and as a result, the number of patients at Waverly Hills consistently declined until the building was officially shut down in 1961.

The Final Years

A short time later, Waverly Hills was reopened as the Woodhaven Geriatric Center. This chapter of the building's history came to an end around 1980 amid whispers of patient cruelty and abuse. Before long, those whispers became full-blown urban legends involving depravities such as electroshock therapy. Not surprisingly, it wasn't long before people started saying that the abandoned, foreboding structure was haunted.

Meet the Ghosts

So who are the ghosts that are said to haunt Waverly Hills? Sadly, the identities of most of them are unknown, but many of them have been encountered. Almost every floor of the building has experienced paranormal activity, such as disembodied voices and ghostly footsteps. Doors have been known to open and close by themselves, and bits of debris have been thrown at unsuspecting visitors. It is said that all one has to do is wait quietly to spot one of the many shadow people that walk down the hallways. Of course, if you're looking for a more interactive ghost encounter, you can always head up to the third floor.

There, you might find the spirit of a young girl in the solarium. If she's not there, check the nearby staircases—apparently she likes to run up and down them.

Waverly Hills is also home to the ghost of a young boy who likes to play with a small ball that sometimes appears on the floor. Not wanting to wait to find the ball, some visitors have resorted to bringing their own, which they leave in a certain spot, only to see it roll away or even vanish before it appears on a different floor altogether.

Welcome to Room 502

Of all the allegedly haunted areas at Waverly Hills, none holds a candle to Room 502. Most of the legends associated with the room center on two nurses, both of whom supposedly committed suicide on the premises. One nurse is said to have killed herself in the room in 1928. Apparently, she was a single woman who discovered that she was pregnant. Feeling that she had nowhere to turn, the young woman chose to slip a rope around her neck and hang herself. The other nurse who worked in Room 502 is said to have killed herself in 1932 by jumping from the roof, although the reason why is unclear. Although no documentation substantiating either of these suicides has been unearthed, that has not stopped visitors to Room 502 from experiencing paranormal activity. Upon entering the room, people often report feeling "heavy" or the sensation of being watched. It is quite common for guests to witness shadow figures darting around the room, and occasionally, a lucky visitor catches a glimpse of a spectral nurse standing by the window.

The Body Chute

When expansion of the building began in the 1920s, a rather morbid (though some would say essential) part of the sanatorium was constructed: the Body Chute—a 500-foot-long underground tunnel leading from the main building to a nearby road and set of railroad tracks. Some believe that the tunnel was created simply for convenience, while others think it was designed to prevent patients from seeing the truth—that many of them were dying. Although it was called a chute, bodies were never dumped into it; rather, they were walked through it on gurneys. The tunnel was even equipped with a motorized cable system to help with transportation.

People walking through the Body Chute have reported hearing disembodied voices, whispering, and even painful groans. Sometimes, shadowy figures are seen wandering through the tunnel. But because the only light down there comes from random air vents, the figures vanish almost as quickly as they appear.

Lights, Camera, Ghosts!

After the TV show *Scariest Places on Earth* featured Waverly Hills in a 2001 episode, numerous programs began filming at the sanatorium. *Ghost Hunters* visited there twice—once in 2006 and again in 2007 as part of its annual live Halloween investigation. *Most Haunted* came all the way from the UK in 2008, and *Ghost Adventures* spent a night locked inside the sanatorium in 2010. But of all the episodes filmed at Waverly Hills, none was more bizarre than that of the short-lived VH1 show *Celebrity Paranormal Project*.

The series' debut episode, which aired in October 2006, was shot at Waverly Hills and featured actor Gary Busey, comedian Hal Sparks, *Survivor* winner Jenna Morasca, model/actress Donna D'Errico, and model Toccara Jones conducting an investigation. The supernatural activity began early in the evening, shortly after Busey and Morasca were sent to Room 418 to investigate. They weren't there long before their thermal-imaging camera picked up shapes moving around the room and even sitting on a bed near them. When Morasca was left in the room alone, she heard all sorts of strange noises and even encountered a small red ball, which wasn't there when the team first entered the room.

When Sparks was in the solarium, he rolled balls across the floor to convince the spirits to play with him. Footage shows what appears to be one of the balls rolling back to him. Around the same time, Sparks saw a small black shape—like that of a child—run past the doorway. Later in the evening, D'Errico reported feeling that someone was following her, an incident that was accompanied by the sound of footsteps. She also heard what sounded like people screaming. She was so frightened that she ran away from the building screaming. Once back in the company of the other investigators, D'Errico said that she saw the figure of a man standing in a hallway.

The night ended with the group attempting to contact the spirits in Room 502. As they asked questions, banging noises and footsteps were heard coming from all around them. When they left the building, they were still hearing noises and encountered a child's ball that seemed to appear out of nowhere.

"Come Join Us"

Waverly Hills Sanatorium is open for tours, both during the day and for overnight ghost hunts. Just be assured that no matter how many ghosts inhabit Waverly Hills, they always have room for more.

Are There Different Types of Ghosts?

Yes, there are. The general belief is that there are two main categories of ghosts and that the way a ghost behaves can help determine which type you are dealing with.

⁕ ⁕ ⁕ ⁕

Residual Hauntings

ONE TYPE OF ghost is known as a residual, which gets its name from the term *residue*, or the idea that something was left behind. Simply put, a residual ghost is believed to be nothing more than energy that is left behind when someone dies. A residual spirit is often explained using the metaphor of an old movie projector that, over time, stores up enough energy that it switches on, plays a short scene, and then shuts back off. For that reason, residuals can be identified by the fact that they always perform the same actions over and over again, never varying what they do. They do not interact with the living.

A residual ghost is believed to be the result of an activity that an individual executed frequently while he or she was alive or the result of a violent, unexpected death—which may explain why so many battlefields are supposedly haunted. In both cases, the release of energy leaves an imprint on the area. A violent death typically results in a sudden release of energy, while a repeated activity results in smaller, more sustained releases. In both cases, it is believed that the released energy is stored in a specific location and somehow replays itself from time to time. For example, the routine of a man who for 50 years would walk from the dining room onto the porch and smoke a pipe every night after dinner might end up causing enough residual energy to linger after the man passes that the action continues to repeat itself. Similarly, a phantom scream that is consistently heard at the same time of day or night could be the result of a residual spirit reenacting its violent death and its last moments among the living.

Intelligent Ghosts

Unlike residuals, intelligent ghosts not only interact with the living, they tend to seek them out—hence the term *intelligent*. And whereas a residual haunting repeats the same action in the same place, intelligent ghosts are free to roam wherever they please. So if you sometimes see a particular ghost late at night in the kitchen and at other times in the attic in broad daylight, you are dealing with an intelligent. These are believed to be the spirits of people who, for whatever reason, simply refused to move on after they passed away. In some cases, it's because they want to remain with the people and places they loved while they were alive; other intelligents seem to have unfinished business on this plane. And since intelligents seem to be aware of the living, they are the ghosts that most often make themselves known to us and to mediums. Unlike residuals, intelligents seem to have the ability to communicate from beyond the grave.

The Black Sheep of the Intelligent Family

Two rather intriguing subcategories of intelligents are demonic entities and poltergeists. Exactly who or what demonic entities are varies depending on your religious beliefs, but they are still defined as intelligents because they appear to understand that the living are nearby. In other words, demonic entities have been known to interact with humans, which means that they are intelligents. The same would be true for nonhuman ghosts, such as those of dogs, cats, and other animals. If spirits acknowledge or interact with the living, they are intelligents.

The categorization of poltergeists is a topic that is hotly debated. Some believe that because humans appear to be the targets of poltergeist activity—such as flying plates and glasses—they are intelligents. But others believe that the flying objects are caused by nothing more than violent releases of energy, which would make poltergeists residual. Finally, some even believe that poltergeist activity is the result of irregular brain waves from the living, which would make them nonghostly. So for now, the jury is still out on poltergeists.

Ghostly Guests Stay for Free at the Hotel del Coronado

If you're like most folks, you love to get extras during a hotel stay: Complimentary breakfast, Wi-Fi, and fancy shampoo are all welcome. But how about extra guests? At the Hotel del Coronado in Coronado, California, you get all the usual amenities plus the chance to share your room with resident ghost Kate Morgan.

✻ ✻ ✻ ✻

Any building that dates back to 1888 is certainly rich with history, and the Hotel del Coronado is no exception. For more than a century, travelers—many with stories and secrets of their own—have passed through its elegant doors. When it was built, this grandiose structure—which is perched right next to the Pacific Ocean—was the largest building outside of New York City with electric lighting. Today, Coronado is a rather affluent town, but in the late 1800s, it was filled with crime and debauchery. "The Del," as the hotel is affectionately known, offered its guests a peaceful escape where they could relax and forget their troubles. The building was named a National Historic Landmark in 1977, and it is still in operation today.

Over the years, the Del became a vacation hot spot for celebrities and politicians. Marilyn Monroe stayed there while filming *Some Like It Hot* (1959), and author L. Frank Baum is said to have written much of *The Wonderful Wizard of Oz* in his room at the famous hotel. In fact, it is believed that the Emerald City was inspired by the Del's architecture. But the hotel's most notable guest may be Kate Morgan—a young woman who checked into the Del in November 1892... and never left.

Many stories have been told about Kate Morgan, but most accounts agree that the 24-year-old woman checked into Room 302 under the alias Lottie Bernard. Strikingly beautiful, she appeared to be either ill or upset, and she had no luggage. She said that she was planning to meet her "brother" for the Thanksgiving holiday, but several days later, she was found on the hotel steps... with a bullet in her head. Her death was ruled a suicide.

The Background

The story behind the story is that Kate and her husband, Tom, had been staging a bit of a con game. They traveled the rails setting up card games that Tom invariably won. While Kate pretended to be Tom's sister, she flirted shamelessly with men who tried to impress her with their card-playing skills. She was impressed all right—to the tune of hundreds of dollars.

Kate finally tired of the scheming and the traveling, and like other young women her age, she longed to settle down in a home and start a family. For a brief time, Tom and Kate lived in Los Angeles, but Tom grew restless and headed back out on the rails. Shortly thereafter, when Kate discovered that she was pregnant, she made the mistake of telling this joyous news to her husband while on a train to San Diego. They quarreled, and he went on to another city, while she continued to her final destination: the Hotel del Coronado.

Evidence suggests that Kate may have tried to abort her baby by drinking large amounts of quinine. When that didn't work, she traveled across the bay to San Diego, where she purchased a gun and some bullets. Those who saw her reported that she seemed pale and sickly; they weren't surprised when she was found dead.

The Spirited Kate

Guests and employees alike have felt Kate's presence in several places around the Del, including her guest room, the beach, and some of the hotel shops. One boutique, known as Established in 1888, has been the site of some particularly unusual activity. A display of Marilyn Monroe memorabilia was often targeted—items literally flew off the shelves. Staff members came to the conclusion that Kate was jealous of the famous starlet. When the Marilyn souvenirs were moved to a corner and replaced with mugs, both areas settled down and no more unusual activity was reported.

An apparition dressed in a long black dress has been seen around the shop and in the hallways. And a maintenance man at the hotel reports that there is one light on the property that will never stay lit: It's the one over the steps where Kate's lifeless body was found.

The most notable haunting, however, is in the room where Kate stayed in 1892. The room number has since changed from 302 to 3312 to 3327. Possibly confused as to which room is hers, Kate also seems to make frequent visits to Room 3502, which is thought to be haunted as well. Odd, unexplained events have occurred in both of these rooms. Guests have reported toilets flushing themselves, lights flickering, curtains blowing when the windows are closed, and a lingering floral scent. Ashtrays have been seen flying through the air, temperatures dip and surge, and televisions blare one minute and are silent the next. Several visitors have also reported seeing a ghostly figure standing by the window of Kate's room, and a strange glow has been observed inside that window from the outside. The screen on that same window has fallen off mysteriously more than once, and hotel guests have reported hearing soft murmurs inside the room. Is it the ocean . . . or the sound of a young woman reliving her distress over and over again?

North Carolina's Train of Terror

North Carolina is rife with haunted houses. In fact, even the Governor's Mansion in Raleigh is said to contain a ghost or two. But one of the Tarheel State's most unusual paranormal events isn't housebound—it takes place on an isolated train trestle known as the Bostian Bridge near the town of Statesville.

* * * *

On August 27, 1891, a passenger train jumped the tracks while crossing the Bostian Bridge, plunging seven railcars 60 to 75 feet to the ground below. Nearly 30 people perished in the tragic accident.

According to local legend, on the anniversary of the catastrophe, the sounds of screeching wheels, screaming passengers, and a thunderous crash can be heard near the Bostian Bridge. The ghostly specter of a uniformed man carrying a gold pocket watch has also been observed lingering nearby.

Another Victim Claimed

Sadly, on August 27, 2010, Christopher Kaiser, a Charlotte-based amateur ghost hunter, was struck and killed by a real-life train that surprised him on the Bostian Bridge.

According to police reports, Kaiser had brought a small group to the trestle in hopes of experiencing the eerie sounds that are said to occur on the anniversary of the 1891 crash. The group was standing on the span when a Norfolk-Southern train turned a corner and headed toward them. With the train rapidly approaching, Kaiser managed to push the woman in front of him off the tracks. His heroic action saved her life but cost him his own.

Other than witnessing this horrific accident, Kaiser's group saw nothing unusual that night. But many others claim to have seen strange phenomena on the Bostian Bridge. On the 50th anniversary of the 1891 tragedy, for example, one woman reportedly watched the wreck occur all over again. More than 150 people gathered near the trestle on the 100th anniversary of the crash in 1991, but nothing supernatural happened that night.

Conjuring Up Spirits

Can someone believe in a ghost so much that one actually appears? A 1972 study in Toronto, Canada, suggests just that.

* * * *

EIGHT PEOPLE INVENTED a ghost named Philip and met weekly in Toronto to make contact with him via séance. None of the group members claimed to have had any previous psychic experience. They simply wanted to see what would happen if they created a ghost and then summoned him.

The group made up a whole story about Philip: He was a wealthy British aristocrat living in the mid-1600s, who supported the king of England during the English Civil War. But when his beloved mistress, Margo, was convicted of witchcraft and burned at the stake, Philip committed suicide. It was a romantic story that included real locations and events, but Philip, his wife, and his mistress were entirely fictional.

When it came time for the group to put its theory to the test, members took turns acting as a medium to contact Philip. But after a full year, nothing had happened.

The Imaginary Ghost Speaks

After many failed attempts at otherworldly contact, the group reviewed a 1960 study conducted by several paranormal researchers, who supposedly made a table weighing nearly 40 pounds rise from the floor—without touching it.

Inspired by the 1960 group, the Toronto group modified its approach. Instead of sitting quietly around a table in dim lighting, they talked normally and used regular light levels, and within three or four meetings, Philip began to manifest. First, he vibrated the table gently, then he rapped on the table loud enough for everyone to hear it.

Using table raps—one for yes and two for no—the group learned to communicate with Philip. They learned more about Philip by asking him questions, and soon the eerie phenomena increased. The table began to slide around the floor during part of each session, and sometimes the table creaked loudly. At other times, lights flickered in response to questions.

The Dancing Table

In later meetings, the Toronto group reportedly used a new, sturdier table that was made entirely of wood, and it behaved even more strangely. At first, the table made rapping noises and slid gently around the room. Then, it rose and stood on one leg. It danced across the room, and at other times, it chased group members. When a new guest arrived, the table would rush over as if to greet the person.

This continued for months. In November 1973, both the ghostly Philip and the table continued to produce phenomena, even while being filmed by the Canadian Broadcasting Company for a talk show. This was followed by a January 1974 documentary, *Philip, The Imaginary Ghost*. On a TV show soon afterward, the table rocked back and forth and climbed three stairs, one at a time. The studio audience gasped in amazement as cameras filmed the strange event.

Beyond Philip

Following the success of the Philip experiment, the Toronto group started a second study with a newly invented ghost named Lilith. As with the first experiment, the group made up a story about Lilith, a fictional woman who, as a member of the French Resistance during World War II, had been captured and shot as a spy.

Lilith also manifested as Philip had. Like Philip, when she told her story during séances, every detail matched the group's made-up tale. The group concluded that "inventing" a ghost is a learned skill, and any group can repeat this experiment.

Later studies, including the Skippy Experiment in Sydney, Australia, and the research of Dr. Michael A. Persinger of Canada's Laurentian University, have demonstrated that ghosts—or the appearance of ghostly phenomena—can be created simply by believing. So, if you believe in ghosts—or want to—be careful what you wish for. You just might have a terrifying encounter with "the other side."

Ghosts of the *Titanic* Exhibit—Just the Tip of the Haunted Iceberg

On April 14, 1912, the supposedly unsinkable RMS Titanic *struck an iceberg in the North Atlantic, and by the end of the next day, approximately 1,500 souls lay in a cold, watery grave. Nestled deep in frigid waters, their resting place remained undisturbed until the wreck was discovered on the ocean floor in 1985. Two years later, salvage divers began to recover artifacts from the wreck. Since then, about 5,900 items have been taken from the site. Many of these objects are included in* Titanic: *The Artifact Exhibition, which has traveled the world since the early 1990s—and a few ghosts have gone along for the ride. A permanent* Titanic *museum is also haunted by the past.*

* * * *

Haunting Atlanta

WHILE TAKING IN the traveling exhibit at the Georgia Aquarium in Atlanta, a visitor initially attributed a sense of being watched and feelings of overwhelming sadness to the items on display. One volunteer who worked at the exhibit said she sensed that lost souls were embedded in the artifacts; she also felt a hand moving over her head and touching her hair. And a four-year-old boy repeatedly asked his mother and grandmother about a lady that he saw in a display case, which—to the adults—held only a dress and a love seat. Another visitor saw a man in a black-and-white suit who seemed out of place amongst the other, more casually dressed people; later, when she felt as if she was being watched, the visitor turned around to see the man in the suit staring at her.

In 2008, when a paranormal investigation team visited the exhibit, its members saw shadowy figures and picked up the voices of spirits on audio recordings. The team concluded that at least three ghosts are attached to the exhibit—those of an older gentleman, an elderly woman, and a young crew member.

Jason Hawes and Grant Wilson and their team from The Atlantic Paranormal Society (TAPS) also conducted an investigation of the exhibit; their results aired on a 2009 episode of *Ghost Hunters*. Their findings were similar to those compiled by other paranormal research groups. In the Iceberg Exhibit (which contains a replica of an iceberg to give visitors a sense of what one feels like), Hawes and Wilson detected a moving cold spot that seemed to be about four to five feet tall and one to two feet wide. Hawes also felt something tug on his shirt. While in the Artifacts Exhibit, both investigators saw a shadowy figure walk into another room, and Hawes felt an unseen hand touch his shoulder. Before they left the room, Hawes asked the spirit to knock on the wall if it wanted them to leave. They heard nothing, but their audio recorder captured an EVP (electronic voice phenomenon) that sounded like a man whispering, "No, please wait," or perhaps even, "Don't leave me." In the end, TAPS concluded that both intelligent and residual hauntings were present at the aquarium, which are likely associated with the exhibit.

Rocking New York City

When the exhibit traveled to New York City, motion-activated security cameras clicked on every night at 3 A.M., even though no living being was present. And one visitor said that as she and her cousin walked through a small hallway, they experienced a rocking sensation, as if they were actually on a ship. They asked an employee if this was some sort of special effect; it was not, but the employee reported that many other people had asked the same question.

Chilling St. Paul

When the artifacts visited St. Paul, Minnesota, one visitor left the Iceberg Exhibit because she suddenly felt dizzy. Then, she felt a hand touching her shoulder, even though no one else was around. Her shoulder felt cold and then hot, as if she had frostbite. A few minutes later, a red mark appeared on her shoulder. Another person reported similar sensations when leaving the

Iceberg Exhibit, including dizziness, shoulder taps, and the queasy sensation of being rocked.

Ghosts in Branson

Even a permanent *Titanic* museum seems to be haunted. The *Titanic* museum attraction in Branson, Missouri (and its sister site in Pigeon Forge, Tennessee), is said to be teeming with ghosts. This re-creation of the legendary ship displays approximately 400 artifacts from the actual *Titanic*—and is thought to be home to at least one ghost-child. Members of the cleaning staff have found child-size fingerprints on the glass separating the Promenade Deck and the Bridge; given that many children visit the museum, this is not unusual. However, the prints reappear after the glass is cleaned when the museum is closed. Even stranger, one museum employee photographed a wet footprint in the shape of a child's bare foot.

Ghosts have also been spotted in other parts of the ship: A man in formal wear has been seen at the top of the Grande Staircase; a gowned figure glides around the First Class Dining Salon and emerges from the area carrying her belongings; and although the museum has a no smoking policy, both staffers and visitors have smelled cigar smoke around the Grande Staircase on numerous occasions.

To determine once and for all if ghosts from the *Titanic* haunt the museum, the owners invited two different teams of paranormal investigators to conduct research overnight. Both teams found high psychic-energy levels that became higher when staff members asked the spirits questions. As museum staffers communicated with two passengers who died on the ship—Robert Douglas Spedden and Mr. Asplund—one staff member became weak and nauseous.

Final Attachments

When the exhibit visited Athens, Greece, employees heard English-speaking voices in the galleries after hours. And in Monterrey, Mexico, several people commented on a man who they described as a "character actor" dressed in a black suit that would have been fashionable in 1912. However, the exhibit did not employ any reenactors.

It makes sense that *Titanic's* victims would follow these artifacts—items that they knew in life and rested with in death for many decades. The artifacts will continue to travel the world as long as there is interest in them, but hopefully, the spirits will eventually separate themselves from their earthly belongings and finally rest in peace.

The Sad Fate of British Airship *R101*

On October 5, 1930, the British airship R101 *crashed during its maiden flight, killing nearly all aboard. Two days later, a woman with absolutely no knowledge of airships explained the incident in highly technical and freakishly accurate detail. Were the ghosts of the tragedy speaking through her?*

* * * *

Foretellers or Frauds

PSYCHICS AFFECT PEOPLE in different ways. Those who believe in concepts such as mental telepathy and extrasensory perception can find validity in a medium's claims. Skeptics, on the other hand, aren't so sure. Psychic researcher Harry Price straddled the fence between the two camps: He deplored fakery but had witnessed enough of the supernatural to believe that there was indeed something to it. At his National Laboratory of Psychical Research, Price worked diligently to separate the wheat from the chaff—the real from the fake.

On October 7, 1930, Price arranged a séance with a promising medium named Eileen J. Garrett. Price's secretary Ethel Beenbarn and reporter Ian D. Coster were enlisted to record the proceedings. Price hoped to contact the recently deceased author Sir Arthur Conan Doyle (of *Sherlock Holmes* fame) and publish an account of the proceedings. Like Price, Doyle held a keen interest in the paranormal. Making contact with him could bring Price the evidence that he sought about the existence of an afterlife.

Strange Contact

Just two days before Price and Garrett met, a horrific tragedy occurred. The British airship *R101* crashed in France, killing 48 of the 55 people on board. Questions about the event arose as quickly as each newspaper went to press. Why had the ship crashed? Who or what was at fault? Were airships inherently unsafe? A Court of Inquiry was assembled to answer these queries, but not before Price and Garrett had their meeting.

At the séance, Garrett fell into a trance and began to speak. In a deep, animated voice, she identified herself as Flight Lieutenant H. Carmichael Irwin, commander of the *R101* (not Sir Arthur Conan Doyle, whom Price had hoped to contact), and began to speak words that were as confusing as they were disjointed:

"I must do something about it. The whole bulk of the dirigible was entirely and absolutely too much for her engine's capacity. Engines too heavy... Oil pipe plugged. Flying too low altitude and never could rise... Severe tension on the fabric, which is chafing... Never reached cruising altitude—same in trial... Almost scraped the roofs of Achy!"

Coster recognized Irwin's name from the recent *R101* tragedy. After the séance, the reporter published highlights from the meeting. Shortly thereafter, a man named Will Charlton contacted Price. Charlton worked as a supply officer at the base where the *R101* was built and was familiar with the airship's construction. He asked the researcher for a transcript of the séance and studied it intently. What he saw astounded him: Garrett—who had no previous knowledge of or interest in the workings of airships—had spoken about one in highly technical terms. Moreover, she seemed to be explaining *why* the *R101* had crashed.

Passing Muster

As details of the crash emerged, Garrett's words proved even more insightful. It was revealed that the airship had passed over the village of Achy so low that it almost scraped a church tower, as Garrett had stated during the séance. Garrett also spoke of an "exorbitant scheme of carbon and hydrogen" as being "completely wrong" for the airship. When Charlton and other airship officials heard this, they were stunned. Only a handful of project team members had been privy to this top-secret information. Parlor tricks, no matter how clever, couldn't possibly account for Garrett's knowledge of this information.

The transcript yielded more than 40 highly technical details related to the airship's final flight. Charlton and his colleagues pronounced it an "amazing document." Before launching an official inquiry, they decided to stage another séance. Major Oliver Villiers of the Ministry of Civil Aviation sat down with Eileen Garrett and observed her as she drifted into a trance. This time, the medium channeled the spirits of others who had perished in the crash. Villiers asked pointed questions regarding the airframe of the *R101*, and the medium responded in startling detail:

Villiers: "What was the trouble? Irwin mentioned the nose."

Garrett: "Yes. Girder trouble and engine."

Villiers: "I must get this right. Can you describe exactly where? We have the long struts labeled from A to G."

Garrett: "The top one is O and then A, B, C, and so on downward. Look at your drawing. It was starboard of 5C. On our second flight, after we had finished, we found the girder had been strained, not cracked, and this caused trouble to the cover..."

Conclusion

When the Court of Inquiry's report was released, Garrett's words matched almost precisely with the findings. The phenomenon so impressed Charlton that he himself became a

Spiritualist. After Garrett's death in 1970, Archie Jarman—a psychic researcher and columnist for the *Psychic News*—revealed that the medium had asked him to dig deeper into the *R101* case: She wished to learn just how close her description of the event was to reality. After six months of dogged research, Jarman concluded that the technical terms expressed so vividly by the medium could only have come from the Other Side. In the end, the goal of contacting Sir Arthur Conan Doyle was not achieved; however, this fantastic development had advanced psychical studies immeasurably. Without question, Price had found his "wheat" and the answers to his questions.

A Voice from Beyond the Grave

Following the murder of Teresita Basa in the late 1970s, another woman began to speak in Basa's voice—recounting things that only Teresita could have known—to help solve the mystery of her murder.

* * * *

IN FEBRUARY 1977, firemen broke into a burning apartment on North Pine Grove Avenue in Chicago. Beneath a pile of burning clothes, they found the naked body of 47-year-old Teresita Basa, a hospital worker who was said to be a member of the Filipino aristocracy. There were bruises on her neck and a kitchen knife embedded in her chest. Her body was in a position that caused the police to suspect that she had been raped.

However, an autopsy revealed that she hadn't been raped; in fact, she was a virgin. Police were left without a single lead: They had no suspects and no apparent motive for the brutal murder. The solution would come from the strangest of all possible sources—a voice from beyond the grave.

"I Am Teresita Basa"

In the nearby suburb of Evanston, shortly after Teresita's death, Remibios Chua started going into trances during which she spoke in Tagalog in a slow, clear voice that said, "I am Teresita Basa." Although Remibios had worked at the same hospital as Teresita, they worked different shifts, and the only time they are known to have even crossed paths was during a new-employee orientation. Remibios's husband, Dr. Jose Chua, had never heard of Basa.

While speaking in Teresita's voice, Remibios's accent changed, and when she awoke from the trances, she remembered very little, if anything, about what she had said. However, while speaking in the mysterious voice, she claimed that Teresita's killer was Allan Showery, an employee at the hospital where both women had worked. She also stated that he had killed her while stealing jewelry for rent money.

Through Remibios's lips, the voice pleaded for them to contact the police. The frightened couple initially resisted, fearing that the authorities would think that *they* should be locked away. But when the voice returned and continued pleading for an investigation, the Chuas finally contacted the Evanston police, who put them in touch with Joe Stachula, a criminal investigator for the Chicago Police Department.

Lacking any other clues, Stachula interviewed the Chuas. During their conversation, Remibios not only named the killer, but she also told Stachula exactly where to find the jewelry that Showery had allegedly stolen from Teresita. Prior to that, the police were not even aware that anything had been taken from the apartment.

Remarkably, when police began investigating Showery, they found his girlfriend in possession of Teresita's jewelry. Although the authorities declined to list the voice from beyond the grave as evidence, Showery was arrested, and he initially confessed to the crime. When his lawyers learned that information leading to his arrest had come from supernatural sources, they advised him to recant his confession.

The Surprise Confession

Not surprisingly, the voice became a focal point of the case when it went to trial in January 1979. The defense called the Chuas to the witness stand in an effort to prove that the entire case against Showery was based on remarks made by a woman who claimed to be possessed—hardly the sort of evidence that would hold up in court.

But the prosecution argued that no matter the origin of the voice, it had turned out to be correct. In his closing remarks, prosecuting attorney Thomas Organ said, "Did Teresita Basa come back from the dead and name Showery? I don't know. I'm a skeptic, but it doesn't matter as to guilt or innocence. What does matter is that the information furnished to police checked out. The jewelry was found where the voice said it would be found, and Showery confessed."

Detective Stachula was later asked if he believed the Chuas: "I would not call anyone a liar," he carefully stated. "... Dr. and Mrs. Chua are educated, intelligent people ... I listened and acted on what they told me ... [and] the case was wrapped up within three hours."

Showery told the jury that he was "just kidding" when he confessed to the crime; he also claimed that the police had coerced him into an admission of guilt. Nevertheless, after 13 hours of deliberation, the jury reported that they were hopelessly deadlocked and a mistrial was declared.

A few weeks later, in a shocking development, Allan Showery changed his plea to "guilty" and was eventually sentenced to 14 years in prison. Some say that Teresita's ghost had visited him and frightened him into confessing.

Obviously shaken by the experience, the Chuas avoided the press as much as possible. In 1980, in her only interview with the press, Remibios noted that during the trial, people were afraid to ride in cars with her, but she said that she was never afraid because the voice said that God would protect her family. Still, she hoped that she would never have to go through such an experience again. "I've done my job," she said. "I don't think I will ever want to go through this same ordeal."

Having attracted national attention, the case quickly became the subject of a best-selling book and countless magazine articles, a TV movie, and a 1990 episode of *Unsolved Mysteries*. The case is often cited as "proof" of psychic phenomena, possession, and ghosts, but it's simply another mystery of the paranormal world. Exactly what it proves is impossible to say; after all, the ghost of Teresita Basa is no longer talking.

Riddles of the Riddle House

While functioning as a cemetery caretaker's home, West Palm Beach's Riddle House was always close to death. Since then, it's been relocated and repurposed, and now it sees its fair share of life—life after death, that is.

* * * *

The "Painted Lady"

BUILT IN 1905 AS a gatekeeper's cottage, this pretty "Painted Lady" seemed incongruent with the cemetery it was constructed to oversee. Cloaked in grand Victorian finery, the house radiated the brightness of life. Perhaps that's what was intended: A cemetery caretaker's duties can be gloomy, so any bit of spirit lifting would likely be welcomed. Or so its builders thought. In the case of this particular house, however, "spirit lifting" took on a whole new meaning.

The first ghost sighted in the area was that of a former cemetery worker named Buck, who was killed during an argument with a townsperson. Shortly thereafter, Buck's ghost was seen doing chores around the cemetery and inside the cottage. Luckily, he seemed more interested in performing his duties than exacting revenge.

In the 1920s, the house received its current name when city manager Karl Riddle purchased it and took on the responsibility of overseeing the cemetery. During his tenure, a despondent employee named Joseph hung himself in the attic. This was the beginning of a cascade of paranormal phenomena inside the house, including the unexplained sounds of rattling chains and disembodied voices.

After Riddle moved out, the reports of paranormal activity slowed down—but such dormancy wouldn't last.

Traveling Spirits

By 1980, the Riddle House had fallen into disrepair and was abandoned. The city planned to demolish the building but instead decided to give it to John Riddle (Karl's nephew). He, in turn, donated it for preservation. The entire structure was moved—lock, stock, and barrel—to Yesteryear Village, a museum devoted to Florida's early years. There, it was placed on permanent display as an attractive token of days long past. There, too, its dark side would return—with a vengeance.

When workers began to reassemble the Riddle House, freshly awakened spirits kicked their antics into high gear. Ladders were tipped over, windows were smashed, and tools were thrown to the ground from the building's third floor. Workers were shocked when an unseen force threw a wooden board across a room, striking a carpenter in the head. The attacks were blamed on the spirit of Joseph, and the situation became so dangerous that work on the structure was halted for six months. After that, however, the Riddle House was restored to its previous glory.

Ghostly Unveiling

During the dedication of the Riddle House in the early 1980s, two unexpected guests showed up for the ceremony. Resplendent in Victorian garb, the couple added authenticity to the time period being celebrated. Many assumed that they were actors who were hired for the occasion; they were not. In fact, no one knew *who* they were. A few weeks later, century-old photos from the Riddle House were put on display. There, in sepia tones, stood the very same couple that guests had encountered during the dedication!

When the *Ghost Adventures* team spent a night locked inside the Riddle House in 2008, a medium warned the investigators that the spirit of Joseph is an evil entity that did not want them there. But that didn't stop investigator Zak Bagans from provoking the spirit. Bagans left a board at the top of the stairs and asked the entity to move it if it didn't want them there. Later, after the team heard footsteps in the room above them, the board fell down several stairs on its own. Throughout the course of the night, the team experienced unexplained noises and objects moving and falling by themselves. In the end, the researchers concluded that the Riddle House is definitely haunted and that whatever resides in the attic does not like men in particular, just as the medium had cautioned.

Ethereal stirrings at the Riddle House continue to this day. Unexplained sightings of a torso hanging in the attic window represent only part of the horror. And if history is any indicator, more supernatural sightings and activity are to come.

McRaven House: The Most Haunted House in Dixie?

Located near Vicksburg, Mississippi, the McRaven House was haunted even before it became a Civil War hospital.

* * * *

THE OLDEST PARTS of the estate known as McRaven House were built in 1797. Over the next 40 years, its owners gradually added to the property until it became a classic southern mansion, standing proudly among the magnolia blossoms and dogwood trees of the Old South.

And like nearly all such mansions, it has its share of resident ghosts. Today, McRaven House is often referred to as "the most haunted house in Mississippi." Some researchers believe that environmental conditions on the property make it particularly susceptible to hauntings: Ghosts that may not be noticeable in drier, less humid climates seem to be more perceptible in the dews of the delta. Of course, it helps that the McRaven House has seen more than its share of tragedy and death during its 200-year history.

The Ghost of Poor Mary

In the early 1860s, the house's supernatural activity seemed to center on an upstairs bedroom in which Mary Elizabeth Howard had died during childbirth in 1836 at age 15. Mary's brown-haired apparition is still seen descending the mansion's grand staircase. Her ghost is blamed for the poltergeist activity—such as pictures falling from the wall—that is often reported in the bedroom where she died. And her wedding shawl, which is occasionally put on display for tourists, is said to emit heat.

Ghosts of the Civil War

Mary Elizabeth's ghost alone would qualify McRaven House as a notably haunted reminder of Mississippi's antebellum past, but she is far from the only spirit residing there, thanks in part to the bloody atrocities of the Civil War.

The Siege of Vicksburg, which took place in 1863, was one of the longest, bloodiest battles of the entire conflict. When General Ulysses S. Grant and his Union forces crossed the Tennessee River into Mississippi, Confederate forces retreated into Vicksburg, which was so well guarded that it was known as a "fortress city." But as more and more Union forces gathered in the forests and swamps around Vicksburg, Confederate General John C. Pemberton was advised to evacuate. Fearing the wrath of the local population if he abandoned them, Pemberton refused.

By the time the siege began in earnest, the Confederate troops were significantly outnumbered. Weary rebel forces surrendered the city of Vicksburg on July 4, 1863, after more than a month of hard fighting. Nearly all of the Confederate soldiers involved in the battle—around 33,000 in all—were captured, wounded, or killed. The Union victory put the entire Mississippi River in northern hands, and combined with the victory at Gettysburg that same week, it signalled the beginning of the end for the Confederacy.

Captain McPherson's Last Report

In the middle of the action stood McRaven House. In the early days of the siege, it served as a Confederate hospital, and, at that time, it was full of the screams of anguished and dying men. Cannons from both armies shot at the mansion, destroying large portions of it.

Later, after Union forces captured the house, it served as the headquarters for General Grant and the Union army. One of the officers put in charge of the house was Captain McPherson, a Vicksburg native who had fled to the North to fight for the Union. At some point during the siege, he disappeared. Soon after, according to legend, McPherson's commanding officer awoke to find the captain in his room. He was furious at the intrusion until he noticed McPherson's mangled, bloody face and torn uniform. The commanding officer then realized that this was not actually McPherson himself—it was his ghost, which had returned to deliver the message that Rebels, who couldn't forgive him for abandoning the South, had murdered him. McPherson's ghost reputedly still wanders the grounds dressed in Union blue with blood oozing from a bullet wound in his forehead.

Other Civil War Ghosts

Nearly a year after the siege ended, John Bobb—the owner of McRaven House at the time—spotted six Union soldiers picking flowers in his garden. Outraged by the trespassers, Bobb threw a brick at them and hit one of the Yankees in the head. After going to the local field commander to report the intruders, Bobb returned home to find 25 Union soldiers waiting for him; they marched him into the nearby bayou and shot him to death. His ghost has been seen roaming the property ever since.

The War Ended, But the Ghosts Kept Coming

Mary Elizabeth and the Civil War-era ghosts aren't the only spirits that haunt McRaven House. In 1882, William Murray purchased the home, and over the next 78 years, five members of his family died on the premises. The most recent death there was that of his daughter Ella, who spent her last years as a recluse in the house, where she reportedly burned furniture to stay warm. After her death in 1960, the mansion was restored, refurbished, and opened for tours and battle reenactments. In the early morning hours, tour groups and staffers have often spotted the ghosts of Ella and the other Murrays who died in the house.

The Most Haunted House in the South?

Today, visitors can tour the McRaven House in all of its antebellum glory. Extensive collections of 19th-century furnishings, artwork, jewelry, and other artifacts are displayed at the mansion, and several ghosts from both sides of the Civil War are believed to share the house with Mary Elizabeth and the other spirits from the mansion's past. Ghost hunters have been conducting investigations at the house since at least the 1980s, and they've frequently photographed mysterious forms outside the building, often around the portion of the property that served as a burial ground for soldiers; some are simply odd blobs of light, but others appear to be human-shaped forms.

Few argue with the claim that McRaven House is "the most haunted house in Mississippi." In fact, some even call it "the most haunted house in Dixie."

The Worried Husband

"You never pay any attention to me!" is a common lament heard in marriages when a person feels neglected by his or her partner. But what about the opposite situation—when a person's concern for his or her spouse extends beyond the grave?

* * * *

IN THE LATE 1940s, Elaine and her husband lived in an apartment in Oskaloosa, Iowa. They shared the floor with a single woman named Patricia, whose husband had died in an industrial accident. The devastated young woman had moved there to try to regroup.

One evening while her husband was working, Elaine decided to take a bath. Just as she was about to turn on the bathroom light, she smelled pipe smoke and then saw a young man with black hair and a horseshoe-shaped scar on his cheek; he was holding a pipe.

After a moment, Elaine realized that the man was not really looking *at* her, he was sort of looking *through* her. She then deduced that the man was a ghost. Elaine watched as he began to move through her apartment. She followed him as he glided down the hall toward Patricia's apartment. When he got to Patricia's door, he vanished.

Uncertain of what she was doing, Elaine turned the doorknob to her neighbor's apartment; it was unlocked, so Elaine went inside. There, she found Patricia lying on her bed, barely alive: She had slashed her wrists, and her lifeblood was quickly draining away. Elaine bandaged Patricia as best she could and called her husband. He raced home with a doctor, who treated Patricia's injuries.

The next day, Patricia thanked Elaine for saving her life. She said that she had been deeply saddened by her husband's death and had turned to the bottle as a result. Overcome by grief, the idea of joining her husband seemed appealing to her, so she had slit her wrists. If it had not been for Elaine, her plan would have succeeded.

Elaine said nothing about why she had entered Patricia's apartment in the first place. But when Patricia showed Elaine a picture of her late husband, everything suddenly made sense: The man in the photo was the same man who Elaine had seen in her apartment.

Bobby Mackey's: Ghosts That Like Country Music

Just over the Ohio River from downtown Cincinnati is the town of Wilder, Kentucky, home of Bobby Mackey's—a country-music nightclub and allegedly one of the most haunted locations in the United States. Over the years, the property is said to have seen such atrocities as a beheading, a poisoning, a suicide, numerous unsolved murders, and even a case of possession. On top of all that, some say there's an entrance to hell in the basement.

✻ ✻ ✻ ✻

Hell's Gate

THE FIRST BUILDING that is believed to have stood on the property now occupied by Bobby Mackey's was a slaughterhouse, which operated from the 1850s until the late 1880s. During that time, it was said to have been so busy that the ground floor was often literally coated with blood. To alleviate that, a well was dug in the basement, which allowed the blood to be washed off the floor and carried out to the nearby river. Needless to say, gallons upon gallons of blood and other assorted matter were dumped into that well. Perhaps that's why legend has it that after the slaughterhouse closed, a satanic cult used the well as part of its rituals. Some even claim that these rituals opened a portal to the Other Side, a portal that—to this day—has yet to be closed.

An Unspeakable Crime

On February 1, 1896, the headless body of Pearl Bryan was found less than two miles from the site of the former slaughterhouse. It was later discovered that Bryan's boyfriend, Scott Jackson, and his friend, Alonzo Walling, had murdered her after a botched abortion attempt. The two men were arrested, but they refused to reveal the location of Bryan's head. Both men were hanged for the crime in March 1897, without ever disclosing the location of Bryan's head. The consensus was that the head was probably thrown into the old slaughterhouse well. Perhaps that's why Pearl Bryan's ghost is seen wandering around inside Bobby Mackey's, both with and without her head. And although Jackson and Walling did not take their last breaths on the property, it is believed that their ghosts are stuck there too; they have both been seen throughout the building, but Jackson's ghost seems to be more active . . . and angry. Those who have encountered his ghost—usually around the well in the basement—say that it is a dark and unhappy spirit.

Gangsters and Unsolved Murders

Shortly after the executions of Jackson and Walling, the former slaughterhouse was torn down, leaving only the well. In the 1920s, the building now known as Bobby Mackey's was built on the property directly over the well. During Prohibition, it functioned as a ruthless speakeasy and gambling den where several people lost their lives. Eventually, the building was shut down and cleared out—presumably of everything except the restless spirits.

In 1933, after Prohibition was lifted, E. A. "Buck" Brady purchased the building and renamed it The Primrose. Brady was competing with powerful gangsters who began showing up at The Primrose trying to scare him into giving them a cut of the profits. But Brady refused to be intimidated and continually turned them down. All this came to a head on August 5, 1946, when Brady and gangster Albert "Red" Masterson were involved in a shootout. After that, Brady decided that he was done. After many years of having to continually (and often forcibly) reject advances by Cincinnati-area gangsters, Brady sold the building. But if the stories are to be believed, as he handed over the keys, he cursed the building, saying that because he couldn't run a successful business there, no one should.

Today, the ghosts of both Buck Brady and Red Masterson are seen inside Bobby Mackey's. Brady's ghost has been identified from photographs taken of him when he was alive. And even though he cursed the building, his ghost seems harmless enough. Masterson's ghost, on the other hand, has been described as "not friendly" and has been blamed for some of the alleged attacks on bar patrons.

Johanna

After Brady sold the building, it reopened as The Latin Quarter. According to legend, Johanna, the daughter of The Latin Quarter's owner, fell in love with (and became pregnant by) Robert Randall, one of the musicians at the nightclub. After Johanna's father found out about the pregnancy, he ordered Randall killed. When Johanna learned of her father's involvement in her boyfriend's death, she first unsuccessfully tried to poison him and then committed suicide in the basement of the building.

Johanna's ghost is seen throughout the building, but it is most often reported on the top floor and in the stairwells, where she will either push or hug people. She is also said to hang out in the Spotlight Room, a secret place in the attic where she allegedly wrote a poem on the wall before committing suicide. Even those who cannot see her apparition can always tell that Johanna is around by the scent of roses.

One of the strangest phenomena attributed to Johanna's ghost is that the turned-off (and unplugged) jukebox sometimes springs to life by playing "The Anniversary Waltz"—despite the fact that the song is not even a selection on the device's menu and the record is not even in the machine.

Bobby Mackey's Music World

In the spring of 1978, musician Bobby Mackey purchased the building, and it has been in operation ever since. Besides operating as a bar, Bobby Mackey's has a stage and a dance floor and has featured performances by many popular country music acts over the years.

Shortly after her husband purchased the building, Janet Mackey was working in the upstairs apartment when she was shoved out of the room toward the stairs while being told to "Get out" by a spirit that she later identified as Alonzo Walling. After that, Janet refused to set foot in the room. So Bobby hired Carl Lawson as a caretaker and allowed him to stay in the apartment. Upon moving in, Lawson reportedly heard strange noises and saw shadowy figures moving around the bar late at night. Believing that the spirits were coming in through the well in the basement, Lawson threw holy water down the hole. As a result, Lawson claimed that he became possessed and was only able to break free from the demon's grasp after an exorcism was performed on him.

In 1993, a man sued Bobby Mackey's alleging that while he was in the bar's men's room, he was punched and kicked by a "dark-haired apparition" wearing a cowboy hat. The victim stated that he might have angered the ghost because he dared it to appear shortly before being attacked. While the suit was thrown out, it did result in the now-famous sign that hangs above the front doors of Bobby Mackey's, which alerts guests to the possibility that the building may be haunted and that they are entering at their own risk.

Prime-Time Ghosts

Bobby Mackey repeatedly turned down requests to have his bar investigated; however, in 2008, he allowed the TV show *Ghost Adventures* to film an episode at his club, which yielded some interesting and controversial footage. Investigators encountered odd cold spots and claimed to have heard the voice of a woman.

While using the men's room, investigator Nick Groff also heard banging noises, which startled him so much that he ran out of the restroom without zipping up his pants. The team also captured some odd video of what appeared to be a man in a cowboy hat moving around in the basement.

But the episode will forever be remembered as the one in which overly dramatic ghost hunter Zak Bagans claimed to have been attacked by a demonic entity after challenging the evil forces in the basement. As proof, Bagans proudly displayed three scratch marks on his back. Bagans was so shaken by the event that he proclaimed it to be one of the scariest things he had ever encountered. But that didn't stop *Ghost Adventures* from returning to Bobby Mackey's in 2010.

Investigators spending time at Bobby Mackey's might be a bit disappointed if they don't experience as much paranormal activity as the *Ghost Adventures* team, but that doesn't mean that the place isn't active. For example, when the organization Ghosts of Ohio visited Bobby Mackey's, the investigation seemed uneventful. But when the team members reviewed their audio afterward, they found that a recorder set up near the infamous well picked up a voice clearly saying, "It hurts."

They're Waiting for You

While Bobby Mackey states that he does not believe in ghosts and doesn't think that his bar is haunted, reports of paranormal activity continue. So should you ever find yourself sitting at the bar at Bobby Mackey's late at night, make sure you take a look around and keep in mind that just because the barstool next to you appears to be empty, you may not be drinking alone.

Europe's Most Haunted Hotels

Many of Europe's haunted hotels are located in Britain and Ireland, where ghosts are often considered as friends or even members of the family, and are given the same respect as any living person—or even more. Other European cultures aren't as comfortable with ghosts—opting to tear down haunted hotels instead of coexisting with spirits—but there are still a few places in Europe where ghost hunters can explore.

* * * *

Comlongon Castle, Dumfries, Scotland

LADY MARION CARRUTHERS haunts Scotland's beautiful Comlongon Castle. On September 25, 1570, Lady Marion leaped to her death from the castle's lookout tower rather than submit to an arranged marriage. Visitors can easily find the exact spot where she landed; for more than 400 years, it's been difficult to grow grass there. Because Lady Marion's death was a suicide, she was denied a Christian burial, and it seems her spirit is unable to rest in peace. Dressed in green, her ghost wanders around the castle and its grounds. In 2007, Comlongon Castle was voted the "Best Haunted Hotel or B&B" in the UK and Ireland.

Ettington Park Hotel, Alderminister, England

You may feel chills when you see the Ettington Park Hotel, where the classic 1963 horror movie *The Haunting* was filmed. It was an apt choice for the movie locale because the hotel features several ghosts.

The Shirley family rebuilt this Victorian Gothic structure in the mid-1800s, and the ghost of the "Lady in Gray" has appeared on the staircase regularly since then. Her identity is unknown, unlike the phantom "Lady in White," who was supposedly a governess named Lady Emma. The voices of crying children are probably the two Shirley children who drowned nearby in the River Stour; they're buried by the church tower.

Watch out for poltergeists in the Library Bar, where books fly across the room. And don't be alarmed if you hear a late-night snooker game when no one is in the room—it's just the ghosts having fun.

Ye Olde Black Bear, Tewkesbury, England

If you're looking for headless ghosts dragging clanking chains, Ye Olde Black Bear is just the place. Built in the early 1300s, the structure is the oldest inn in Gloucestershire. The hotel's headless ghost may be one individual or several—without a head, it's difficult to tell. However, the ghost's uniform suggests that he was a soldier killed in a battle around the 1470s. Those who've seen the figure at the hotel suspect he doesn't realize he's dead—Ye Olde Black Bear was supposedly a favorite hangout for soldiers during his era.

Renvyle House Hotel, Galway, Ireland

Renvyle House Hotel is not old by haunted hotel standards. The site has been built on, destroyed, built again, destroyed again—once by a fire set by the IRA—and so on, until the current hotel was erected in the 1930s. But its ghosts have an impressive pedigree, dating back to a 16th-century Irish pirate queen, Gráinne O'Malley. A redheaded boy is a more recent spirit, possibly a son of the Blake family who owned the site in the 19th century. The hotel is haunted by so many spirits that it was regularly visited by celebrities, such as poet W. B. Yeats, who conducted séances there. Today, Renvyle House Hotel is still a favorite destination for ghost hunters, and it is included in many "haunted hotel" tours.

Royal Lion Hotel, Lyme Regis, England

The Royal Lion Hotel was built in 1601 as a coaching inn, but some of its ghosts may visit from across the street, where executions allegedly took place. Other ghostly figures around the hotel may be the spirits of pirates who sailed into the port, or they could be some of the rebels who were hung and quartered on the nearby beach after trying to overthrow King James II in

1685. Waterfront hotels are often haunted due to their association with pirates and wrecked ships. However, with several dozen different spirits, this site reports more ghosts than most.

Dragsholm Slot Hotel, Nekselø Bay, Denmark

In Danish, the word slot means "castle," and the Dragsholm is one of the world's great haunted castle hotels. According to legend, Dragsholm's "Gray Lady"—a 12th-century maid who loved working at the hotel—visits on most nights. She silently checks on guests to be sure they are comfortable. The "White Lady" haunts the corridors nightly. She may be the young woman who was allegedly walled up inside the castle; her ancient corpse was found during 19th-century renovations.

James Hepburn, the Fourth Earl of Bothwell, is the castle's most famous ghost. Hepburn became the third husband of Mary, Queen of Scots, after he helped murder her previous spouse. For his role in that crime, Bothwell spent the last ten years of his life chained to a pillar in Dragsholm. If you think you've seen his ghostly apparition, you can compare it to his mummified body in a nearby church in Faarevejle.

Hotel Scandinavia, Venice, Italy

The Hotel Scandinavia is in a building dating back to the year 1000, and it's surrounded by stories of ghosts and apparitions. In the 15th century, the apparition of a wealthy (and rather buxom) Madonna first appeared close to the hotel's palazzo. Witnesses report hearing sounds from the sorrowful ghosts of condemned prisoners who long ago crossed the nearby Bridge of Sighs. This famous bridge was where convicts caught a final glimpse of Venice before being imprisoned. These spirits apparently visit the hotel, and their voices are most often heard in the lobby. Because of the location's unique ghosts and how often they're heard, the Hotel Scandinavia is consistently ranked as one of the world's top five haunted hotels.

The Haunted Toy Store

The ghost at a California Toys"R"Us is just as playful as the customers.

✻ ✻ ✻ ✻

WITH A CHEERFUL name like Sunnyvale, this mid-sized town in California's Silicon Valley may seem like an odd place for a haunting, but odder still is the location being haunted: a popular Toys"R"Us store. And yet, according to employees, something unseen routinely wreaks havoc there after the staff has left for the night.

Indeed, the actions of the mischievous spirit seem almost like things that a spoiled child would do: Books are tossed on the floor and roller skates are scattered about, even though everything was put away when employees locked up the night before.

Sometimes the ghost gets more personal. More than one employee has reported being tapped on the shoulder only to find no one there, and several female employees have complained of feeling unseen hands stroke their hair. And there was the time that a group of employees, including a manager, rolled down a metal door in the store and then heard someone yelling and pounding from the other side. When they rolled the door back up, no one was there.

Psychic Encounter

In 1978, renowned psychic Sylvia Brown visited the store, intent on identifying the silly spook. Brown said that she saw a tall, thin man wearing a coat and that his name was Johnny Johnson. During their chat, Johnny informed Brown that she should move or her feet would get wet. An examination of county records later revealed that there had once been a well where Brown had been standing.

The store has done nothing to get rid of Johnny Johnson's spirit. In fact, most of the employees are fond—even protective—of him, and very few say that they feel scared or threatened. And whether he's just a little clumsy or he simply wants to have fun, it seems that he's in the right place. After all, what better place is there to have some fun than a toy store?

Haunted Objects: One Person's Treasure Is Another's Torment

Many people would be frightened to encounter a haunted object. The idea is just a little creepy, whether the object in question is a doll, a painting, or a hairbrush. But some people actually scour estate sales and surf the Web searching for haunted objects. To those people we say, "Let the buyer beware."

* * * *

What Is a Haunted Object?

A HAUNTED OBJECT IS an item that seems to give off a certain energy or vibe. Paranormal occurrences accompany the object itself and begin after the object is acquired. Sometimes, human characteristics—such as breathing or tapping sounds—are associated with the item. In other cases, a person can place a haunted object in one place only to find that it mysteriously moves while he or she is absent from the room, is sleeping, or is away from home.

Becoming Haunted

No one knows for sure what causes an object to become haunted. Some people think that the items are possessed. Renowned psychic Sylvia Browne says that oftentimes a spirit has a "lingering fondness" for an object and may just stop by to visit it. She stresses that all items are capable of holding imprints, which are not always pleasant.

Another explanation is that certain objects are cursed, but that doesn't seem as likely. Most experts feel that a "haunting" comes from residual energy associated with the people or places connected to the item. For example, a beloved doll or stuffed animal may retain some energy from its human owner. This is especially likely to be the case with an item that was near—or even involved in—a violent event such as a murder, the death of a child, or even a heated argument. The "haunting" occurs when the residual energy plays back or reenacts the traumatic event. Like other residual phenomena, haunted objects can't communicate or interact with humans.

When people experience a paranormal event, they often assume that the building in which the incident occurs is haunted, but sometimes it's just one item. Here's a look at some objects that are reportedly haunted.

An Especially Evil Ouija Board

Many people avoid Ouija boards because they may connect us with the Other Side or evil entities. This certainly seemed to be the case with the board Abner Williams loaned to a group of El Paso "Goths." In mid-2000, after the board was returned to him, Williams complained of scratching noises coming from the board, along with a man's voice addressing him and the sound of children chanting nursery rhymes at his window. When Williams tried to throw the board away, it mysteriously reappeared in his house. A paranormal investigator borrowed the board, and a hooded figure appeared from out of nowhere and growled at his son.

When a paranormal research team investigated the Ouija board, they found spots of blood on the front of it and a coating of blood on the back. They measured cold spots over the board, and photos revealed a strange ectoplasm rising from it. The board was eventually sent to a new owner, who did not want it cleared of negative energy. That person has remained silent about subsequent activity surrounding the board.

Although this is an unusually well-documented haunted Ouija board, it is not an uncommon tale. Many psychics warn that if you ask a spirit to communicate with you through a Ouija board, it's like opening a door between the worlds. You never know what kinds of spirits—good or evil—will use that Ouija board to visit you. Therefore, it's wise to be cautious with "spirit boards" of any kind.

Haunted Painting

Actor John Marley purchased a painting titled *The Hands Resist Him* after he saw it at a Los Angeles art show. Many years later, the piece of art—which Bill Stoneham painted in 1972—was found in a trash bin behind a brewery, and in strict accordance with "finder's keepers" rules, the person who found it took it home.

Unfortunately, it soon became clear why the artwork had been abandoned. The finder's four-year-old daughter claimed that she saw the children in the painting fighting. And sure enough, a webcam that recorded the painting for several nights confirmed that the figures were indeed moving. The artist didn't have any insight as to why this particular painting might be haunted, but he did remember that both the gallery owner and a Los Angeles art critic died soon after that show. Coincidence? Maybe. Nevertheless, the family listed the painting and its bizarre story on eBay and came away $1,025 richer.

Robert the Doll

When artist Robert Eugene "Gene" Otto was a young boy growing up in Florida in the early 1900s, he owned a doll, which he named Robert. He took this doll with him everywhere and liked to talk to it. The problem was that the doll talked back—and this was long before the days of Chatty Cathy and other "talking" dolls. It wasn't just the young boy's imagination either—servants and other family members also witnessed the phenomenon. Neighbors were surprised to see the doll moving by itself, and when Otto's parents found their son's bedroom trashed, Gene said that Robert the doll did it. Did it? Maybe so, at least according to the daughter of the family that bought the house in 1972: She was terrified when she discovered the doll in the attic. She said that it wanted to kill her. Her parents had no intention of finding out if this was true, so they gave the doll to a museum in Key West. Visitors to the museum are advised to ask permission before they snap a photo of the famous doll. A tilt of his head means yes, but if you don't get the OK, don't even think about taking a picture, or you'll be cursed.

Nathaniel Hawthorne and the Haunted Chair

You may have seen a creepy old chair or two, but when author Nathaniel Hawthorne encountered one that actually seemed to be haunted, he wrote a short story about it: "The Ghost of Dr. Harris." According to Hawthorne, Dr. Harris sat and read the newspaper in the same chair at the Boston Athenaeum each morning. When Harris died, his ghost continued to visit, and Hawthorne, who was researching at the library, saw it daily. The author said that the spirit had a "melancholy look of helplessness" that lingered for several seconds, and then vanished. So if you visit the Boston Athenaeum, be careful where you sit: Dr. Harris may be in that "empty" chair.

Annabelle and the Haunted Doll

Raggedy Ann and Andy dolls have been popular for decades. But after a young woman named Donna received a Raggedy Ann doll in the 1970s, she didn't have such a warm and fuzzy experience. The doll would often change positions on its own: Once, it was found kneeling—a position that was impossible for Donna and her roommate Angie to create due to the soft and floppy nature of the doll's body. The girls also found mysterious notes written in a child's hand. Worried, the roomates called in a medium, who told them that their apartment building was once the home of a young girl named Annabelle. But after the doll attacked Angie's boyfriend, the girls called in demonologists Ed and Lorraine Warren, who determined that "Annabelle" was not the friendly, playful spirit of a young girl, but instead was a demonic entity. The doll went to live with the Warrens, who knew how to handle its antics, and it now resides in a glass case at the Warren Occult Museum in Connecticut.

⁕ Chapter 4

The Civil War: The Brilliant and the Bungled

That Devil Forrest

A master tactician and a ruthless leader, Nathan Bedford Forrest saw his influence last long after the war ended.

⁕ ⁕ ⁕ ⁕

THE ESSENCE OF Confederate General Nathan Bedford Forrest's military strategy was to "get there first with the most men." It may sound elementary enough, but General Forrest, who fought against many better-educated generals during the Civil War, won nearly all of his battles. It's said that during the war he came under fire 179 times and captured 31,000 prisoners.

Rising from the Ranks

Unlike many famous Civil War generals, Forrest wasn't a product of West Point. His education came from his own experience. His father died when Forrest was still a teenager in northern Mississippi. As a result, he left school to support his mother and eight siblings. Initially, he worked on his family's farm, but he soon expanded into trading cotton, livestock, real estate, and slaves. By 1860, he was running a highly successful business and was the owner of a 3,000-acre plantation in Memphis, Tennessee.

At the outbreak of war, he enlisted in the Confederate army as a private, but he soon received permission to raise his own cavalry unit. Spending his own money, he outfitted a cavalry battalion of 600 and earned the rank of lieutenant colonel. Shortly thereafter, his exploits as a cavalry leader became the stuff of legend.

A Ruthless Reputation

A number of notorious tales and legends surround Forrest. He had a reputation throughout most of the war for being tough and ruthless. This may be summed up in one of the most famous quotes attributed to him: "War means fighting, and fighting means killing."

One of Forrest's earliest exploits that started to build his reputation is his cavalry regiment's escape from besieged Fort Donelson in Tennessee. Ulysses Grant's army had won the day, and Confederate commanders were getting ready to surrender the fort to him the next morning. Forrest would have none of it, declaring, "I did not come here for the purpose of surrendering my command." A few officers planned to slip away, Forrest among them. He announced to his troops: "Boys, these people are talking about surrendering, and I am going out of this place before they do or bust hell wide open." His unit disappeared into the night.

While covering the Confederate retreat from the Battle of Shiloh two months later, Forrest was wounded. He was far ahead of his troops when he got shot, but he grabbed a Union soldier and used him as a shield to make his escape back to Confederate lines.

Raiding Behind Enemy Lines

When he recovered, Forrest was promoted to brigadier general and began a series of successful cavalry raids against Union supply lines. In one memorable raid into Tennessee in 1863, his regiment of fewer than 1,000 attacked a Union garrison twice that size at Murfreesboro. His horse soldiers trounced the 2,000 bluecoats there and captured all survivors, including Union commanding officer General Thomas T. Crittenden.

In an attempt to disrupt General Sherman's campaign on Atlanta, Forrest and his unit went behind enemy lines to attack a Union supply depot at Johnsonville, Tennessee. The Union forces were caught by surprise, and Forrest was able to destroy a gunboat fleet and several million dollars' worth of Union supplies. Sherman wrote to Grant, "That devil Forrest was down about Johnsonville and was making havoc about the gun-boats and transfers." Still, it had no effect on his maneuvers in Georgia. Years later in his memoirs, Sherman had to admit a grudging respect, calling the incident "a feat of arms which, I confess, excited my admiration."

The Fort Pillow Massacre

Forrest cemented his reputation as a ruthless general at Fort Pillow, Tennessee, in April 1864. He led a successful attack against the fort and offered fair treatment to the defenders if they surrendered. When Union commanders refused, Forrest's army brutally overran the fort and its defenders, reportedly shooting many men—primarily black soldiers—as they attempted to surrender. Forrest denied these accusations, but he later said, "The river was dyed with the blood of the slaughtered for 200 yards." The incident, which came to be known as the Fort Pillow Massacre, became a rallying cry for black Union soldiers.

This made Forrest a bit of a target himself, leading Sherman to wish that the army would "go out and follow Forrest to the death, if it cost 10,000 lives and breaks the Treasury." Four times Union generals sent forces out specifically to defeat Forrest; each time they failed. The second of these attempts led to the Battle of Brice's Crossroads in Mississippi, where Forrest's troops routed a federal force that was twice their size. He was finally beaten in the spring of 1865 and forced to surrender in Gainesville, Alabama, that May. This was after both

General Lee and General Johnston had surrendered their own forces at Appomattox Court House and Durham Station.

A Lasting Legacy

After the war, Forrest returned to planting cotton and speculating on railroads. He certainly didn't abandon his Confederate ideals, though. Forrest, known as the Wizard of the Saddle during the war, became the first Grand Wizard of the Ku Klux Klan in 1867. The Klan formed to terrorize blacks, Northerners, and Republicans. Forrest died on October 29, 1877, at the age of 56.

The Slow Slide of General Rosecrans

Every Civil War general had his ups and downs, but few started so high and ended so low as William S. Rosecrans.

* * * *

UNION GENERAL WILLIAM S. Rosecrans came from a family of patriots. His mother, the former Jemima Hopkins, was reputed to be related to Stephen Hopkins, the governor of the Rhode Island colony and a signer of the Declaration of Independence. His father, Crandell Rosecrans, had fought against the British in the War of 1812. It was only natural that young William Starke Rosecrans would follow in the family tradition and seek to attend West Point.

Rosecrans graduated fifth in the West Point class of 1842. While at the academy, he once came to the defense of a young Ulysses S. Grant, who was a year behind "Old Rosy," as he came to be called. Rosecrans saved Grant from being hazed by older students. After graduation, Rosecrans went on to a career in army engineering, but he left the service in 1854 to seek more success in civilian life as an architect and civil engineer. His new venture was quite successful—Rosecrans created a number of inventions and made profitable mining discoveries.

When the Civil War broke out, Rosecrans joined the staff of General George McClellan, who was commanding Ohio's state forces at the time. Rosecrans took command of the 23rd Ohio Volunteer Infantry and soon earned a general's appointment. His talent at battle planning led to a victory over the Confederates at Rich Mountain, Virginia, but, as he was still on McClellan's staff, most of the credit and glory for the victory went to the superior officer. McClellan rode this success to eventual command of the entire U.S. Army. Rosecrans, on the other hand, felt cheated and requested a transfer to the West.

In the West, he engineered even more Union successes, initially at the Battle of Corinth, Mississippi, and later at Stones River and Tullahoma in Tennessee. Yet despite these successes, his relationships with his superior officers continued to decline. General Grant was unhappy that he hadn't pursued the Confederates after the victory at Corinth and believed that Rosecrans took too long to commit to battle.

Rosecrans's command fully broke down at the Battle of Chickamauga. His army had been skirmishing with Braxton Bragg's forces with neither getting the upper hand. At one point, Rosecrans mistakenly believed he had a hole in his line and ordered one of his subordinates to fill it. Although it was clear that rearranging the troops would create a hole rather than fill one, General Thomas Wood followed the order. The Confederates took advantage of the opportunity and pushed the Union army from the field. This finally provided Grant with a reason to fire Rosecrans from his command. Rosecrans was given meaningless jobs in Missouri thereafter, and he concluded a promising Civil War career in disappointment.

Regardless of this, Rosecrans enjoyed a respected career in public service after the war. He was appointed minister to Mexico and then elected to Congress from California. President Grover Cleveland made him register of the Treasury in 1885, a post he held until his retirement in 1893. He died five years later.

Supreme Bunglers: Bad Generalship in the Civil War

Oh, how they botched it. Suicidal attacks, failures of military intelligence, lack of human intelligence, even cowardice. Some of the generals were admirable leaders who just had bad days; others should never have worn the bars or the stars.

✻ ✻ ✻ ✻

Nathaniel Banks, U.S.A. Banks's many losses included Shenandoah Valley and Cedar Mountain (Virginia, 1862), plus Port Hudson (Louisiana, 1863). He didn't get the nickname "Commissary" for his superior logistical skill; rebel troops gave him that nickname in appreciation for the supplies they captured from him. His crowning blunder was the Red River Expedition (Louisiana, 1864), in which his troops floundered toward Shreveport without achieving anything.

Braxton Bragg, C.S.A. You might want discipline-stickler Bragg running a basic training center, but not your army. The Bragg cycle would start when he'd botch in some way: mishandle a battle, pick the wrong terrain, or fail to exploit a win. His subordinates would either defy him or write home to Richmond begging reprieve. Bragg would learn of this, punish them for insubordination, then commit a new error to restart the process. Eventually, the Confederate government put Bragg to work on the supply-and-draft pipeline, something he was actually good at.

Ambrose Burnside, U.S.A. There's bad, and then there's Burnside's version of bad. His problems began when this competent brigadier gained a second star. Burnside—a modest man—didn't think he had what it took to command the Army of the Potomac, but his superiors felt otherwise. Robert E. Lee spanked Burnside at Fredericksburg (Virginia, 1862), showing that Ambrose's one brilliant command judgment had been that he shouldn't command.

Burnside's Homeric error came at Petersburg (Virginia, 1864). Union engineers tunneled beneath Confederate positions and detonated explosives, breaching the lines with a big crater. Burnside sent a force commanded by a drunkard (James Ledlie) into the crater, where they halted long enough for the rebels to contain and mow them down. Burnside's solution: Send more men into the hole to die with the first group! The Battle of the Crater was one of the war's most tragic episodes.

James Ledlie, U.S.A. The charge was GWI (generaling while intoxicated). Ledlie nominally commanded one of the divisions Burnside squandered in the Crater at Petersburg. Instead of leading his men into danger, he stayed behind and got crocked. Drunkenness is one thing in a leader—a number of Civil War generals took a drink—but few deliberately ducked combat. Even incompetent Ambrose Burnside was neither a drunkard nor a coward; Ledlie was both.

Nathan Evans, C.S.A. As drunkards went, Evans had a pretty decent generaling career, except for one incredible blunder at Kinston (North Carolina, 1862). Ordering a fighting withdrawal across a river under heavy Federal assault, Evans burned a bridge behind him. That strategy would have worked well had he not accidentally left half his force on the far bank. Observing the scene from a safe distance, he mistook his forsaken troops' gunsmoke for Union fire and ordered his artillery to shell his own men.

Ulysses S. Grant, U.S.A. Even great generals botch now and then. Until mid-1863, the rebel fortress guns of Vicksburg dominated Mississippi River traffic; no Union vessel could safely pass. In 1862, Grant decided to dig a long ditch on the Louisiana side of the Big Muddy. He figured that the water would rise, scour out a generous channel, and let Union warships sail upriver outside Confederate gunnery range. After months of hard labor, the river leaked into the ditch—then silted it up. Grant's canal was useless.

Joseph Hooker, U.S.A. Many officers command well at one level but fail at another. "Fighting Joe" was a heck of a corps commander: fierce, brave, and popular. Promoted to army command, he managed to lose to Lee at Chancellorsville (Virginia, 1863) despite outnumbering the Southerners two to one. After this debacle, Hooker went back to corps command, where he resumed his competent ways.

George McClellan, U.S.A. McClellan might have been elected president in 1864 had he led the Army of the Potomac better in 1862. He was reportedly Caesar, Napoleon, and Hannibal rolled into one, posturing for public approval, dodging battle despite superior numbers, and whining for more troops. It's hard to avoid thinking that McClellan feared a career-spiking defeat. Abe Lincoln summed it up: "If General McClellan does not want to use the army, I would like to borrow it for a time."

George Pickett, C.S.A. No, this isn't for his Charge at Gettysburg, in which he was simply obeying Lee's direct orders with the only fresh division on the field. On April Fools' Day in 1865, Pickett's force suffered a terrible defeat at Five Forks (Virginia). For that he might be forgiven, except that he and some generaling cronies had decided to attend a fish fry and missed the entire battle.

Gideon Pillow, C.S.A. This doesn't seem difficult: When you blow a hole in your enemy's lines, you exploit the hard-won breach. You don't do what Pillow did at Fort Donelson (Tennessee, 1862)—march your guys back to their trenches, giving someone as smart as Ulysses S. Grant time to patch his lines and regroup. While Pillow's 12,000-man garrison hauled down its flag, he fled across the river, thus helping the Union again by avoiding captivity.

David Porter, U.S.A. Admirals, the seagoing version of generals, must also take their medicine. Like Grant, Porter was an outstanding officer who only once opened a jug—but Porter drank the entire thing. In 1865, the Union was trying to capture the tough Fort Fisher (North Carolina). Porter decided that Navy sailors and U.S. Marines armed with swords and pistols could "board" Fort Fisher and capture it, not realizing that the defenders' long-rifled muskets made this a suicidal proposition for the Yankee tars.

Daniel Sickles, U.S.A. His movement of III Corps to the Peach Orchard at Gettysburg—leaving a big gap in the line—has long been debated. "Boob" or "savior"? Most say the former, since he disobeyed orders and got his corps pounded doing so. Somehow the Union lines held anyway. Sickles might have been court-martialed, but he lost a leg in the fight, so he was just chewed out by George Meade, commander of the Army of the Potomac at the time.

A Quick Victory

As America started to divide, most people believed the war would be a simple, short-lived skirmish.

* * * *

BEFORE FIGHTING BEGAN, many throughout the North anticipated a quick Union victory. Shortly before blue-clad Federals marched from the outskirts of Washington to meet the rebels outside of Manassas Junction, Virginia, publisher Horace Greeley's New York Tribune had published the following notice: "Forward to Richmond! Forward to Richmond! Forward to Richmond! The Rebel Congress must not be allowed to meet there on the 20th of July. By That Date the Place Must be Held by the National Army!"

Excitement over this opening match at the First Battle of Bull Run quickly dissipated, however, as the vanquished Union army fled from the battlefield in panic. Greeley, whose paper helped to push the North into battle, wrote the following in a letter to Lincoln: "On every brow sits sullen, scorching, black despair... If it is best for the country and for mankind that we make peace with the rebels at once, and on their own terms, do not shrink even from that."

Thimbles and Handkerchiefs

As states began seceding from the Union, many in the South also felt the war would be quick and virtually bloodless. Some Southerners assumed the conflict would end with their victory at the First Battle of Bull Run, and once they saw the Northern troops fleeing, many thought the war was over. Confederate supporter Edmund Ruffin saw in "this hard-fought battle virtually the close of the war."

General William Sherman, who had been the superintendent at a college in Louisiana before the war, wrote about the time in his memoirs: "In the South, the people were earnest, fierce and angry, and were evidently organizing for action; whereas, in Illinois, Indiana, and Ohio, I saw not the least sign of preparation. It certainly looked to me as though the people of the North would tamely submit to a disruption of the Union, and the orators of the South used, openly and constantly, the expressions that there would be no war, and that a lady's thimble would hold all the blood to be shed." James Chesnut, husband of diarist Mary Chesnut, suggested that so little blood would be shed over secession that he could drink it all. Lucius Lamar, who wrote the bill of secession for Mississippi, made the same offer.

As the number of states joining the Confederacy increased, Leroy Pope Walker, the Confederate secretary of war, had a hard time generating enthusiasm for war preparation. The most common sentiment he encountered was that war could be avoided or, if fought, would end quickly, so there was no need to get ready for a long struggle. "At that time," Walker later wrote, "I, like everybody else, believed there would be no war. In fact, I had gone about the state advising people to secede, and promising to wipe with my pocket-handkerchief all the blood that would be shed."

The Confederate attack on Fort Sumter galvanized the North in its resolve to defeat the Confederates and forcibly bring them back into the Union. Most Northerners saw the initial defeat at Bull Run as a temporary setback. Lincoln ignored the calls for peace—the dismal battle strengthened his resolve to win the war. It didn't take very long before Southerners realized that thimbles and handkerchiefs wouldn't come close to mopping up the blood yet to be spilled.

The Man in Grant's Shadow

General George Meade was the victor at the Battle of Gettysburg. No hero with the newspapers, however, he was easily eclipsed by his superior officer.

* * * *

THE FIFTH MAN to lead the most important Union army in the East, General George Gordon Meade is regarded by many to have been one of the finest Union generals of the Civil War. But Meade was certainly not subject to universal praise. Why? One reason is that, like many generals, he didn't trust journalists, and those journalists took revenge by writing little about him. He also may have been in the wrong place at the wrong time.

Meade graduated from West Point in 1835. He served in the Mexican War and the Second Seminole War, and he was a captain in the topographical engineer corps when the Civil War broke out. Completing a survey of the Great Lakes at the time, he was immediately given a position as a brigadier general in the Pennsylvania Reserves. This unit joined the Army of the Potomac after training and fought in battles at Beaver Dam Creek and Gaines' Mill.

General Meade was severely wounded in a battle at Glendale, Virginia, in June 1862. However, he recovered soon enough to lead his troops into the Second Battle of Bull Run and at

Antietam. He served well in those conflicts and was promoted to major general. More success eventually led him to be named commander of the Army of the Potomac after Joseph Hooker resigned. That appointment came just three days before the Battle of Gettysburg, the largest confrontation of the war, so he had little time to prepare. Nevertheless, he masterfully commanded his forces to a major victory that turned the tide for the Union.

Yet Meade never seemed to get the attention that he so obviously deserved. Journalists and the newspapers and magazines they wrote for played a dominant role in how key events and people were remembered. This fact caused the cautious Meade to be suspicious of them, and he was known to be cruel to the journalists who crossed his path—including those who traveled with the troops to report on the war. One anecdote has him forcing a newspaper reporter to leave camp riding backwards on a mule. Major newspapers in the North retaliated by infrequently mentioning Meade's name—usually in articles about defeats.

Although Meade remained in command of the Army of the Potomac through the end of the war, he faded into the background when General Grant came from the Western Theater to become General-in-Chief of all armies. Grant organized his own headquarters with the Army of the Potomac, which resulted in the spotlight passing from Meade to Grant. The war ended with Meade as a mere footnote to the more charismatic General Grant.

Fortunately, Grant himself understood and appreciated Meade's abilities, and he gladly helped advance Meade's career. After the Civil War Meade remained with the Army and commanded Southern reconstruction projects. He died in Philadelphia in 1872.

Undercover Raiders

A few clever Confederate operatives attempted raids along the Union's northern border or far behind enemy lines. Some succeeded, and some didn't, but all went down in history for their daring efforts.

* * * *

Island Dwellers

THE UNION HOUSED Confederate soldiers and generals in several island prisoner-of-war camps. Such locations were remote, difficult to reach, and disconnected from any methods of easy access—ideal conditions if the purpose is to prevent prisoner escapes. Johnson's Island in Ohio was extremely secluded in the middle of Lake Erie. Many considered it an inescapable prison. The Confederate army, however, saw an opportunity. Hundreds of Southern soldiers and officers occupied the stockade on the island. If a small force could sneak through and free these prisoners, the Confederacy would suddenly have a significant military presence deep within Union territory and the possibility of creating a new front to the war.

Since it was far from the front lines of battle, Lake Erie was guarded by only one major gunboat, the USS *Michigan*. Confederate Captains Charles Cole and John Beall came up with a plan to commandeer the *Michigan*, which would allow them to take control of the lake without opposition. If they could achieve that, they assumed the guards at Johnson's Island would surrender. Beall commandeered the Philo Parsons, a Lake Erie steamer, and prepared an attack on the *Michigan*. He was waiting for a signal from Cole, who'd made his way onto the ship. That signal never came, however, because Cole was found out and captured by the *Michigan* crew. Beall's Parsons crew, on the other hand, quickly became concerned when the signal failed to appear and forced Beall to take them to safety in Canada on the other side of the lake, aborting the mission.

Charles Cole did ultimately reach Johnson's Island. He was imprisoned there for the rest of the war. Beall was later caught and executed, and any thoughts of forming an army of prisoners died with him.

An Unsuspecting Town

Believe it or not, the planned assault on Johnson's Island was not the most outlandish border raid of the Civil War. That honor is perhaps held by a group of Southerners who trekked all the way to the quiet town of St. Albans, Vermont, looking to make their mark. It was 1864, and the South was getting more and more desperate. Lieutenant Bennett Young, a brash rebel soldier, was stationed in Canada and had been put in charge of formulating secret sabotage missions into Union territory. On October 19, 1864, he finally put his plans into action.

Young gathered 20 conspirators and silently slipped across the border and into St. Albans. He broke the relatively calm mood of the town by pulling a .38-caliber revolver from his coat and declaring they were taking possession of the town for the Confederate States of America. Initially, many St. Albanians didn't take the situation seriously, assuming that Young was drunk. They watched, bemused, as he and his crew robbed the town's bank and attempted to burn down buildings with crude incendiary bombs (not a single one of which worked). All in all, the attack lasted 20 minutes, although it did claim one life, that of Elinius Morrison, who attempted to confront the gang.

With more than $200,000 dollars, Young and his followers slipped back across the border into Canada, where they were protected by friends and supporters. After handing the money over to a Confederate agent who turned around and smuggled it back out of the country and into the South, Young was banned from entering the United States until 1868. He used the time to study law in Europe. When he was permitted to return home, he became a well-respected attorney in Kentucky and a minor celebrity with his fellow former rebels.

Confederates in the Big City

In the fall of 1864, the Confederacy began to hatch another assault on the Deep North, this time in the populous and important city of New York. Again, the plot involved an elaborate setup. Coded messages were published in the Richmond Whig, a Virginia newspaper, and secretly reprinted in many New York papers. These messages were placed by Confederate agents in Canada, who, like Bennett Young, had been ordered by President Davis to undertake secret espionage missions. One agent implicated in planning the New York assault was Jacob Thompson, a former House representative for Mississippi and U.S. secretary of the interior turned provocateur. The plan was to set a number of large fires in order to incite chaos in the city and ultimately capture it. Like the Johnson's Island raid, the New York plotters hoped to free prisoners of war from local camps and use them to sack the city.

The hotels were to be set ablaze by Confederate agents who checked in, lit their rooms on fire, and left, locking the door behind them. They actually succeeded in starting small fires in all the major hotels of New York City, as well as in P. T. Barnum's American Museum, but the flames were quickly put out by the city's experienced fire department and did not have the effect they desired. In the end, the Confederates escaped the city under cover of darkness. Only one, Robert Cobb Kennedy, was ultimately captured. He refused to sell out his co-conspirators and went to the gallows by himself in March 1865.

Preserving the Union

As the Civil War got started, the U.S. Congress wanted everyone to know what they were—and were not—fighting about.

* * * *

ONE COMMON MISCONCEPTION about the Civil War is that it was fought to abolish slavery. In reality, the issue was more complicated. In fact, some people wanted to take the argument over slavery entirely out of the discussion.

Early in the war, Congress tried to do just that. Representative John J. Crittenden of Kentucky (who had just finished his term in the Senate before running for a seat in the House) and Senator Andrew Johnson of Tennessee presented a resolution to state the North's war aims clearly. It was passed on July 25, 1861—a little more than three months after the war began.

The Official Reason for Fighting

The Crittenden-Johnson Resolution stated the war with the Southern states was being fought neither to abolish slavery nor for "over throwing or interfering with the rights or established institutions of those States," but only to "defend and maintain the supremacy of the Constitution and to preserve the Union."

Claiming slavery had nothing to do with the war didn't sit well with abolitionists. A primary opponent of slavery in the House, Pennsylvania's Thaddeus Stevens fought against the measure. He couldn't stop it from passing, but he got it repealed in December 1861.

By that time, the resolution had served its main purpose: to discourage the slave states of Delaware, Kentucky, Maryland, and Missouri from seceding. The losses of these border states would have been devastating. If Maryland had seceded, Washington, D.C., would have been completely surrounded by Confederate territory. But by assuring residents that their slaves were not at risk, the Union gave itself a bit more protection.

A Little Foreign Intrigue

Prince and Princess Salm-Salm were a unique duo, making contributions to the battlefield and impressions in high society.

* * * *

Bad Habits

BORN IN 1828 IN Westphalia, Prussia, Prince Salm-Salm attended military school in Berlin and entered the Prussian army in 1846 as a second lieutenant—just in time to serve in the Schleswig-Holstein War. He later moved over to the Austrian army. During European conflicts, the prince proved his skills on the battlefield, but his heroic efforts did not mask his extravagant spending habits and card playing. Gambling debts—and possibly his attempts to avoid them—resulted in his discharge from the Austrian army for a lifestyle unbecoming an officer. Salm-Salm's father purchased his son's way to the United States so he could escape that tainted status and live a soldier's life elsewhere. With the sectional conflict underway in his new home, Salm-Salm headed for the Union to suit up for a new fight. The U.S. Army was happy to accept help where they could get it, mercenary or not.

Immigrant Regiment

Salm-Salm joined the many immigrant recruits under General Louis Blenker. He soon met his bride, Agnes Joy, a charming and ambitious girl, who, according to one biographer, would always overshadow her soldier husband. Agnes had a somewhat mysterious past: She worked in circus shows as a ropedancer and horse rider under the name Agnes Leclerq. In Washington, D.C., where she had apparently charmed her way into local society, she was known for daring morning rides on her wild mustang and for attending parties in influential circles. Felix Salm-Salm took note of the dashing Agnes and set up a situation in which he could meet her at a capital reception. Their romance was quick—they married in August 1862.

Salm-Salm had gained a commission in command of the 8th New York, but soon after the wedding, as a result of intended military reorganization, that position became insecure. His record in battle both abroad and in the United States was respectable enough, but by April 1863, the 8th New York was getting ready to muster out of the service. Even The New York Times questioned the "absurd and impolitic regulation of red-tape" that brought Salm-Salm and his troops home after a two-year tour. "The men . . . have proven themselves worthy patriots," the Times declared.

A New Troop

Princess Salm-Salm was able to secure her husband a new command. She had a certain reputation in political circles, but in hindsight, her degree of influence at the national level and her Washington connections were likely questionable and exaggerated, by herself as much as anyone else. She sometimes showed up at parties uninvited, and other women, including Mary Todd Lincoln, looked down on her. Despite this, she did have influence with the New York governor, and this connection landed her husband the command of the 68th New York.

On June 8, 1864, Salm-Salm became colonel of this outfit and took his troops to Nashville, where they guarded the Nashville and Chattanooga Railroad. The 68th New York then spent the next few months on an island in the Tennessee River near Bridgeport, Alabama, guarding against Confederate raids on pontoon and railroad bridges. Salm-Salm, however, had a deep desire to fight rather than simply to defend points of interest, so after hearing about Confederate General John Bell Hood's campaign into Tennessee, he convinced Union General James Steedman to include him on his staff. Reporting on one skirmish, the colonel relayed, "My men poured two well directed volleys into [the outlaw guerrillas] and they skedaddled as quick as they came." The successes in Tennessee and Alabama netted Salm-Salm a brevet promotion to brigadier general in April 1865.

"Princessly" Duties

For much of Salm-Salm's tour in the latter half of the war, he was accompanied by his wife, who didn't want to be away from him. Although the high-society women in Washington didn't like the princess, she earned respect from the troops in the field. A strong advocate for rank-and-file soldiers, she often took supplies and clothing intended for officers and gave them to the wounded enlisted. She even stole her husband's bedsheets and tore them into bandages.

Off to Mexico

After the Civil War, Prince Salm-Salm, a warrior who wasn't comfortable living in peacetime, began looking for his next fight. He and Princess Agnes traveled to Mexico, where they joined the effort to defend the regime of Emperor Maximilian in 1867. Maximilian was supported by France, and he was under attack from Mexican forces seeking to restore President Benito Juarez. Salm-Salm nearly lost his life after he was captured, but his wife saved his neck. Through tireless effort, she established an accord with the reinstated President Juarez and his generals that secured her husband's release. Salm-Salm would have been executed had it not been for her.

After fleeing Mexico, the couple headed for Europe, and the prince joined yet another fight in the Franco-Prussian War. He was killed in the Battle of Gravelotte in August 1870. His princess bride, a widow at 30, lived on to publish her diary, which details some of her adventures. She died in 1912.

Captured at Sea

Already embroiled in a devastating Civil War, the United States almost went to war with the entire British Empire over four men at sea.

❋ ❋ ❋ ❋

TOWARD THE END of 1861, Jefferson Davis hoped to gain some aid from Britain and France to help the South break the Union blockade that was slowly cutting off his shipping lanes. Accomplishing this would require convincing those two countries to recognize the Confederate States of America as a nation. The South had little chance of building a fleet capable of breaking the blockade, but both European nations had fleets at the ready for such a task. Factors that shored up Confederate chances of convincing these countries to intercede were the facts that the French and British were major consumers of Southern cotton, and that France had its eye on Mexico, which bordered Confederate Texas. Thus, the mills and banking industry of Britain and the pro-expansion political forces in France favored intervention in the American conflict, but sentiment among the public did not. In Britain, especially, the fact that the South's economy was based on slavery was not popular.

Southern Ambassadors

Davis chose former U.S. Senators James Mason and John Slidell as emissaries to those countries to negotiate recognition. The diplomats were smuggled out of Charleston by a blockade runner on October 12 and sailed to Havana, where they transferred to a British mail steamer, the *Trent*. The *Trent* left Havana on November 7 and was intercepted at sea by a U.S. vessel, the *San Jacinto*, the next day. In violation of international law, the *San Jacinto* fired warning shots across the bow of the *Trent* and forced its captain to hand over Mason, Slidell, and their two secretaries. Because Britain was not at war with the United States, seizing these men was a violation of the same laws that Britain had violated to start the War of 1812. In short, Captain Charles Wilkes of the *San Jacinto* had just committed an act of war against Britain.

Armed and Ready

When the *San Jacinto* reached port with its prisoners, its captain was treated like a hero by the Union press and public. Initially, the official American response was also favorable, with Congress sending its official thanks to Wilkes. President Lincoln, a lawyer, quickly saw the legal problem he faced, however. On the flip side, when the *Trent* reached port with word of Wilkes's actions, the British public and government were outraged. And Britain was not alone—France was also up in arms over the incident and indicated that if Britain joined the American conflict, France would quickly follow.

Lincoln was presented with a demand for the release of the four men, and British troops were sent to Canada, from where they could attack southward if need be. Plans were drawn up detailing how many British ships would be needed to destroy the American navy, and weapons and equipment were stockpiled for the coming war against the United States. The debates in Britain became heated, with conservative newspapers, businesspeople, and politicians wanting war, while major figures ranging from Charles Darwin to Albert, the Prince Consort, argued for peace.

In late December, Lincoln decided that he needed to keep to "one war at a time." He ordered that a formal note of "apology" be sent to the British, disavowing the illegal actions of Captain Wilkes, and he then released Mason, Slidell, and their secretaries. Thousands of British troops arrived off the coast of America, only to find they weren't going to fight after all.

Mason and Slidell resumed their journeys to Britain and France, respectively. The Confederates had finally reached Europe, but Britain and France remained neutral.

Beaten on the Senate Floor

After Massachusetts Senator Charles Sumner's speech got personal, a South Carolina representative took matters into his own hands—literally.

* * * *

RAW EMOTION AND angry exchanges can often come to dominate a political debate, and at times it can be downright dangerous. In the tension-filled years that led up to the secession of the Southern states, the U.S. Senate was a frequent setting for passionate, and often bitter, oratory in which the adversaries were North versus South, slaves states versus free.

An Acrid Address

It can be argued that no speech in the history of the Senate has been more provocative than the address made by Massachusetts Senator Charles Sumner on May 19 and 20, 1856. For two days, the fervent abolitionist delivered his blistering "Crime Against Kansas" speech. It was an attack on proslavery forces in the Kansas territory and beyond that was loaded with personal invective directed toward the supporters of that peculiar institution. Sumner made a number of personal insults against senators who advocated slavery, calling them imbeciles and immoral. He referred to Illinois Senator Stephen Douglas, one of his political enemies on the slavery question, as a nameless animal unfit to be an American senator.

Sumner also specifically singled out South Carolina Senator Andrew Butler for scorn. Butler was ill at the time and therefore not present in the Senate chamber to hear the address himself. Sumner identified Butler as a prominent example of hypocrisy on the slavery question and repeatedly disparaged South Carolina as a state whose contributions to the Union were so historically insignificant that it would be immediately overshadowed by the admission of a new territory such as Kansas. He also made a number of pointed personal attacks against Butler, the most graphic of which was his repeated suggestion that Butler kept a "harlot, Slavery."

A Personal Attack

The fallout from this incendiary speech was far more than merely political. The response from Butler's camp came two days later when Preston Brooks, a member of the House of Representatives from South Carolina and the nephew of Senator Butler, offered his reply. He had heard the Sumner speech from the Senate gallery and, incensed by its tone, was determined to avenge the honor of both his uncle and his state.

Brooks entered the Senate chamber during a recess. Seeing Sumner sitting at his desk, Brooks began to denounce the senator for the insults he'd made two days earlier. Before Sumner could reply, Brooks suddenly struck him with his hollow, metal-tipped cane, delivering a series of vicious blows to Sumner's head and body that left the senator battered and semiconscious. When his cane broke after 30 blows, Brooks walked away from the crippled Sumner and returned to his business in the House of Representatives. In many contemporary reports of the attack, Southern senators present in the chamber were said to have laughed as Sumner was beaten. It took three years for Sumner to fully recover from the injuries he received.

Although a motion to expel Brooks from the House of Representatives for dishonorable conduct was defeated when representatives from the South refused to support it, the House did vote to censure him. He resigned from office, but his South Carolina district elected him right back into Congress. Brooks became a hero in the South, a symbol of a man prepared to defend principle and honor when challenged. Sumner became a heroic figure in the North, where he was characterized as both a victim of Southern brutality and an example of how proslavery forces attempted to suppress free speech.

Breakfast at Shiloh

The Union soldiers turned the tide in the brutal Battle of Shiloh when the Confederates eyed their breakfast.

✻ ✻ ✻ ✻

IN APRIL 1862, President Lincoln's grand plan of splitting the Confederacy in two by controlling the Mississippi River was nearing reality. The battle-tested armies of Union generals Ulysses S. Grant and Don Carlos Buell were marching unopposed toward Corinth, Mississippi, where the South's rail lines converged. If Grant and Buell took Corinth, not only would the Confederacy be cut in two along the western rivers, but Southern rail transportation would be crippled, as well.

In a Holding Pattern

Grant's army rolled into the town of Pittsburg Landing, Tennessee, north of Corinth on the west bank of the Tennessee River, on April 6. He planned to wait until Buell's force joined him there and then to march together into Corinth. But Grant, and his subordinate General William Tecumseh Sherman, made a critical tactical error. They set up camp with their backs to the wide and deep river.

Confederate General Albert Sidney Johnston recognized an opportunity. He took 44,000 of his troops who were guarding Corinth and marched them right up to the edge of the Union camp. At dawn, the rebels sprang upon the unsuspecting bluecoats, completely surprising the Union soldiers and driving them quickly to the bank of the river.

Hmm, That Smells Like Breakfast!

Once the enemy was on the run, however, the hungry Confederates' stomachs betrayed them. The federal soldiers had been making breakfast when the battle started, and they'd left the food behind in the rush of the battle. Some of the rebels slowed their attack to gobble down a meal.

This gave Grant's army a chance to pull themselves together. Hundreds of them made a stand in a sunken road known as "The Hornet's Nest." Their defense was bolstered further when General Johnston was mortally wounded. A bullet—possibly from one of his soldiers—had torn through an artery in his leg.

Fighting Back

Union forces clearly lost the battle that day, but they weren't crushed. Grant acknowledged the defeat but was quick with a comeback—"Whip 'em tomorrow, though." Not long after that conversation, Buell arrived. On April 7, 1862, the combined Union armies retook the ground lost the previous day. In two days of fighting, Shiloh had become the bloodiest battle of the war so far, with more than 23,000 combined casualties.

Jeb Stuart Rides

The success and leadership qualities of this Confederate cavalry commander place him among some of the finest leaders of Civil War history.

* * *

CONFEDERATE CALVARY GENERAL James Ewell Brown "Jeb" Stuart was one of the most charismatic and daring leaders on either side of the war. A Virginian, Stuart possessed an impeccable military pedigree that included graduation from West Point, a U.S. Army commission as a cavalry officer, and marriage to the daughter of renowned cavalry officer Colonel Philip Cooke. Before the war, Stuart played a key role in the capture of John Brown and his rebels at Harpers Ferry in 1859. He seemed destined for both honor and high rank in the American military.

Leading a Local Crew

When Virginia seceded in 1861, Stuart resigned his commission in the U.S. Army and took command of a state cavalry unit as its colonel. As a local commander leading local troops who rode their own mounts over their own countryside, Stuart possessed significant advantages in cavalry warfare. The Virginia riders knew the subtleties of their home terrain and were aware of how their horses would perform under the pressure of combat. This familiarity with their geography often meant the difference between military success and failure. Knowledge of the safest places to cross creeks and rivers, where to find watering holes and forage spots for their mounts, and the best hiding places in the densely forested Virginia hills were each critically important to Stuart's success. It also often provided security from Union ambushes.

Long List of Achievements

The list of engagements in which Stuart and his Virginians were prominent in Confederate battle movements is remarkable. Stuart's unit, referenced in various battle reports as the "Black Horse Cavalry," provided support to the main Confederate army at the First Battle of Bull Run in July 1861, and as a result, Stuart was promoted to general. Stuart built an impressive and flamboyant reputation for himself, even going so far as to wear an ostrich plume in his hat. His rebel horse soldiers followed their dashing leader enthusiastically.

Stuart achieved enduring fame in 1862's Peninsula Campaign in northern Virginia, as Union forces led by General George McClellan sought to advance on Richmond. Directed by General Robert E. Lee to find the right flank of the Union army, Stuart boldly took a brigade of cavalry around the entire 80,000-strong Union force, allowing him personally to provide his reconnaissance to Lee at Richmond. Stuart replicated this feat in August 1862 during the Second Bull Run campaign, taking 1,500 soldiers around the rear of Union General John Pope's army to attack his supply line and a Union gunboat.

The Black Horse Cavalry executed a third surprise penetration of Union defenses after Antietam on September 17, 1862, sweeping around the much larger and robust Northern forces.

Covering 80 miles in just 27 hours, Stuart's Virginians wreaked havoc on Union positions and supply depots. Other key successes included prominent roles in the crucial Confederate defense of Fredericksburg in December 1862 and at the Battle of Chancellorsville in April 1863. Ultimately promoted to the rank of major general, Stuart possessed talent and an ability to lead that placed him alongside Jackson and Lee in the pantheon of great Confederate commanders.

His Final Battle

Like Jackson, however, Stuart did not survive the war. In May 1864, General Philip Sheridan was given orders to destroy Stuart's cavalry. Sheridan's force took on the Confederate horse soldiers at the Battle of Yellow Tavern. Although it had worked to his advantage so far, Stuart's plumed hat proved too effective a target, and the general was shot and killed. Lee was particularly saddened by the death. "He never brought me a piece of false information," General Lee said in tribute.

The Fox of Harpers Ferry

A Southern colonel fighting for the Union uses his heritage against the rebels.

* * * *

THE FEDERAL TROOPS garrisoned at Harpers Ferry in September 1862 were grimly hunkered down, surrounded by eight regiments of Confederates under General Lafayette McLaws. Southern artillery had been lobbing shells into town most of the day, and morale was low. Garrison commander Colonel Dixon S. Miles stood useless, stunned by fear and indecision. When some Union regimental commanders began discussing mutiny, Colonel Benjamin Franklin "Grimes" Davis of the 8th New York Cavalry knew he had to act.

Born in Alabama and raised in Mississippi, Davis was a rare Mississippian who stayed with the U.S. Army when the conflict erupted. It apparently never occurred to him to break his oath to the Union.

Getting Out

At Harpers Ferry, Davis went to Colonel Miles and informed him that he and his troops would not be captured in Harpers Ferry because they were breaking out. Miles objected vehemently, but in the end, he reluctantly agreed to let them go.

After dark, Davis organized his 8th New York Cavalry, the 12th Illinois Cavalry, and smaller units of Maryland and Rhode Island horse troops into two columns. They planned to go through Confederate lines with two locals who knew the area. At 9:00 P.M., moving as quietly as cavalry could, they crossed the pontoon bridge over the Potomac and turned west. Davis didn't know that General McLaws had withdrawn seven of his eight regiments to defend Crampton's Gap from Union attack, so he was astonished to find the roads unguarded. In the wee hours near Sharpsburg, his troopers slipped quietly past groups of retreating, disorganized, Confederate soldiers.

A Little Subterfuge

Just before dawn, they came upon a Confederate wagon train. Davis was unwilling to plunge his tired force into a fight and had an idea of how to avoid it. Going to the lead wagon, he informed the wagon master in his best back-home Mississippi accent that Union cavalry were on the road ahead and that the train needed to detour—down a road that would take them north into Pennsylvania. The Illinois cavalry so smoothly ran off the rebel escort following the train that the wagon drivers suspected nothing until the sun came up.

"What outfit you with?" asked a driver suspiciously.

"The 8th New York," a trooper replied.

"The hell you say."

It was too late to resist. Davis's horse soldiers had the situation well in hand. At 9:00 A.M., they reached Greencastle, Pennsylvania, and Davis took stock of what he'd captured. It was General Longstreet's reserve ammunition train: more than 40 mule-drawn wagons and 200 prisoners. The citizens of Greencastle turned out with food and drink, delighted finally to hear good news about their cavalry instead of the usual exploits of General Jeb Stuart and his Confederates.

Meanwhile, Back at Harpers Ferry

By this time, the remaining Yankees at Harpers Ferry, pounded by 50 Confederate guns, had surrendered to a shabbily attired Stonewall Jackson. One observing Union soldier said, "Boys, he is not much for looks, but if he had been in command of us we would not have been caught in this trap."

This was the Union's biggest surrender of the war: 70 cannons, 13,000 rifles and muskets, and 12,500 soldiers. Colonel Miles himself barely lived to see it. He was killed by a shell after he laid down his arms. The only sour note for Confederates was that the cavalry had escaped. There would be no fresh mounts for Jeb Stuart. Colonel Davis would soldier on until he was killed the following summer in the largest cavalry action of the war, the Battle of Brandy Station, which occurred just before Gettysburg.

What Could Have Been

After years spent teaching the leaders of the Civil War, General Charles Ferguson Smith became a hero in his own right.

* * * *

UNION GENERAL WILLIAM Tecumseh Sherman was a protégé and friend of Ulysses Grant, but there was another officer who Sherman respected even more. "Had C. F. Smith lived," Sherman stated, "Grant would have disappeared to history after Donelson."

Charles Ferguson Smith entered West Point in 1820 at 14, graduating in 1825. Smith returned to West Point as an assistant instructor of infantry tactics in 1829, ultimately rising to commandant of cadets. He held that position while Grant, Sherman, and various other Civil War figures were cadets.

Smith left that post in 1842 and, a few years later, made a name for himself in the Mexican War. He won three brevet promotions, rising to the rank of lieutenant colonel. When the Civil War started, he was promoted to brigadier general of volunteers. He was transferred to Grant's Army of the Mississippi in January 1862 and made a division commander during the campaign against Forts Henry and Donaldson. When the Confederates attempted to break through Union lines outside of Fort Donaldson, he led a counterattack that sealed the fate of the rebels. When Confederate General Simon Buckner—another former student—requested surrender terms, Smith advised Grant to accept no less than unconditional surrender. Grant agreed and was forever after known as Unconditional Surrender Grant. Smith was promoted to major general. "I owe my success at Donaldson emphatically to him," Grant wrote.

A short time later, however, Smith jumped into a rowboat and fell and scraped his leg. The wound became infected and killed him on April 25, 1862. To hear Sherman tell it, Smith was arguably Grant's superior. Had he lived, could he have replaced Grant and the public career we know today?

Confederates Have a Riot in Baltimore

Federal troops heading to Washington, D.C., were met by hostile civilians in Baltimore. At the end of the altercation, 17 were dead.

✻ ✻ ✻ ✻

On April 14, 1861, two days after the cannons fired on Fort Sumter, Abraham Lincoln called for 75,000 volunteer troops. Some of these would be assigned to defend Washington, D.C.—the federal army numbered only 13,000 at the time, with most of the soldiers stationed out West. Almost all that stood between the Capitol and the Confederates were clerks, quartermasters, and engineers.

The 6th Steps Up

The first to respond to the President's call was the 6th Massachusetts. The new troops marched through Boston in long frock coats and tall caps while the city erupted with cheers. As they passed through other towns on their way south to Washington, they were met with waving handkerchiefs, fireworks, and marching bands.

The Mason-Dixon Line—traditionally the dividing line between North and South—is the boundary between Pennsylvania, a free state, and Maryland, a slave state. Crossing that boundary, the soldiers started to notice a difference in how they were treated. By the time the 6th arrived in Baltimore, all enthusiasm had stopped. Instead, unsmiling locals greeted their train as it chugged to a stop at the President Street Station. The soldiers had a second train to catch at Camden Street Station, a mile away. Meeting that connecting train would be a battle the troops had not bargained for.

Hostile Hellos

Military planners had seriously underestimated the hostility awaiting the troops in Maryland. About the time the 6th had boarded the train in Boston, Virginia had announced its secession in protest of the president's proclamation. By the time they arrived in the Old Line State, rumors were flying that Maryland's state government would stop the advance of the Yankees to the nation's capital.

Connections between the two Baltimore stations were usually made by hooking train cars to horses and carting them from one station to the other. Seven military companies made the journey safely, but after livid Baltimore residents barricaded the roads, the remaining 260 volunteers were forced to get out and walk to the connection. Rocks, bricks, cobblestones, knives, and fists flew at the 6th as they marched and were escorted by Baltimore Mayor George William Brown, who was trying to keep the peace in his city.

Under Attack

In his official report, Colonel Edward Jones of the 6th Massachusetts recalled that the remaining troops "were furiously attacked by a shower of missiles, which came faster as they advanced." He reported that the volunteers picked up their pace, which seemed only to make the mob angrier. Jones reported pistol shots being fired at the volunteers, killing one soldier.

Although he wrote that at this point soldiers were ordered to stop and fire into the crowd to protect themselves, it is certain that some of the soldiers had already started shooting. Several in the mob were hit, and the troops again picked up their pace and continued toward the station. Mayor Brown, still marching with the Massachusetts volunteers, tried to calm the situation between them and the people of Baltimore, but as Jones wrote, "The mayor's patience was soon exhausted." Brown took a soldier's gun and fired on a citizen himself, as did a Baltimore police officer.

The scrambling soldiers who made it to the second train left behind their marching band, as well as five slain soldiers and at least a dozen dead Marylanders. Baltimore had now become a crisis. Lincoln was determined to keep the route clear for troops to pass through to Washington and even more determined to save Maryland for the Union, so he suspended the writ of habeas corpus and declared martial law. A number of citizens were jailed for inciting a riot, and even Mayor Brown and other state and city officials ultimately found themselves behind bars.

The "Battle" of New Orleans

No shots were fired nor blood shed on the soil of New Orleans in the ultimate lightning-fast takeover.

⁜ ⁜ ⁜ ⁜

THE CONFEDERATES NEVER imagined losing New Orleans. They believed its defense was impenetrable. On the Mississippi 70 miles south of the city, two brick forts, Jackson and St. Philip, were armed with 126 large guns that would surely sink any invasion fleet. "Nothing afloat could pass the forts," one New Orleans citizen proclaimed. "Nothing that walked could get through our swamps." That thinking hit a brick wall in April 1862. Led by Captain David G. Farragut, a Union squadron of 24 wooden vessels and 19 mortar schooners broke the heavy chain defense cables that had been stretched across the river near the forts and immediately blitzed and bombarded the forts. The shocked Confederates defended the forts valiantly, even using tugboats to push flaming fire rafts in the direction of the Yankee ships. This created a huge fireworks display but couldn't stop the Union fleet from sweeping past.

The Element of Surprise

The attack so thoroughly shocked Confederate General Mansfield Lovell, that he evacuated his 3,000 troops from the city. Though the mayor refused to surrender and mobs of citizens took to the streets, New Orleans fell easily on April 29 when marines entered the city to raise the American flag. Not one shot was fired.

Were the Confederates overconfident? Complacent? Historians will forever debate the circumstances that surrounded the "Battle" of New Orleans. But what's incontrovertible is that the loss of geographically crucial New Orleans—the seceded states' most populous city and busiest port—proved to be a major blow to the Confederacy.

A Step Too Far

The Union occupation of New Orleans turned ugly when the city's women started their own resistance.

❋ ❋ ❋ ❋

When Union General Benjamin Butler and Captain David Farragut captured New Orleans, federal troops occupied the city under Butler's command as military governor. The New Orleans citizenry was not happy with this turn of events, to say the least.

Butler took a draconian approach to governing the city. New Orleans residents despised the federal occupation and Butler's tactics, but men—particularly those who had fought against the Union—couldn't openly express their disdain for Butler and his soldiers without facing drastic punishment. But in the chivalrous mood of the era, women sometimes had more flexibility. If fathers, husbands, and brothers couldn't express their feelings, wives, sisters, and daughters could.

When Union soldiers passed on the streets, the usually genteel Southern ladies contemptuously crossed streets to shun them, issued insulting remarks, and gathered their skirts as if walking through mud. Some Union troops suggested that the women wanted to lure federals into an unpleasant exchange that might rally local men to retaliate. At one point, in response to a group of ladies on a balcony who had turned their backs to him, General Butler said, "Those women evidently know which end of them looks best."

The Last Straw

The situation had gone too far when one woman in the French Quarter upended a chamber pot onto Captain Farragut's head. Butler took action beyond snide remarks. On May 15, 1862, he issued General Order 28, which declared: "When any female shall by word, gesture, or movement insult or show contempt for any officer or soldier of the United States she shall be regarded and held liable to be treated as a woman of the town plying her avocation." In other words, he authorized his soldiers to treat them as nothing more than prostitutes.

For the most part, as one might imagine, the insults ceased. Women who continued to be aggressive were sent to Ship Island on the Gulf Coast. This included one lady who laughed loudly when a Union soldier's funeral procession passed. The response from locals and Southern newspapers, however, highlighted the fact that Butler had already been despised for his harsh rule of the city. Empowering his soldiers to treat disrespectful women as prostitutes struck at Southern womanhood and angered those beyond Louisiana's borders. He was soon dubbed "Beast Butler." Jefferson Davis branded Butler and his lot "outlaws" and promised to hang the general if he were captured. The Daily Mississippian offered a reward for Butler's head, and others joined the cause. "I will be good for $5,000," one reader submitted. "Let the money go to the family of the party who succeeds in the undertaking, if he should forfeit his life in so doing."

Butler's tenure in New Orleans remained controversial, and not surprisingly, his regard among the populace never improved. President Lincoln, of course, was faced with the challenge of bringing New Orleans and its residents back into the Union. Before the year was out, he had relieved Butler of his command of the city. General Order 28, however, was not rescinded.

Not Much to Bragg About

This most despised of Confederate generals was one of the most ineffective, as well.

* * * *

IRRITABLE, TEMPERAMENTAL, AND highly contentious, the hard-driving Braxton Bragg routinely antagonized his officers and alienated his troops. He never listened to the constructive advice of others, and he routinely blamed his blunders on everyone else. He was quick to execute soldiers who didn't meet his standards of discipline. Meanwhile, many officers openly rebelled against the Confederate general, his personality, and his erratic decision making. General Nathan Bedford Forrest in particular minced no words. He told Bragg: "I have stood your meanness as long as I intend to. You have played the part of a damned scoundrel . . . If you ever again try to interfere with me or cross my path it will be at the peril of your life."

No Winning Record

Some generals, though they lose popularity contests in the ranks, still manage to win battles on the field. That group doesn't include the spindly, frizzy-haired Bragg. Historians put the number of his battle victories at exactly one. Bragg's sole success occurred at Chickamauga, Georgia, in September 1863, when General James Longstreet's corps poured through a hole in the Union line. Bragg failed to turn that victory into the rout that it could have been when he allowed the Federals to retreat to Chattanooga. That misstep later cost the general and his Confederates dearly when the surrounded Union army broke out of Chattanooga and soundly defeated Bragg in one of the most pivotal battles of the war. Bragg's losses and strategic failures would also include Pensacola, Corinth, Perryville, Stones River, Dalton, and finally Wilmington in 1865.

Bragg wasn't totally lacking in good qualities. In addition to being a highly organized administrator, he apparently possessed some level of self-awareness. In a letter to Confederate President Jefferson Davis regarding Chattanooga, Bragg admitted, "I fear we both erred in the conclusion for me to retain command here after the clamor raised against me."

Friends in High Places

The general's biggest problem may have been that Davis himself was his only friend. The President always stood by his friends, no matter what level of competence they may have displayed. Thus, when he finally removed Bragg from his position as commander of the Army of Tennessee, Davis made the fatal mistake of appointing the general to be his military advisor. Bragg suggested nothing to Davis that would work. Later, put back in the field to stop General Sherman's army in North Carolina toward the end of the war, Bragg failed again. He drew vital reinforcements from where they were desperately needed by rebel General Joseph Johnston at the Battle of Bentonville. It would be the last time he was used in the war—and even he couldn't argue with that.

Quaker Guns Stall the Union

A clever Confederate ruse spelled humiliation for General George McClellan.

* * * *

In deciding how best to attack the Confederate capital of Richmond, Virginia, Union General George B. McClellan decided to transport his army by ship to Fort Monroe near Norfolk. From there, they would rapidly march up the Virginia Peninsula and capture Richmond before the Confederates could construct strong defenses. This plan collapsed in part due to a clever ruse: McClellan had been tricked by a small Confederate force using Quaker guns.

On April 4, 1862, McClellan advanced the 24 miles up the peninsula from Fort Monroe to Yorktown, flanked by approximately 50,000 soldiers. Meanwhile, Confederate General John B. Magruder's small force was ordered to hold off the federal army as long as possible while the bulk of the Confederate army, commanded by General Joseph E. Johnston, strengthened the Richmond defenses. Magruder used a series of deceptions to give McClellan the impression that he faced a force twice as large as was really there. It looked as though the area was heavily fortified with plenty of cannons and artillery. In fact, only 17,000 Southern soldiers held the 13 miles of the Yorktown defenses along the Warwick River.

McClellan decided to use siege operations to avoid a direct assault on the "heavily defended" positions. Union engineers constructed field works along the federal line, but it was a month before McClellan felt that he had enough artillery to begin action of any sort.

McClellan finally fired on Yorktown at midnight on May 5. Federals met no resistance in entering the Confederate works. Magruder had fallen back that evening, leaving behind a number of Quaker guns—logs painted black and mounted on wheels to look like cannons. McClellan had easily been deceived, which marked a poor beginning to what would be a disastrous Peninsula Campaign.

Nothing Could Keep Sheridan Down

Unlike many whose careers seesawed throughout the war, Philip Sheridan enjoyed an almost vertical rise to success.

* * * *

To look at Philip Sheridan's unimpressive early career, no one would have expected greatness. It took him five years to graduate from West Point. Eight years later, when the Civil War broke out, he still held the lowest officer grade of second lieutenant.

As a quartermaster in the Army of Missouri, Sheridan lobbied for a combat command and was finally promoted to colonel and assigned to lead a cavalry unit. Within days he took his troops into a successful battle at Booneville, Mississippi, getting himself a quick brigadier general's star and command of a division in Kentucky and Tennessee. He led that division to success in three out of four major battles.

To the Shenandoah Valley

Sheridan was moved east in 1864, getting the job of clearing Confederate General Early's army from the Shenandoah Valley. It was here where he made his most famous stand. On October 18, Sheridan was away, meeting with Grant. The next morning, Early staged an initially successful surprise attack at Cedar Creek. Returning from his meeting, Sheridan heard the fighting. Galloping swiftly to the front, he found his army retreating in panic. He quickly reorganized them and took the offensive, turning the tide. The Confederates fled.

He later helped Grant take Petersburg and Richmond, preventing Lee's escape near Appomattox, which encouraged Lee's decision to surrender.

Sheridan ended the war as a major general, an amazing climb from second lieutenant. Remaining in the Army, he died as a full general in 1888.

The Union Dress Parade

General Grant used a clever ruse to distract the Confederate army and win a major Union victory.

⁂ ⁂ ⁂ ⁂

IN THE AFTERNOON of November 23, 1863, two divisions of Union troops left their line around Chattanooga, for a plain near Missionary Ridge, an area occupied by Confederate infantry. For about an hour, the Union troops marched back and forth as if on dress parade.

The Confederates left their rifle pits to watch the military pageant. "It was an inspiring sight," observed a federal colonel. "Flags were flying, the quick, earnest steps of thousands beat equal time... The ringing notes of the bugles, companies wheeling and countermarching and regiments getting into line—all looked like preparations for a peaceful pageant, rather than the bloody work of death."

It's All a Trick

At 1:30 P.M., a cannon fired. This was a signal to the Union troops to break formation and charge the Confederate position a few hundred yards away on Orchard Knob, which was a 100-foot-high hill near the base of Missionary Ridge. Joined by three additional divisions, the determined Union soldiers stormed the rebel line, overrunning the surprised enemy defenders. The "pageant" had been a ruse to catch the Confederates off guard. It succeeded, driving the Confederates—those who could escape—to the fortifications along Missionary Ridge.

What began as a reconnaissance of the forward enemy position became a full-scale assault that exceeded General Grant's expectations. He had used the dress parade trick because the ground in front of Orchard Knob was relatively flat and treeless and provided little cover for attacking troops. The "parade" allowed the Union infantry to move closer to the enemy position without raising their suspicions. Little did Grant and his subordinates realize that this was just the opening act of a spectacle rarely seen in warfare.

On November 25, Grant ordered the troops under Generals Sherman and Hooker to attack both flanks of Bragg's line on Missionary Ridge. When each of those attacks stalled, Grant ordered General Thomas's Army of the Cumberland to seize the rifle pits at the base of the ridge. The Confederates looked to be too formidable for a direct attack, so this was primarily intended as a distraction.

An Unexpected Charge

Before the smoke from the signal guns had cleared, Thomas's force moved forward. "Fifteen to twenty thousand men," a Union lieutenant described the charge, "in well-aligned formation, with colors waving in the breeze, almost shaking the earth with their cadenced tread."

A hail of Confederate artillery and rifle fire pelted the charging federal troops, but this didn't deter them. "A terrific cheer rolls along the line," wrote the Union lieutenant. "The quick step has been changed to the 'double quick.' Another cheer, and the enemy's first line of works . . . is ours."

Without orders, the massive Union battle line quickly regrouped at the foot of the ridge and surged forward. In his memoir, Grant recorded how "our troops went to the second line of works; over that and on for the crest." Confederate fire was fierce but generally inaccurate. Afraid of hitting their own troops who were retreating from the first two Confederate lines, Bragg's gunners on the crest fired too high. Many Southern soldiers surrendered after being caught in the crossfire.

Bragg, pressured by Sherman on his right flank and Hooker on his left, watched helplessly as the center of his line crumbled. Barely escaping capture, Bragg left nearly 6,700 of his troops behind. What started out as a "harmless" dress parade ended in a major Union victory.

Miles of Mud

Downpours and thick mud put an end to Union General Ambrose Burnside's hopes of a winter attack.

* * * *

WINTER AND EARLY spring were difficult times for a Civil War army to fight. Instead of marching and battling, troops hunkered down in their camps to await the end of the rainy season. Storms that ruined gunpowder and turned roads into muddy traps for wooden wheels usually started to diminish by mid-spring.

Retaliation Time

In the middle of January 1863, General Ambrose Burnside was on thin ice with President Lincoln, having been stung by his disastrous defeat at Fredericksburg, Virginia, the previous month. In that botched assault on the Southern army entrenched on the opposite side of the Rappahannock River, Burnside crossed the river only to have Robert E. Lee's soldiers cut down thousands of his federal troops. Desperate to turn the tide of the war and repair his damaged reputation, once he felt a warming in the weather, Burnside ordered his Army of the Potomac to strike "a great and mortal blow to the rebellion."

Bogged Down

Burnside's second plan for crossing the river singled out Banks' Ford, a bit upriver from Fredericksburg, as the spot from where he would launch a surprise flank attack on Lee. The plan was to build five floating pontoon bridges to cross the Rappahannock and come up beside and behind Lee's force. On January 20, as Burnside's army of 100,000 blue-clad soldiers began their march west along the river, the ground was dry underneath their feet. But soon dark clouds formed, and before long it would not stop drizzling. The rainfall increased throughout the day, and the ground became muddier and muddier. After very slow progress—one regiment reported marching only a mile and a half for the whole day—the troops set up camp. Rain continued all night, pelting the tents and turning the landscape to muddy goo.

All the activity had not been ignored on the Confederates' side of the river. Lee had been monitoring Burnside's moves, speculating that the Federals might try to cross the Rappahannock in such a way.

Second Time's A Charm?

The next morning, the sun returned, the ground dried out a bit, and Burnside's army set out anew. But then the sky opened again. Muddy confusion reigned—in addition to the rain itself, the thousands of marching feet and the heavy wagons churned the mud to unmanageable levels. Military formations gave way to chaos as troops simply tried to make any progress at all. Wheeled weaponry and supply wagons sank to their wheel hubs in the muck. With the rain driving at them, determined teams of soldiers and mules strained to pull the wagons and cannons along the roads but couldn't make them budge. More teams of horses and mules were added to the wagon trains but could get no additional traction. Stopping in the afternoon of the second day, Burnside resolved to try again tomorrow.

As a new day dawned and the rain continued, determination gave way to a mood of wild despair. Supplies were bogged down, and engineers who had been able to get to Banks' Ford didn't have enough supplies to build even one pontoon bridge, let alone five. Ordering a whisky ration for his troops to raise morale, Burnside—himself caked with mud—strained to push a cannon into position. The alcohol simply added to the frustration, so not only did Burnside have troops that were stuck, many were now drunk, too. Tempers already at a fever pitch began to boil even more. There were reports of a massive fistfight between regiments and of a drunken artillery captain cutting down a sergeant with his sword. Across the river, the rebels did what they could to ridicule the Union effort, planting mocking signs that read, "This way to Richmond" and "Burnside stuck in the mud." By this time, soldiers were getting sick and hundreds of horses and mules had died.

Sloppy Defeat

On the afternoon of the third day, Burnside finally recognized the futility of the situation and ordered everybody back to where they had started. Conditions weren't any better on the way back, of course, so the Union army still had to withstand a great deal of suffering to return to the place they'd left a few days earlier. Over two days, the Union leader's troops slogged away to return to their original positions, defeated without a fight. The loss of equipment was as terrible as if they had been in a real lost battle. Illness and desertions made the march's casualties as high as if the army had experienced combat. To Union soldiers, the Mud March embodied the incompetence of its commanders. "Hell with Burnsides," one trooper said. To others, it became a symbol of the army's will to press on, whatever the elements.

Lee might have seized the moment to crush the federal army, already shattered without a battle, but he was wary of following Burnside's lead and rushing into the same morass himself. He later reported that the weather turned cold a week after the Mud March's first day, and two days later he was in six inches of snow. By this time, however, he was no longer facing Burnside. Before the end of January, Burnside had already been sacked and replaced by General "Fighting Joe" Hooker.

General Sherman Marched Straight into History

A man who couldn't settle down before the war gained national recognition for keeping his army on the move.

* * * *

WILLIAM TECUMSEH SHERMAN seemed to come alive on the battlefield. Before the war and at its beginning, he was viewed as a man unsure of himself, but before the war was over, he had evolved into a general full of bravado, one of the most fearsome warriors this country has ever seen.

A Disappointing Beginning

Sherman was born in 1820 in Lancaster, Ohio. When he was only nine years old, his father died, and his mother, overwhelmed, put him in the care of Thomas Ewing. Ewing later became a U.S. senator and secretary of the interior and used his influence to get Sherman into West Point at age 16. Sherman graduated sixth in his class in 1840. His military career thereafter was lackluster, however. He served mostly in Southern states, and he regretted missing the action of the Mexican War. He wrote to his future wife, "I feel ashamed having passed through a war without smelling gunpowder."

Seeing little future in an army career, Sherman quit and became a banker in 1853, running the San Francisco branch office of Lucas, Turner and Company of St. Louis. Accounts differ concerning his success; some historians say he was cautious and prudent in the role, while others call him a failure. In any case, he soon left banking to become a lawyer, a field in which he didn't gain much more success: He lost his only case. Tired of moving from one place to another and being too often separated from his wife and children, Sherman once wrote, "I am doomed to be a vagabond, and shall no longer struggle against my fate." In 1859, he finally settled for a short time into the job of superintendent of Louisiana State Seminary of Learning & Military Academy (which later changed its name to Louisiana State University). He proved himself an efficient administrator there, and he became a popular storyteller among young professors and students.

Taking a Stand with the Union

Though Sherman enjoyed his job and loved the South, he knew that he could not follow Louisiana out of the Union if it came to that. When Louisiana seceded in 1861, he went north and took a position with a street railway company in St. Louis. Two months later, when he realized war was inevitable, he volunteered to return to his Army uniform.

Sherman continued his mediocre career performance through the first half of the Civil War. He was made a colonel and saw his brigade routed along with everyone else in the Union defeat at Bull Run, although he himself was said to have performed well. Transferred to Kentucky, he blundered politically by stating that it would take a force of 60,000 to hold that state and another of 200,000 to open the Mississippi Valley. Newspapers and Northern politicians called him insane for these estimates. By the end of the war, however, his estimates had been proven right.

Sherman had backed himself into a corner and was relieved of his Kentucky post, but he'd made an important friend while in the West: General Grant. In Sherman's next few battles—Vicksburg, Jackson, and Chattanooga—he had minor success but was praised by Grant. His career was on the upswing.

Sherman Steps into His Own

When Grant was called to take command of all military operations for the war, Sherman took over command of the West. As was the case for many Union generals, his early, less successful battles had served as a training ground. In his new position, Sherman understood that his objective went beyond the military force opposing his vast army. "War is cruelty and you cannot refine it," he wrote, "and those who brought war into our country deserve all the curses and maledictions a people can pour out." Militarily, Sherman led his army against that of Confederate General Joseph E. Johnston. He pushed Johnston all the way to Atlanta and crushed Johnston's successor, John Bell Hood, in three battles outside that pivotal city. Then began his famous March to the Sea, which cemented his reputation and his place in American history. Sherman finished the war marching north through the Carolinas until hostilities ended.

He accepted the surrender of Johnson's army a little more than two weeks after Lee surrendered to Grant.

After the War

Sherman remained in the army after the fighting was finished, but his activity in the last year of the war had made him a high-profile political figure. Because he had spent so much of his time in the South before the war, he had a lot of friends in the areas undergoing Reconstruction. He always advocated a light Reconstruction policy, falling on the side of those who preferred to "welcome back" seceded states. When Ulysses Grant was elected to the office of president, Sherman took his place as general of the army, the top military officer of the day. He continued to hold that position until his retirement in 1883.

✳ Chapter 5

The Other WWII

The Bizarre Story of Rudolf Hess

On May 10, 1941, Deputy Führer of the Third Reich Rudolf Hess flew a Messerschmitt Bf-110 long-range fighter to Scotland, presumably to negotiate peace with Britain. But he wasn't acting on Hitler's orders. In fact, everyone seemed confused by the mission, including Hess himself.

✳ ✳ ✳ ✳

REICHSMINISTER RUDOLF HESS was an early lieutenant of Hitler's and had helped him create Mein Kampf in Landsberg Prison. In 1941, Hess flew to Britain without authorization, hoping to broker peace; he left his wife and son behind in Germany, perhaps expecting to return.

From the moment his parachute caught the Scottish air until his asphyxiation in Spandau Prison at age 93, Hess's life was shrouded in secrets, hidden files, conspiracy theories, and wild claims. The British and German governments both seemed shocked by Hess's bizarre flight. Given his actions in prison, many believed he'd lost his mind.

Was Either Side Really Surprised?

Perhaps not. Hess had spent four months preparing for the flight, which would have been impossible in the Nazi police state without Hitler's assent. The most surprised person was Hess, who expected dignitary treatment (however discreet) followed by a quick return home. Instead he found himself taken prisoner by a Scottish farmer.

Hess claimed that Germany would let Britain be, provided it didn't interfere with German interests in Europe, which meant Germany had no intention of giving up the land it had taken. Most likely, the Nazis wanted to negotiate peace with Britain so they wouldn't have to fight a two-front war—Germany hadn't yet invaded the Soviet Union, but it was preparing to do so.

However, Churchill knew that Hitler had a history of breaking agreements. Should Hitler defeat the Soviets, the British could logically expect him to turn on them. Churchill ordered that Hess be imprisoned for the duration of the war.

How did he handle it? He acted depressed and paranoid. He attempted suicide and claimed there were drugs in his food and drink (he made others taste them before he would eat). Some believed he was mentally unstable, others thought he may have been putting on an act since at times he seemed lucid. Some of his postwar letters while imprisoned in Spandau boasted of duping his captors into thinking him insane—but such boasts do not prove him sane.

Possibly Drugged

Early in his British captivity, Hess described symptoms that resemble being given sodium pentothal, sometimes called truth serum. In hopes of learning about Nazi plans, his captors may have dosed him with this. Even so, Hess's guards treated him quite kindly.

The Nuremberg Court Sentence

Hess had long been an eccentric, fussy hypochondriac. The Nuremberg psychiatrists pronounced him sane enough for trial, which does not necessarily mean normal. He often appeared uncomprehending or apathetic. Some thought he was faking.

Hess acted as he had in British captivity, only he seemed more paranoid, depressed, and withdrawn. He annoyed the other prisoners, wouldn't cooperate with staff, and refused to seek release or even see his family for many years. From 1966 until his death in 1987, Hess was Spandau's only inmate.

Hess had lucid times all his life, even in his feeble final years touched with senility, but those grew progressively rarer. No one knew for sure whether he'd crossed the threshold to insanity, whether faking insanity became his norm, or whether he drifted back and forth between moments of sanity and insanity.

Hess died on August 17, 1987, at age 93. Officially, he hanged himself with electric cord. Dissenters argue that he was too feeble to hang himself, that there were signs of struggle, that his suicide note looks fake, and that his second autopsy suggests manual strangulation.

Spandau was in the British sector of what was then West Berlin. An inter-Allied commission consisting of France, the United States, Britain, and the Soviet Union operated the prison. They did not apply a high standard of forensic investigation. Soon after Hess's death, Spandau was razed, every brick ground to powder and the rubble dumped in the North Sea.

Was It Murder?

Not necessarily. Hess's death meant closure of a prison, a four-power Cold War diplomatic issue, and a major expense for the West German government. For 21 years, Spandau had no purpose but to house a single ancient Nazi. At 93, likely close to a natural death, he seemed to have killed himself.

Theories of Death in a Crash and Replacement by a Double

A British flying boat did crash under odd circumstances (i.e., a flying boat crashed on land) in Scotland on August 25, 1942, 15 months after Hess's Bf-110 flight. Some allege Hess was aboard the second flight, and a double had replaced him in the

Bf-110. Major questions regarding the prisoner included loss of lifelong habits and knowledge, as well as major discrepancies, including the prisoner's lack of a major chest scar Hess received during World War II. Yet Hess's wife and son believed him genuine. Their view is not easily dismissed. Also, photos of Hess at Spandau look quite like elderly versions of earlier photos of him. Hess had a distinct look, with brambly eyebrows and a square jaw.

We May Never Know

In 2017, more British archives are slated for declassification. That doesn't mean they'll be intact. Some information is on permanent loan to the Royal Archives, where only the monarch's direct order can reveal it. This permanent loan raises suspicion that the lent files might humiliate the House of Windsor, which may be why many of them remained secret.

Operation Pastorius: Nazi Saboteurs in New York

In the seventeenth century, a group of German immigrants to the New World arrived under the care of their leader, Franz Daniel Pastorius, and founded what would become Germantown, Pennsylvania. Nearly 300 years later, a man claiming to be Franz Daniel Pastorius called the FBI from a hotel room in New York and confessed that he had led a new group of Germans to American shores. This time, however, the recently arrived Germans were saboteurs who had come to America intending to sabotage key industrial targets.

* * * *

RECRUITED AND TRAINED by the Nazi Abwehr ("Intelligence 2"), eight men reached the United States via submarine in June 1942. Split into two groups of four, they landed on beaches in Long Island, New York, and in northern Florida.

The New York group, led by George Dasch, who had lived in the United States for nearly 20 years before the war, did not fare well. While still on the dark beach, the men were discovered by a U.S. Coast Guard patrolman who doubted that they were the stranded fishermen they claimed to be. According to FBI records, Dasch offered the guardsman a $300 bribe and threatened to kill him unless he "forgot" about the event. Outnumbered on the desolate beach, the guardsman accepted the bribe and then reported the incident to his superiors. When they investigated the site, they discovered discarded German uniforms and bomb-making equipment and saw evidence of a German submarine slipping beneath the waves offshore.

The saboteurs, meanwhile, had escaped to New York City, where Dasch and another man named Ernest Burger checked into a hotel. Soon thereafter, perhaps convinced that the operation was doomed to failure, Dasch resolved to surrender. The FBI, however, had not yet learned of the Long Island incident and thought the man claiming to be Pastorius was playing a joke. Nevertheless, Dasch traveled to the Mayflower hotel in Washington to surrender in person and the FBI, having learned of the landing, immediately took him into custody. The FBI arrested the other saboteurs—the Florida group had made it as far as Chicago and New York.

All eight men were quickly tried and convicted by a secret military court ordered by President Franklin Roosevelt and condoned by the Supreme Court. Six of the men were executed, but at the request of FBI Director J. Edgar Hoover, President Roosevelt commuted Dasch and Burger's sentences to long prison terms in return for their cooperation. Both men were granted clemency by President Truman in 1948 and subsequently deported to Germany.

During their stay, the New York contingent had spent more than $600 on clothing, meals, lodging, and travel. Besides helping the local economy, Operation Pastorius had no effect upon America's war effort, and with the exception of one other occasion in 1944 that ended similarly to Pastorius, no other German agents landed on the shores of the United States during the war.

The Sullivan Brothers

The tragic loss of five brothers who volunteered and served in the South Pacific epitomized America's commitment to winning the war.

* * * *

ON FEBRUARY 14, 1942, the crew of the light cruiser USS *Juneau* stood proud as their ship was commissioned in a ceremony held at the New York Navy Yard. Members of the commissioning crew included the five Sullivan brothers of Waterloo, Iowa: Albert, Francis, George, Joseph, and Madison.

The brothers, whose motto was "We Stick Together," stipulated as a condition of their enlistment that they be allowed to serve aboard the same ship. Despite accepted navy policy, which discouraged but did not forbid family members from serving together, their request was approved.

After the *Juneau*'s Atlantic christening, the cruiser was assigned to the South Pacific. On November 13, 1942, the *Juneau* participated in an intense naval battle near Guadalcanal. One torpedo launched from Japanese submarine I-26 scored a direct hit on *Juneau*'s magazine. The resulting explosion cut the cruiser in half. *Juneau* sank within minutes, taking many of the trapped crew to watery deaths.

While dozens of *Juneau*'s crew were killed in the initial explosion, approximately 115 sailors survived the attack. All but ten perished in the following days as they awaited rescue. Survivor accounts indicated four of the Sullivan brothers died in the initial explosion, while the fifth, George, survived the attack but died in the water five days later.

The Sullivan brothers were survived by their parents, sister, and one spouse. News of the brothers' deaths was exploited for the war effort. A navy press release dated February 9, 1943, stated that the Sullivans' parents visited "war production plants urging employees to work harder."

President Roosevelt wrote a letter to Mrs. Sullivan after learning her sons were listed as missing in action: "...the entire nation shares your sorrow. I offer you the condolence and gratitude of our country. We, who remain to carry on the fight, must maintain the spirit in the knowledge that such sacrifice is not in vain. The Navy Department has informed me of the expressed desire of your sons...to serve on the same ship. I am sure that we all take pride in the knowledge that they fought side by side."

Although an act forbidding family members from serving in the same military unit was proposed after the death of the Sullivan brothers, no such act was ever passed by Congress. In addition, no president has ever issued an executive order forbidding family members from serving in the same unit or on the same ship.

Two U.S. Navy ships were named in honor of the Sullivan brothers: USS *The Sullivans* (DD-537; a destroyer) was commissioned in 1943, while USS *The Sullivans* (DDG-68; a guided missile destroyer) was commissioned in 1997.

Fugitive Nazis in South America

Nazi leaders were sentenced to death after the war, but a few managed to escape the grasp of Nuremberg and sneak into South America.

* * * *

IN JULY 1972, an old man complaining of intense abdominal pain checked himself into a hospital in Sao Paulo, Brazil. The admitting physician noted that the man, obviously a foreigner, looked much older than the 46 years listed on his identity card but accepted the excuse that the date was misprinted. The man had an intestinal blockage: He had a nervous habit of chewing the ends of his walrus mustache, and over the years the bits of hair had accumulated in his digestive tract. The man had good reason to be nervous. He was Josef Mengele, a Nazi war criminal, who had been hiding in South America since the end of the Second World War.

After the war, many Nazi war criminals escaped to start new lives in nonextraditing countries like Argentina, Brazil, Uruguay, Chile, and Paraguay. They lived in constant danger of discovery, though in several cases, friendly dictatorships turned a blind eye in exchange for services. Many were found and kidnapped by the Israeli Mossad, tried, and sentenced for their crimes. Some, including Mengele, the cruel concentration camp doctor, eluded capture until their deaths. Mengele died by accidental drowning in 1979.

South America was a welcome refuge from the net of international justice that swept Europe at the end of the war. Several of the countries had enjoyed friendly prewar relations with the Reich—Argentina, in particular, was sympathetic to the plight of the fugitive Nazi war criminals. Many Nazis found positions in Argentina's Fascist government, controlled by Juan Perón. His wife, Evita, later traveled to Europe, ostensibly on a goodwill tour. However, she was actually raising funds for the safe passage of war criminals.

In Argentina, Uruguay, Chile, Bolivia, and Paraguay, scores of Nazis found familiar traditions of elitism and militarism. Perhaps most importantly, corruption in those countries made it easy to obtain false identification papers through bribery.

Nazis Who Slipped Overseas

* Walter Rauff invented the Auschwitz "death trucks" that killed more than half a million prisoners. Rauff was arrested in Santiago, Chile, in 1963, but the Chilean authorities released him after three months in jail. Many believe that

Rauff designed the concentration camps where Chilean political prisoners were killed under the Pinochet regime.

* Paul Schäfer served in the Hitler Youth. Years later, he was arrested for pedophile activities while working at an orphanage in postwar Germany. Schaefer fled to Chile where he established a bizarre settlement known as Colonia Dignida, in which abuse, torture, and drugs were used to control cult followers. The adults were taught to call Schaefer "The Führer"; the colony's children called him "Uncle." Schaefer sexually abused hundreds of children in the decades that followed. The colony prospered economically and, despite information from escaped members, it remained in operation until 1993 when Schaeffer went into hiding. He managed to escape justice until 2006 when he was sentenced to 20 years in prison.

* Klaus Barbie, the notorious "Butcher of Lyon," became a counter-intelligence officer for the U.S. military which helped him to flee to Bolivia soon after the war. In 1971, he assisted in a military coup that brought the brutal General Hugo Banzar to power. Barbie went on to head the South American cocaine ring called Amadeus that generated funds for political activity friendly to U.S. interests in the region. When a more moderate government came to power in 1983, Barbie was deported to France, where he was tried and convicted of war crimes. He died of cancer after spending four years in prison.

Nueva Germania

German colonists had settled in South America long before the start of the Second World War. One of the more outlandish attempts occurred in the late-nineteenth century when Elisabeth Nietzsche-Forster and her husband Bernhard Forster led a group of 14 families to Paraguay. They founded a utopian village dedicated to anti-Semitism and an "authentic rebirth of racial feeling." They called their remote settlement Nueva

Germania. Within two years, many of the colonists had died of disease, Bernhard drank himself to death, and Elisabeth returned to Europe where she successfully worked to have her husband's writings adopted as the favorite philosophy of the Nazi Party. After the Second World War, many fugitive Nazis were rumored to have sheltered in Nueva Germania, including Josef Mengele.

Submarines Full of Nazi Gold

Despite evidence suggesting that Hitler's secretary, Martin Bormann, died in the streets of Berlin in 1945, rumors persist about his escape to South America. According to the same accounts, Bormann witnessed his Führer's suicide and then followed prearranged orders to escape to Argentina, where he had been transporting large amounts of Reich gold and money using submarines. In the final days of the war, go the stories, Bormann and dozens of other Nazi officials left Germany in ten submarines, five of which arrived safely in Argentina. Waiting for them was a friendly government and the stolen wealth of Europe, safely deposited by the Peróns in Swiss bank accounts. Some estimates place the accumulated wealth at $800 million.

Code Name "Diane"

During stints for both British and U.S. intelligence, Virginia Hall so excelled at her duties that she became a marked woman by the Gestapo and ultimately was awarded the U.S. Distinguished Service Cross.

* * * *

"The woman who limps is one of the most dangerous Allied agents in France," proclaimed Gestapo wanted posters, showing a young brunette American. "We must find and destroy her." So dangerous was Virginia Hall's position that even her wooden leg was given a code name, "Cuthbert." Escaping France by crossing the Pyrenees on foot in November

1942, Hall cabled London that "Cuthbert is giving me trouble, but I can cope." Misunderstanding that Cuthbert was another agent, a Special Operations Executive (SOE) officer cabled back, "If Cuthbert is giving you trouble have him eliminated."

Born in Baltimore, educated at Radcliffe and Barnard colleges, and fluent in French and German, Hall had aspired to a Foreign Service career and worked at the U.S. Embassy in Warsaw in 1931. Her hopes were dashed a year later when she accidentally shot herself during a hunting trip in Turkey and her left leg was amputated. In Paris at the outbreak of World War II, Hall volunteered for the French Ambulance Service Unit. When France fell to Germans in June 1940, Hall trekked to London and volunteered for British intelligence.

During 15 months of SOE service, Hall was instrumental in Britain's effort to aid the French resistance. Working from Vichy, she posed as an American journalist while securing safe houses, setting up parachute drop zones, and helping rescue downed Allied airmen. After the United States entered the war, Hall went underground. Her position became untenable when German troops occupied Vichy following Rommel's defeat in North Africa and she barely escaped to Spain.

Back in Britain, Hall volunteered for the U.S. Office of Strategic Services (OSS) and trained in Morse code and wireless radio operation. Unable to parachute because of her leg, she landed in Brittany by British patrol boat prior to the D-Day invasion. Code-named "Diane," she contacted the French Resistance in central France and helped prepare attacks supporting the Normandy landings.

Still hunted by the Gestapo, Hall adopted an elaborate disguise as a French milkmaid, layering her fit physique with heavy woolen skirts that hid her limp. Peddling goat cheese in city markets, she listened in on the conversations of German soldiers to learn the disposition of their units. Hall helped train three battalions of partisan fighters that waged a guerrilla campaign against the Germans and continued sending a valuable stream of intelligence until Allied troops reached her position in September 1944.

After the war, President Truman awarded Hall the Distinguished Service Cross, though she turned down a public presentation to protect her cover for future intelligence assignments. In 1951, she joined the CIA as an intelligence analyst and retired in 1966. She died in Rockville, Maryland, in 1982.

Major Richard I. Bong: America's "Ace of Aces"

Calvin Coolidge, wet laundry, and a plane named Marge helped create America's highest-scoring ace of all time.

* * * *

THE SCENE WAS a farm on the outskirts of Poplar, Wisconsin. The year was 1928, and the event was the daily flyover by a plane carrying mail to President Calvin Coolidge's summer home in Superior. An eight-year-old boy watched with rapt fascination as the plane passed directly over his house. Even at that tender age, Dick Bong knew he wanted to be a pilot. He grew up a typical American boy of the period—fishing, sports, church, and chores—but he never lost his dream of becoming a pilot. While attending Superior State Teacher's College, Dick became a certified pilot through the government-sponsored Civilian Pilot Training Program and soon afterward joined the Army Air Corps. He never looked back.

Dick Bong's instructors, including future presidential candidate Barry Goldwater, noticed the young man's natural abilities. Bong's penchant for antics while stationed at Hamilton Field near San Francisco, however, raised the indignation of his commanding officer, Major General George C. Kenney. He'd been willing to tolerate loops around the Golden Gate Bridge and low-level flights through the city's business district, but reached his breaking point when a lady called the base to complain that one of the pilots had flown so low over her house that all the laundry had blown off her line. Bong was ordered to go to the woman's home and hang whatever laundry needed to be dried.

In October 1942 Bong was assigned to the 9th Fighter Squadron in the Southwest Pacific. He soon began racking up kills in his P-38 Lightning, which he named Marge after his fiancée. He developed a dogfighting technique in which he would swoop down on enemy planes and engage them at extremely close range. By 1943, he was an ace (five kills), and on April 2, 1944, he shot down his 27th aircraft, surpassing Eddie Rickenbacker's World War I record. He was sent home on leave but returned in time for the South Pacific campaign, in which his tally reached 40 confirmed kills. Douglas MacArthur personally awarded Bong the Congressional Medal of Honor, and noted that the pilot had "ruled the air from New Guinea to the Philippines." General Kenney pulled Bong out of combat and sent him home with orders to marry Marjorie and start a family. In the meantime, his services promoting war bonds were in constant demand.

At home Bong was a hero. The international press covered his wedding and his accomplishments continued to be celebrated. However, he wanted to be a test pilot. Soon after their wedding, Dick and Marjorie moved to Dayton, Ohio, where new aircraft were being tested at Wright Field. On the day that the atomic bomb was dropped on Hiroshima, Dick Bong was tragically killed testing a Lockheed P-80 Shooting Star jet plane.

Mavericks with Wings: The Flying Tigers

The ragtag group of volunteer pilots and outdated planes may not have seemed impressive at first, but the Flying Tigers earned a legendary reputation for defending the Burma Road.

❋ ❋ ❋ ❋

IN 1937, CAPTAIN Claire L. Chennault of the U.S. Army Air Corps was 44 years old and suffering from partial deafness. He had served in the First World War, had spent time as the Chief of Pursuit Training, leading the development of combat tactics, and had been part of the three-man Air Corps' precision flying team, "Three Men on a Flying Trapeze," but he was fed up with military bureaucracy. He had argued frequently with his superiors. Chennault retired from the Air Tactical School in 1937, and after receiving an offer from Madame Chiang, wife of Generalissimo Chiang Kai-shek, to train China's fledgling air force, he moved to China.

The Chinese pilots considered instruction demeaning, and Chennault watched helplessly as the nation's air force was reduced to almost nothing by the superior Japanese pilots. During this time, however, Chennault was able to observe the Japanese pilots' tactics firsthand, and he amassed a collection of training manuals and aircraft specs from downed enemy planes.

In 1940, he returned to the United States with the intention of recruiting American pilots and planes to form a volunteer air force, paid by the Chinese but ostensibly operating under the aegis of the Central Aircraft Manufacturing Company (CAMCO, an American company that also assembled planes in China). Despite the protests of numerous military officials, President Roosevelt gave the plan his approval. The company soon recruited 300 men (100 pilots and 200 ground crew) to form the core of the American Volunteer Group (AVG) in

China. The pilots traveled on the Dutch ship Jaigersfontaine, and carried passports that listed occupations as diverse as metalworker, teacher, and farmer.

Birth of the Tigers

Once in China, the group occupied an old British air base, and Chennault began relentlessly drilling his men on the geography of their new home and the flying tactics employed by the Japanese pilots. Chennault managed to procure 100 Curtiss P-40B Tomahawk fighters—ill-equipped for combat, these planes lacked gun sights, bomb racks, and provisions for attaching auxiliary fuel tanks. Nevertheless, the intrepid ground crews learned how to fashion crude ring-and-post gun sights for the planes and improvised using spare parts and scrap metal to make up for their other deficiencies. Though slower and heavier than the Japanese fighters, the P-40Bs were exceptionally tough and could dive much faster than the lighter enemy craft. Chennault taught his pilots how to use these traits to their advantage.

Soon after arriving in China, some of the pilots noticed a magazine illustration of a P-40 in North Africa emblazoned with a grinning shark face on its nose. It made an impression, and soon all of the AVG planes were painted with the characteristic fierce smiles. Walt Disney Studios designed many of the U.S. squadron logos, including a winged tiger flying through a victory "V," and the group quickly earned the nickname the "Flying Tigers."

In case they were shot down in friendly territory, each pilot was issued a silk blood chit scarf with the free China flag, which read: "This foreign person has come to China to help in the war effort. Soldiers and civilians, one and all, should rescue, protect, and provide him medical care." Some of the pilots sewed the chits on the back of their flight jackets, others simply stuffed them in their pockets.

The Tigers Prove Their Mettle

The only supply route to interior China without access to China's ports in the fall of 1941 was the tortuous Burma Road. Beginning at the port of Rangoon, supplies were transported to Lashio in northern Burma and finally carried over the 717-mile road to Kunming in Southwest China. At the behest of the British defenders of Rangoon, Chennault sent one of his three AVG squadrons to Rangoon. On December 25, 1941, the AVG shot down 23 Japanese planes in their first and biggest single-day victory. On January 23, the Japanese launched an all-out attack on the city, losing 21 more planes to AVG. In 10 weeks of intense battles over Rangoon, the Flying Tigers shot down 217 aircraft with the loss of only 4 American pilots and 16 aircraft. The British pilots barely broke even against the Japanese, while the Americans established a 15–1 kill ratio.

The Legend Lives On

During the war, the AVG shot down 286 enemy aircraft and kept the port of Rangoon open for two and a half crucial months, saving China from collapse. Moreover, its victories provided a much-needed morale boost for Chinese and Americans in the darkest days of the war.

Winston Churchill praised the efforts of the AVG in defending Rangoon. "The victories of these Americans over the rice paddies of Burma," he wrote, "are comparable in character, if not in scope, with those won by the RAF over the hop fields of Kent in the Battle of Britain."

The AVG disbanded on July 4, 1942, and was absorbed into the U.S. Army Air Force. By that time, many of the original pilots had returned to their respective services; the techniques they had learned from Chennault spread quickly through the American squadrons. Many of the war's top aces saw their first combat as Tigers, including Tex Hill, Charlie Bond, John Petach, and Gregory "Pappy" Boyington, who went on to form the Black Sheep squadron.

Rommel's Way

Led by "Desert Fox" Erwin Rommel, Germany's Afrika Korps nearly took the Suez Canal. Was it leadership? Tactics? Luck? Enemy bungling? All of the above? How on earth did Rommel do it?

* * * *

GERMAN GENERAL (LATER Field Marshal) Erwin Rommel fought for Africa's Mediterranean coast from 1941 to 1943. He took over a disintegrated Italian position in western Libya, stormed all the way to Egypt, was forced to retreat to his starting point, and rebounded again to within 60 miles of Alexandria—the gateway to the Suez Canal. In the end, even with his forces caught in Anglo-American pliers, he fended off his enemies in Tunisia for months.

Conditions in Africa

* The German Army was low on supplies near Tripoli, a problem that grew worse every mile his men advanced from that main base.

* The German panzers in Africa were adequate, but rarely superior.

* Rommel's men came to Africa unschooled in desert warfare.

* His Italian allies interfered, and disappointed him in battle.

* British code breaking enabled the Allies to read Rommel's reports to and directives from Germany. (Although Rommel also had access to decoded British messages.)

* In the end, it took a devastating blow followed by a crushing pincer movement to run Rommel and his Afrika Korps out of Africa.

* The Allied divisions in Africa proved tough adversaries for the Afrika Korps. Among the esteemed units were the Coldstream Guards, the 9th Australian Division, the Long-

Range Desert Group, the Royal Gurkha Rifles, General Freyberg's New Zealanders, Klopper's dour South Africans, the Free Frenchmen of Bir Hacheim, and the Seaforth Highlanders—all proven warriors. The Allied forces generally had enough food, water, and fuel. They had air superiority. While some of their gear was unreliable, in the desert nothing was reliable for anyone. While their commanders made mistakes, so did Rommel; in any event, their leaders were battle-tested career military men with guts of iron. And unlike Rommel's superiors, Churchill told his leaders to be aggressive—to march to the sound of their guns.

Rommel's Rules

The Afrika Korps' special élan emanated from its famous chieftain. Rommel's policies embedded themselves in his soldiers' minds. Soon his officers and men did things his way by reflex. If not, Rommel found out, and someone was sorry.

One of his rules taught itself by necessity: Don't waste. Rommel would not squander supplies or equipment, nor did his men. They couldn't afford to, for the Afrika Korps lived in a state of chronic scarcity. Getting Axis supplies past British-held Malta was like climbing a razor-wire fence: He who succeeds also bleeds. If his officers asked Rommel for more supplies, he would order them—without sarcasm—to steal from the Allies, who had plenty. Often they would.

Like most outstanding commanders, Rommel knew that soldiers will nearly always obey two commands: "Do as I do!" and "Follow me!" Afrika Korps troopers knew that Rommel might show up anywhere. He piloted his own reconnaissance plane over enemy territory. In one famous incident, he dropped a weighted note to his troops telling them to get moving or he would come down there. If Rommel's vehicle got stuck, Rommel would help push it free. He ate what his men ate; he took time to thank his troops for work well done. He set the example and expected his officers to follow.

Rommel stressed independence and flexibility. To understand why, it helps to grasp the low troop density of the theater. Virtually the whole campaign was fought along or near a single key road stretching over 1,000 miles, with an average of five to eight effective divisions on both sides until the end in Tunisia. (For comparison, the Russian front had hundreds of divisions. The day it was attacked in 1940, even the Royal Dutch Army had ten divisions.) This meant a scattered battlefield with units spread widely apart in open desert, requiring junior officers to act independently and with initiative. In far-flung desert warfare, his units reorganized themselves as situations developed. Rommel's standing order was "In the absence of orders, go find something and kill it." They usually did.

Near the end, with his forces pushed back in Tunisia, Rommel's health forced him back to Germany. Hitler refused Rommel permission to return to Tunisia and fight on with his troops, but their leader's spirit survived. As they prepared to surrender to the Allies, some of his last Afrika Korps troopers sent Germany a radio message worth remembering: "Ammunition expended. Weapons and equipment destroyed. Obedient to its orders, the Afrika Korps has fought until it can fight no more. The German Afrika Korps will rise again! Heia Safari!"

Had they done any less, they might well have worried that Rommel himself would come to their POW camps under a flag of truce to give them a piece of his formidable mind.

The Most Decorated Unit

From internment camps to combat in Italy, the men of the 442nd "Go for Broke" regiment of mostly Japanese-Americans proved themselves tenacious and loyal soldiers.

* * * *

When the Japanese attacked Pearl Harbor, future U.S. Senator Daniel Inouye, a Japanese-American high school student on Oahu, raced on his bicycle to help out at a first-aid station. An elderly Japanese man grabbed the handlebars of his bike, asking the 17-year-old Inouye, "Who did it? Was it the Germans? It must have been the Germans."

"I shook my head, unable to speak," recalled Inouye, a second-generation American, or Nisei. The Japanese-Americans "had worked so hard. They had wanted so desperately to be accepted, to be good Americans. Now, in a few cataclysmic minutes, it was all undone."

Uprooted After Pearl Harbor

For months, rumors flew around Hawaii that Japanese-Americans had sabotaged power plants and cut a giant arrow in the mountains pointing at Pearl Harbor. There were, in fact, some Japanese-American spies. One ring in Honolulu monitored U.S. warships entering and leaving Pearl Harbor. The West Coast Tachibana ring collected data on weapon factories, dams, and army and navy bases. But, after the attack on Pearl Harbor, roundups on the Hawaiian Islands of 1,800 first-generation Issei extended to businessmen, judo instructors, and Japanese language teachers. Every Buddhist monk, whose religion was linked to Japan, was taken into custody.

On the mainland, 110,000 Japanese-American residents, mostly from Washington, Oregon, and California, were ordered into internment camps. All Americans of Japanese descent were suspended from military service.

In Hawaii, however, 40 percent of the population was Japanese-American, and internment would have wrecked the economy. The small Hawaiian National Guard, largely Nisei in composition, was a different matter. The commander of the U.S. Army in Hawaii, General Delos Emmons, didn't trust them. He recommended sending the Guard's 298th and 299th Infantry Regiments to the mainland for reorganization. In June 1942, they shipped out.

Uncertain Loyalties

"When we landed in Oakland on June 12," recalled the 298th's Raymond Nosaka, "everyone thought we were prisoners." The Hawaiian regiments were formed into the 100th Infantry Battalion. Their ranks swelled in February 1943 when President Roosevelt rescinded the ban on Japanese-Americans in the military.

Still, prospective GIs were handed a loyalty questionnaire. Question 28 asked, "Will you . . . forswear any form of allegiance or obedience to the Japanese emperor, or any other foreign government, power, or organization?" Question 27 asked, "Are you willing to serve in the armed forces of the United States on combat duty, wherever ordered?" About a quarter of Nisei men, especially those with families interned on the mainland, answered "No" or didn't answer. Eventually, about 3,000 Japanese-Americans from Hawaii and 800 from the States joined for service.

Ronald Oba of the 100th Battalion was 17 when a Honolulu policeman spoke to his high school, telling the students, "You must volunteer, to prove your loyalty." Oba recalled, "I didn't take kindly to his words, because I was born an American . . . I didn't need to prove my loyalty." But Oba volunteered for the army for patriotism's sake. "My country needed me," he said.

Training the 100th Infantry Battalion

Combat training began at Camp McCoy, Wisconsin, continuing with more drills at Camp Shelby, Mississippi. There, remembered Oba, "Our superiors wouldn't let us off the trains until after dark, to allay the fears of Americans who hadn't seen Japanese before." He added, "[The] next morning, a newspaper headline read, 'Japs Invade Hattisburg.'"

Every officer, from company commander up, was white. Protestant chaplains took the place of Buddhist monks. There were also tensions at first between enlistees from the mainland (known as "katonks"—from the sound of a coconut falling on a head) and those from Hawaii (nicknamed "buddhaheads"—from Japanese-English slang, "buta," or "pig-headed").

Some recruits were sent to Camp Savage, Minnesota, for training as translators and interrogators of Japanese POWs. Others went to a secret training program at Cat Island, Mississippi, to act as "guinea pigs" for military Doberman pinschers and German shepherds. The army was attempting to train guard dogs to recognize the scent of Japanese soldiers. But "officials realized," said Nosaka, "that Japanese blood doesn't smell any different than American blood."

Action in Italy

As Japanese-Americans were barred from fighting in the Pacific, the 100th fought in North Africa, Italy, and France. It formed part of the 34th, then later the 36th Division, of the U.S. 7th Army.

In Italy, the men saw heavy action at Monte Cassino, the bombed-out Benedictine monastery that blocked the way to Rome. The fighting there in early 1944 helped the 100th earn its nicknames, "the Purple Heart Battalion" and "the little iron men." In January, in an attack along the flooded Rapido River, the battalion faced the crack German 1st Parachute Division; of three assaulting American companies of 187 men, 173 were casualties. The next month, the battalion fought halfway to the monastery, but had to retreat when the units on its flanks couldn't keep up. One platoon lost 35 men out of 40. From an initial strength of 1,300, the 100th was reduced to 521 effectives.

In another Italian battle near San Terenzo, now-Second Lieutenant Inouye won a Distinguished Service Cross. "With complete disregard for his personal safety," reads his citation, Inouye "crawled up the treacherous slope to within five yards of the nearest machine gun and hurled two grenades, destroying the emplacement . . . Although wounded by a sniper's bullet, he continued to engage other hostile positions at close range until an exploding grenade shattered his right arm. Despite the intense pain, he refused evacuation and continued to direct his platoon." Inouye lost his right arm; in 2000, his award was upgraded to the Medal of Honor.

"Go for Broke!"

The 100th Battalion merged into the 442nd Regiment in June 1944, and in August joined the invasion of southern France. The unit's motto was "Go for Broke." Its most memorable action occurred that October in the frigid, thickly forested Vosges Mountains near the German border. The Wehrmacht had cut off 211 men of the 141st Regiment, "the Lost Battalion," on a ridge and surrounded the approach with mines, tanks, machine guns, and mortars. The U.S. commander for the region told the 442nd, "There's a battalion about to die up there and we've got to reach them."

For many terrible days the regiment attacked the ridge. "We were charging uphill all the time," recalled 442nd veteran Henry Arao. The Germans "were just sitting waiting for us with machine guns. They had the hills loaded with mines. If you walked in the wrong spot, you'd get your leg blown off." Stung by frostbite and trench foot, soldiers cut their boots off of their swollen feet and slept in the snow. "The daytime sun doesn't penetrate there; it's dark as hell," recalled the 442nd's Tom Goto.

Surrounded for six days, the San Antonio-based Lost Battalion, nicknamed "the Alamo Regiment," was losing hope. One of its sergeants, Bill Hull, remembered, "I thought I was going to die." So "I kept shooting, not worrying about saving my ammunition. Then suddenly there was a lot of noise behind me . . . when I turned to look I saw this little Japanese-American soldier jumping into the dugout." After four days, the 442nd had broken through, but at horrific cost. At the start, the outfit had 2,943 men. But its casualties were at least 800, with more than 200 killed or missing.

The Purple Heart Battalion had a full strength of just 4,500 men. Yet in three years of combat, as men funneled in and out, it received 9,486 Purple Hearts and 18,142 decorations for bravery. These included 4,000 Bronze Stars with 1,200 Oak Leaf Clusters, more than 500 Silver Stars, 52 Distinguished Service Crosses, and 20 Medals of Honor. Seven hundred and forty-seven men were killed or MIA. In proportion to its size, it was the most highly decorated unit in the annals of the U.S. military.

The 442nd's Stanley Akita became "struck by the fact that other American units . . . wanted soldiers of the 442nd next to them in combat." Akita recalled the samurai role models he enthusiastically admired as a child. He noted, "We were taught from a very early age not to shame the family name, or embarrass the family by being a coward."

The Real River Kwai

For most people, the "bridge on the river Kwai" conjures memories of the Oscar-winning movie about British prisoners of war made to build a bridge in Japanese-occupied Southeast Asia. They might recall the film's compliant British commander, the firm-but-fair Japanese leader, the daring commando raid on the bridge. Yet in most respects, the movie got it wrong. In fact, the prison camp there was part of a system that resulted in some of Japan's worst war crimes.

* * * *

Building the "Death Railway"

THE KHWAE YAI, or Kwai, river bridge was part of a massive construction project, the Thailand–Burma Railway, to supply Japanese troops fighting in Burma (now Myanmar) and support an invasion of British India. The Allies were blocking Japanese freighters from sailing around the Malaysian Peninsula, and Tokyo needed a land-based route to transport 3,000 tons of supplies daily.

From September 1942 to October 1943, 61,000 Allied POWs and more than 200,000 native laborers constructed more than 600 bridges and laid track through deep mountain passes and thick jungle along the 257-mile line. The railway, built from track and other materials looted from Malaysian and Indonesian railroads, stretched from terminals near Bangkok to the Burmese coast. Decades earlier, British surveyors had deemed such a project almost impossible, requiring a minimum of five years to complete. The Japanese succeeded in 16 months—due to the ruthless use of forced labor for the "Death Railway."

The nickname was apt. Of the 61,000 POWs, about 16,000 died. Of those, 6,318 were British, 4,377 American, 2,815 Australian, and 2,490 Dutch. The death rate was higher among the native conscripts, mostly Thais, along with Burmese and Chinese—perhaps 100,000 perished. They endured brutal treatment; were poorly fed; and suffered from beriberi, dysentery, and other tropical diseases. Work shifts lasted 12 to 18 hours. Laborers at the mountainous "Hellfire Pass" were forced to lug 110-pound sacks of rice and share the burden of 600-pound buckets of concrete.

The Nightmare of the Khwae Yai River

One of the major spans constructed along the route was Bridge 277, the concrete-and-steel Khwae Yai bridge, immortalized in The Bridge On the River Kwai. It was built by British POWs at the nearby prison work camp of Tha Maa Kham, or Tamarkan, in Thailand. The 1,200-foot-long construction was composed of 11 arches laid atop pylons painstakingly pulled and then pounded into the riverbed. A POW who survived described the toil:

"[The] Jap standing on the riverbank [shouted] through a megaphone the required rhythm . . . You pulled in unison, you let go in unison. 'Ichi, ni, san, si, ichi, ni, san, si' . . . Hour after hour, day in, day out, from dawn to dusk. On returning to camp it was often difficult to raise the spoon to eat the slop issued to us. Your arms protested in pain, often preventing you from snatching some precious sleep. And yet, come dawn you repeated the misery of the previous day."

In the movie, the bridge is made of wood, but only a temporary span at Khwae Yai was made of that material. The movie also falsely portrays Tamarkan's real-life senior officer, British Lieutenant Colonel Philip Toosey, played in the movie by Alec Guinness as Lieutenant Colonel Nicholson. The real Philip Toosey refused to kowtow to the Japanese, but tried to foil them and protect his men. Under his charge, the POWs poured inferior concrete for the bridge and collected white ants to eat wooden constructions. Toosey complained about his men's mistreatment (although he was beaten for doing so), helped several prisoners escape, and contacted a daring Thai merchant, Boonpong Sirivejjabhandu, to smuggle in medicine and food. Refusing special treatment, he had all officers reside and eat with the enlisted men; his own weight dropped from 175 to 105 pounds during internment. Toosey was later knighted and made a brigadier general.

The film's portrait of Toosey enraged ex-POWs. Indeed, the movie was based on a Pierre Boulle novel Le Pont de la Rivière Kwai, which was based on the author's memory of French, not British, collaborationists.

There was no mistaking the kind of prison life later described by former POW Fred Seiker. Ordered up-country from Tamarkan to work on railway embankments, he recalled:

"You carried a basket from the digging area to the top of the embankment, emptied it, and down again to be filled for your next trip up the hill . . . The slopes of the embankments consisted of loose earth . . . very tiring on thigh muscles and painful, often resulting in crippling cramp. You just had to stop . . . Whenever this occurred the Japs were on you with their heavy sticks, and beat the living daylights out of you."

Caught snatching a container of fruit at another camp, Seiker was tied to a tree with barbed wire, beaten, left overnight, and told, falsely, he was scheduled for execution.

The dangers of disease surpassed the perils of work. Tropical ulcers disfigured limbs, sometimes requiring amputation—without anesthesia. Cholera, transmitted by water and through mud, was prevalent in the rainy season. During monsoons, recalled POW Rod Allanson, "We slept on boards within inches of black slimy mud ... We would return from the railway covered in mud and extremely tired ... We would go out to work each day stepping over the bodies of dead Tamils, and trudge four miles through more mud to the railway." In some camps, cholera killed almost all the native workers.

With barbed wire, the Japanese quarantined cholera-afflicted camps, ordering the corpses burned. During epidemics, inmates piled bodies around the clock onto smoking pyres. "It was particularly macabre ... during nighttime," remembered a POW. "Bodies would suddenly sit up, or an arm or legs extend jerkily."

Camp doctors held "sick parades" late into the night to count the number of ill, trying to convince guards to give the afflicted days off. To survive, men resorted to ingenuity and tricks. Short of intravenous tubes, orderlies cut up stethoscopes to make drips. One overworked physician drew on his experience to later devise medical breakthroughs: After the war, Dr. James Pantridge, a camp doctor, codeveloped cardiopulmonary resuscitation (CPR) and invented the portable defibrillator.

The savagery of camp life ended suddenly with war's end, as fleeing Japanese left the sick and malnourished inmates for rescue by Red Cross trains steaming up the railway the prisoners had built. After war crimes trials, more than 200 Japanese involved in the Death Railway were hanged.

As for the critical Khwae Yai bridge, it was destroyed, not by commandos, as in the movie, but by bombs. Allied flyers attacked the bridge eight times. In April 1945, U.S. pilots flying just above it hit the bulls-eye with four 1,000-pound bombs, destroying two spans, rendering the bridge impassable.

The destroyed sections were rebuilt by Japan as war reparations to Thailand. The bridge, and a lengthy section of the railway, are in use today—along with museums commemorating the tens of thousands lying in mass graves nearby.

Pearl Harbor II: The Improbable Boat Plane Raids

On March 4, 1942, the Japanese public heard news of a second air attack on Pearl Harbor. Supposedly quoting a broadcast from Los Angeles, radio and newspaper reports claimed that considerable damage was done to military installations, with the deaths of 30 sailors and civilians. Propaganda aside, two seaplanes did reach Pearl Harbor for a second encounter in 1942.

* * * *

AFTER DECEMBER 7, 1941, the Japanese Imperial Navy searched for new ways to tie down U.S. forces defending the Hawaiian Islands. A bombing raid was planned using new H8K "Emily" flying boats. Code-named Operation K, the plan called for two planes to fly from the Marshall Islands to a rendezvous with Japanese refueling submarines at French Frigate Shoals, an atoll halfway between Midway and Hawaii. Relying on a full moon for visibility, each Emily would carry four 550-pound bombs, with the intended target of docks at Pearl Harbor.

Lieutenant Hisao Hashizume, an expert seaplane pilot proficient in open-ocean navigation, was chosen as commander. Taking off from the Marshalls on the night of March 3, the two boat planes met two submarines at French Frigate Shoals without incident and completed the dangerous operation of refueling at sea. Closing in on Hawaii on the early morning of March

4, Lieutenant Hashizume was unaware that U.S. radar stations on the islands had already spotted them as potential bogies. Flying at 15,000 feet, cloud cover had obscured the raiders' visibility by 80 percent.

At Pearl Harbor, air-raid alarms were sounded and fighters scrambled to intercept the planes. With the lighthouse at Kaena Point his only visible reference point, Lieutenant Hashizume made his best guess and dropped his bombs on the slopes of Mount Tantalus; the second plane dropped its payload into the ocean at the entrance to Pearl Harbor. U.S. fighter pilots, radar operators, and ground controllers failed to find the Japanese aircraft as they banked toward the west on their flight back to the Marshall Islands. Taking 35 hours and covering 4,750 miles of the Pacific Ocean, the raid was the longest air mission flown by any nation in the war and proved the feasibility of using submarines to extend the range of seaplanes. The second raid was canceled because a delay meant inadequate moonlight for visibility. After its defeat at Midway, the Imperial Navy never considered launching a similar raid.

In the end, Operation K proved much more valuable to the United States. The raid demonstrated the need for airborne radar and fighter pilots trained in night interception, which proved devastating to Japanese fliers in the final years of the war.

The Emperor's Savage Samurai

The widespread German and Soviet atrocities on Europe's eastern front were approximately matched, horror story for horror story, in the Pacific Theater, where the Japanese soldier's ideology and training produced a particularly brutal killing machine.

* * * *

THE VIOLENT IDEOLOGY of the Imperial Japanese Army soldier was cultivated even before his induction into the

Emperor's service. The popular concept of yamato damashi, or "spirit of ancient Japan," was a cultural moral code stressing absolute loyalty to the Emperor, as heaven's regent of the Sacred Islands. The combination of this nationalist spirit with Buddhist and Shinto religious beliefs (which preached reincarnation and the Emperor's divinity) produced a warrior's creed expressed as the modern bushido, a chivalric code originally applied to the samurai class as far back as the twelfth century.

Modern bushido indoctrinated Japanese recruits with simple, inflexible precepts: "Honor is everything." "A soldier never surrenders." "Death in the Emperor's service is an honor." Widespread belief in reincarnation made the thought of death more palatable for Japanese conscripts than for many of their Western counterparts, who generally considered death a necessary evil—but one to be avoided by smart tactics.

Part of the brutal nature of the Japanese rifle-company soldier stemmed from tactical doctrines, which favored cold steel when attacking. Because of a relative inferiority of Japanese firearms, Japanese soldiers spent more time than their Allied counterparts on bayonet practice, which encouraged close-in combat with a distinctively personal element of violence.

Another ingredient of a deliberately brutal soldier was physical punishment. The army's absolute code of obedience allowed superiors to beat and otherwise abuse inferiors. Enlisted men and auxiliary troops (such as Koreans in Japanese service) came in for the worst beatings.

After the invasion of China, one former sergeant major recalled: "In training, we were forced to engage in actual charges in order to kill live humans. They talked about POWs, but in the Japanese Army, any Chinese we caught were called POWs. Didn't matter if they were peasants or what. There were three men tied to stakes, all ready for us. I had applied to be a petty officer, so I got to go first. My squad leader signaled me, shouting, "Forward, forward, thrust!" But it was by no means easy to thrust my bayonet into a living person. I'd never killed anyone before. Still, if I was clumsy at it, it would affect my whole military career. So I showed no mercy or leniency."

Japanese brutality entered its full flower in China, where some 20 million ethnic Chinese died at the hands of the Imperial Japanese Army. As one Japanese officer later conceded: "The major means of getting intelligence was to extract information by interrogating prisoners. Torture was an unavoidable necessity. Murdering and burying them follows naturally. You do it so you won't be found out. I believed and acted this way because I was convinced of what I was doing. We carried out our duty as instructed by our masters. We did it for the sake of our country. From our filial obligation to our ancestors. On the battlefield, we never really considered the Chinese humans. When you're winning, the losers look really miserable. We concluded that the Yamato race was superior."

When Japan widened the war to the United States and the British Commonwealth, its soldiers treated their new foes with greater brutality than their German counterparts did. One historian estimated that a Western POW stood a 4 percent chance of dying before the war's end in a German camp, while the same POW had a 30 percent chance of dying in Japanese hands. It was a miserable, trickle-down effect of the martial spirit that animated the Emperor's soldiers.

The Nazis and the Freemasons

Although Hitler admired its organizational structure, the group that inspired America's Founding Fathers to enshrine the rights to life, liberty, and the pursuit of happiness found no favor in Nazi Germany, and suffered horribly during the war.

* * * *

ALTHOUGH MOST PEOPLE are familiar with the Nazi persecution of the Jews, many are not aware that the

Freemasons also suffered in Hitler's Germany. Adolf Hitler's anti-Mason sentiments were no secret. In *Mein Kampf*, the Nazi dictator had stated that the pacifistic philosophy of the Freemasons encouraged a "paralysis of the national instinct of self-preservation." He believed its members had succumbed to the Jews, who were using the organization as a means to spread their alleged agenda of subverting the upper strata of society.

The reasons Hitler feared the Freemasons and perceived them as a threat to the Nazis are complicated. Surely the Nazi regime could list many prominent Masons among its enemies, including British Prime Minister Winston Churchill, American President Franklin Delano Roosevelt, and General John Pershing, who had directed the American Army against the Germans in the First World War. But, those figures were not the only reason for the Nazis' determination to wipe out Freemasonry in Europe.

Hitler was more afraid of the basic tenets of Freemasonry, which decree that personal political freedom and human dignity are the rightful foundations of all human society. The Nazis perceived Masons as a threat because the group defined human freedom in terms that superseded feelings of nationalism or racial allegiance. Many Freemasons were known as leading figures of the Enlightenment. Men such as George Washington, Benjamin Franklin, and Simón Bolívar led revolutions against aristocratic power. It was natural, then, for a totalitarian regime such as Nazi Germany to want to destroy the organization. As U.S. Supreme Court Justice Robert H. Jackson, the chief prosecutor at the Nuremburg trial, observed, "... many persecutions undertaken by every modern dictatorship are those directed against the Free Masons [sic]."

Justice Jackson understood, as did Hitler, that members of such an organization were not likely to cooperate with the oppressive police states necessary for the survival of a dictatorship. As such, Freemasons were regularly put to death by the SS.

The Secret Stash of the Bomb Project's Foreign Agent

U.S. officials working on the atomic bomb project frantically searched the world for sources of uranium. However, they were unaware that enough nuclear material to make several bombs lay for the taking at a New York City warehouse.

❋ ❋ ❋ ❋

U.S. ARMY COLONEL Kenneth Nichols was engaged in what he thought was a nearly impossible task. With the war at its height in September 1942, U.S. researchers were beginning work on the Manhattan Project. Yet the United States possessed no uranium, the essential building block for the atomic bomb. The Manhattan Project's chief, Army General Leslie Groves, had ordered Nichols to obtain some of the precious material.

Nichols paid a visit to the New York City office of Edgar Sengier, a Belgian tycoon. Sengier was director of the Union Minière du Haut Katanga, a company that owned the world's most precious uranium mine, located in the Belgian Congo's Katanga province.

Nichols asked Sengier if his company could procure uranium for the U.S. government. He added that he was well aware this was an unusual request that could not be swiftly fulfilled.

Sengier carefully checked Nichols's credentials. Then he replied, "You can have the ore now. It is here in New York, a thousand tons of it. I was waiting for your visit."

As it turned out, Sengier had long been aware, far more than most U.S. officials, of the critical importance of his firm's product. As war broke out in Europe, Sengier learned from British and French scientists that it might be possible to devise a superweapon out of processed uranium. In May 1939, he met with Henry Tizard, chief science adviser to the British government. That month, he also consulted with Jean Frédéric Joliot, the cowinner with wife Irène Joliot-Curie—daughter of Marie Curie—of the Nobel Prize in Chemistry. Joliot had taken out patents on path-breaking work to induce chain reactions in atomic piles using uranium and heavy water (deuterium oxide).

Shifting the Stash

In fact, Sengier negotiated with the French to supply them with 55 tons of uranium ore. When the Germans invaded France in May 1940, Union Minière shipped some eight tons of the promised uranium oxide to Joliot's research team. The material, hidden in French Morocco during five years of German occupation, became the basis for France's postwar nuclear program.

Sengier also sought to aid other nations in the fight against the Nazis. He reestablished his firm's headquarters in the safe harbor of New York and had 1,250 tons of uranium ore shipped there. The consignment cleared customs and was unloaded at Staten Island. It was stored—unguarded and unnoticed for nearly two years—at a warehouse there.

Jesse Johnson, the Atomic Energy Commission's director of raw materials, later remarked, "M. Sengier told members of the State Department and other government officials about the shipment, but the secret of the atomic bomb was so closely guarded that no one with whom he talked recognized the importance of the information."

The shipment was a true mother lode. The only other source of uranium at the time, with a uranium content of only .02 percent, was at distant Great Bear Lake in northern Canada. The Katanga ore, in contrast, was up to 65 percent uranium.

After Colonel Nichols's visit, the army bought and transferred Sengier's stockpile. The ore, said Johnson, supplied "the bulk of the uranium for the [Manhattan Project's] development work and early production of fissionable material for bombs."

The army, in the meantime, in one of the war's little-known and most far-reaching operations, acted to exploit the Union Minière's other valuable holdings.

The Katanga mine, Shinkolobwe, had been closed since 1939 because its mineshafts had flooded. The Army Corps of Engineers sent units to restore the mine, to construct a port on the Congo River, and to renovate an airport in Léopoldville. The United States, with British backing, negotiated a ten-year deal for exclusive rights to the Shinkolobwe uranium with the Belgian government-in-exile. Sengier stayed on to manage the mine, and before the war's end, the army had bought up 12 times Sengier's original hoard—some 30,000 tons.

Sengier was also awarded the Belgian Ordre de la Couronne, the French Légion d'Honneur, and the title of Knight Commander in the Order of the British Empire. Sengierite, a radioactive crystal, is named after him. In 1946, General Groves, in President Truman's presence, presented Sengier with the prestigious Medal for Merit—he was the only non-U.S. citizen to receive the accolade up to that time. Details of the Manhattan Project were still hush-hush, so the citation was cloaked in vague language that lauded the Belgian's "wartime services in the realm of raw materials."

The Sleeping Giant Awakens

During World War II, Canada transformed itself from a junior player in the Allied camp to one of the world's leading military and economic powers.

* * * *

On the eve of World War II, Canada was not ready for war. A nation of only 11 million people in 1939, it had a tiny, under-equipped army, an obsolete air force, and a tin-pot navy. Its economy was still in the stranglehold of the Great Depression.

Yet once the war started, Canada quickly mobilized its people and resources. More than one million Canadians would volunteer for service in the armed forces as Canada transformed its military from weakling to juggernaut. Unemployment would virtually disappear as Canada's basket-case economy became an industrial dynamo. By the war's end, Canada would make the fourth largest contribution to the Allied war effort and earn recognition as a major world power.

A Titan on the High Seas

At the outbreak of war, the Royal Canadian Navy (RCN) was perhaps the most war-ready branch of Canada's armed forces. That wasn't saying much—the RCN consisted of just over 1,800 officers and sailors and a fleet of 15 ships. But most RCN personnel were well-trained and experienced from prewar exchanges with the British Navy and would form the core of a world-class naval corps.

Canada immediately embarked on a massive shipbuilding program; the country's moribund shipyards increased production tenfold by 1940. The RCN also began a recruitment campaign for men to crew the new ships, which in 1942 was expanded to include women for shore duties so that more men could be freed for ship duty.

In time, the RCN built a "small-ship" fleet of destroyers, corvettes, frigates and minesweepers that became indispensable in the Battle of the Atlantic. Initially limited to defending Canada's Atlantic coast, the RCN played a leading role in protecting Allied convoys crossing the Atlantic and evolved into a killer antisubmarine force. In 1944, RCN vessels were redeployed to the Pacific to fight with the British Pacific fleet.

By the war's end, the RCN ranked as the world's third-largest navy. More than 89,000 men and 6,000 women served in the RCN aboard 471 warships and smaller fighting vessels, most of which were Canadian-built. Approximately 2,000 Canadian sailors died in action.

Just as impressive was the expansion of Canada's merchant navy, which played an equally vital role in keeping the trans-atlantic lifeline to Britain intact. During the war, Canadian shipyards built some 400 merchant ships. Twelve thousand Canadians served in the merchant navy, which was arguably a more perilous duty than the regular navy: More than 1,600 merchant sailors perished at sea.

Air Canada

In September 1939, the Royal Canadian Air Force (RCAF) was barely an air force. The RCAF had 4,061 personnel, of which only 235 were pilots—fewer than the number of Canadian pilots in the British air force. Only 19 of the RCAF's collection of 275 aircraft were suitable for modern warfare.

Given the state of the RCAF, it was decided that Canada's initial contribution to the Allied air war effort would take place on the ground. Canada established and operated the British Commonwealth Air Training Plan (BCATP) which trained 131,553 aircrew personnel in Canada, including 49,507 badly needed pilots. More than 70,000 BCATP graduates were Canadian; others came from Britain, Australia, New Zealand, France, Poland, and Norway. Recognizing the success of the BCATP, U.S. President Franklin Roosevelt praised Canada as "the aerodrome of democracy."

Canada did eventually make a significant contribution in the skies. Aggressive recruitment and a burgeoning Canadian aircraft industry led to a rapid expansion of the RCAF, which counted 86 squadrons at its peak. Canadian fighter squadrons fought in the epic Battle of Britain and later, Canadian fighter-bombers supported Allied ground offensives in Western Europe. An all-Canadian heavy bomber group participated in the Allies' strategic bombing campaign against Germany, while Canadian maritime patrol bombers fought German submarines. Canadians also flew combat, reconnaissance, and transport missions in Asia and the Pacific. By 1945, Canada boasted the world's fourth-largest air force. Some 232,000 men and 17,000 women served in the RCAF during the war. More than 17,000 Canadian air force members lost their lives.

On the March in Europe

The largest branch of Canada's armed forces in September 1939 was the army. By the end of the month, after calling up reserves, the army numbered 55,500 soldiers. But they were far from ready: They lacked equipment, weapons, uniforms, and boots.

Again, with rapid mobilization of Canadian industry and aggressive recruitment, Canada's army ballooned. By mid-1942, the Canadian Army numbered 400,000 plus men and women in infantry, armored, artillery, and auxiliary units.

The Canadian Army went on the offensive in July 1943 with the invasion of Sicily. The 1st Canadian Infantry Division and 1st Canadian Armored Brigade fought all the way up the Italian peninsula as part of the British 8th Army, engaging in some of the toughest battles of the campaign. The 3rd Canadian Infantry Division and 2nd Canadian Armored Brigade also emerged as formidable fighting forces, commencing with the landing of Canadian troops on Juno Beach on D-Day. Canadians faced and defeated some of the German Army's most fanatical troops as they fought their way through northern France and Belgium. In February 1945, Canadian forces in Italy transferred to northern Europe and a unified 1st Canadian Army, under independent Canadian command, liberated Holland. Over the course of the war, more than 730,000 men and women enlisted in the Canadian Army. More than 45,000 Canadian soldiers found their final resting places in Europe.

An Economic Powerhouse

Once at war, the Canadian government controlled virtually every aspect of the country's economy and mobilized it for war production. During wartime, Canada's economy would recover from the Depression and develop a modern, highly industrialized infrastructure. By the war's end, Canada's economic contribution to the Allied war effort had had as great an impact as its military contribution.

Canada morphed into an industrial powerhouse during the war. It was the world's second-largest producer of military vehicles behind the United States—its production of more than 800,000 trucks outstripped that of the three Axis powers combined. Canada also became a leading aircraft manufacturer, producing more than 16,000 planes of all types. Booming Canadian shipyards, ammunition plants, and textile factories also supplied the Allied war effort.

Canada's agricultural sector experienced unprecedented growth and production levels during the war years. The country's prairie regions produced record wheat harvests, and production of raw materials skyrocketed, especially metals. During the war, Canada accounted for 40 percent of the Allies' aluminum production and a whopping 95 percent of its nickel.

As in the United States, Canada implemented a strict wartime ration program for everyday commodities such as meats, sugar, dairy products, coffee, rubber, and gasoline. More than one million Canadian women entered the workforce to help keep the factories and farms at full production.

A New World Power

Canada emerged from World War I as a nation; it came out of World War II as major world power. By the end of the war, Canada possessed the world's fourth-largest military and was its third-largest industrial producer.

After the war, Canada willingly descended in rank to a middle power and adopted a quintessentially Canadian role in the world as a "peacekeeper." Most Canadians are justifiably proud of their country's unique identity among the world's nations.

Agent 488: Carl Jung and the OSS

World War II called individuals from all walks of life to duty—even famous scientists had parts to play. In America, Einstein wrote Roosevelt to suggest the possibility of an atomic bomb. In Germany, Wernher von Braun raced to build rockets for his country. And in Switzerland, Carl Jung used his own talents to offer the world a glimpse into the psyche of a madman.

⁂ ⁂ ⁂ ⁂

Services in Demand

In the 1930s, Germany had a reputation as a hotbed of scientific innovation across a variety of disciplines. The Nazis were quick to take advantage of much of this research in the form of technologically advanced weapons, but they did view one particular field with suspicion: psychology. The most well-known figure in this science was Sigmund Freud, who was Jewish—and although they were interested in the topic itself, the Nazis couldn't very well make use of a Jewish scientist.

One alternative was Carl Jung. He was Swiss, of German extraction, and well respected in the field. Moreover, he had written about the psychological differences inherent in groups of people, and anything that supported tribal or racial divisions was of interest to the Nazis. They went so far as to suggest that Jung relocate, and even discussed arresting him on a visit to Berlin just to keep him in Germany. He declined the invitation, as he also did with an offer to migrate to the United States, preferring to remain in touch with his roots. Unfortunately, the German invitation gave rise to a rumor that he was a Nazi sympathizer, an accusation he called "an infamous lie."

His accusers needn't have worried. He "despised politics whole-heartedly" and had little use for Hitler or any other leader except, perhaps, as a study in the psychology of power. His independence may have contributed to a request he received in October 1939 from Nazi doctors, who asked if he would be willing to examine the Führer; Hitler had been behaving erratically, and his personal doctors were worried. Jung turned down the invitation, already convinced that the German leader was at least half crazy. He went on to offer the opinion that both Hitler and the German people were possessed, but his invectives were directed at both sides—he once referred to Roosevelt as the "limping messenger of the apocalypse" and believed that the president had all the makings of a dictator.

Recruiting Agent 488

Regardless of his apolitical views, Jung did contribute to the Allied cause. In 1942, American agent Allen Dulles (who would go on to head the CIA) arrived in Berne, Switzerland, which he used as a base to monitor German activity. He enlisted the aid of an American expatriate of some local notoriety, Mary Bancroft, who eventually also became Dulles's lover. Bancroft had been a patient of Jung's, and the two had formed a mutual attachment. Knowing of their association, Dulles began posing questions to the famous psychologist through Bancroft, soliciting his opinion on "a weekly, if not daily basis."

The two eventually met, and Dulles continued to rely on Jung to analyze events. Jung's responses found their way into Dulles's reports, with Jung listed as the source under the code name "Agent 488." Of particular interest was Jung's insight into the personality of the Nazi leaders, especially of Hitler. Dulles urged the OSS to pay particular attention to Jung's analysis of the Führer, which included the opinion that Hitler was capable of anything up until the very end, at which point Jung could not "exclude the possibility of suicide in a desperate moment."

The analysis proved prophetic: Hitler became increasingly debilitated and unstable as the war progressed, eventually taking his own life. After the conflict, the records detailing Jung's full involvement were classified, but Allen Dulles offered one evaluation: "Nobody will probably ever know how much Professor Jung contributed to the Allied cause during the war."

Architects Go to War

Two of the world's most prominent architects helped build German and Japanese working-class villages in the Utah desert. For good reason, no one ever inhabited the buildings.

⁕ ⁕ ⁕ ⁕

IN 1942, WITH war raging in the Pacific and in Europe, the U.S. military and the Standard Oil Company constructed two villages in the Utah desert. The first was modeled after a typical Japanese worker's village; the second was designed to the specifications of German housing for low-wage workers.

The military spared no expense in creating the replica villages. They hired prominent architects to design the structures. Prison laborers constructed the edifices in record time. In the case of the German village, wood was imported from Russia, and the buildings were hosed down to replicate the wet Prussian weather. Finally, designers from RKO Picture Studios furnished the buildings in exact detail, from the bedding, linen, and drapes to the paper wall partitions typical of Japanese interior design and bulky furniture characteristic of German working families' homes.

When the buildings were finally just right, the army destroyed them using napalm, gas, anthrax, and incendiary bombs.

Then they rebuilt them and blew them up again . . . and again.

The site was the Dugway Proving Grounds weapons-testing area 90 miles southwest of Salt Lake City. Created at the request of President Roosevelt, the program's goal was to measure, as accurately as possible, the potential damage from anticipated massive aerial bombing campaigns on enemy cities—in particular, the effects upon the manufacturing segment of the populations. To further this end, the army staff hired two influential architects whose unique knowledge of their respective subjects made them ideal candidates to design mock structures for the tests. Certainly neither man had ever contemplated designing a structure so that it could be immediately destroyed.

Erich Mendelsohn

Erich Mendelsohn was a Russian-Polish Jew from Germany who had interrupted his architecture studies to drive a Red Cross ambulance for the Kaiser's army during the First World War. His bold modernist designs marked him as a master of steel and concrete structures that utilized space to create a unique identity for each project. Paramount among his early work is the Einstein Tower, which still stands in Potsdam near Berlin. Persecuted by the Nazis, Mendelsohn moved first to London and then to Palestine. Eventually, Mendelsohn relocated to the United States where he addressed and befriended peers such as Frank Lloyd Wright. Perhaps because of his persecution by the Nazis, Mendelsohn was eager to assist in the design of German worker housing for the Dugway tests. He was particularly useful in the design of the roofs, which would receive the brunt of any bomb damage. When Germany surrendered in May of 1945, Allied raids had eliminated 45 percent of German housing.

Antonin Raymond

Antonin Raymond was a native Bohemian who had emigrated to the United States in 1910. In 1916, he worked as an associate in Frank Lloyd Wright's studio Taliesin, in Wisconsin, and aided in the design of the Imperial Hotel in Tokyo. While in Tokyo, Raymond developed a deep fondness for Japanese design—a passion that would last his entire life. Nevertheless, he was amicable when the U.S. Army asked him to design the Japanese village at Dugway. Constructed almost entirely of wood, the mock village burned to the ground after a few tests. Its fate predicted that of real Japanese cities that were bombed during the war. In a single raid upon the paper-and-wood city of Asakua, for instance, 334 B-29 bombers dropped 2,000 tons of napalm and destroyed an unprecedented 265,171 buildings.

German POWs Enjoy the War Stateside

World War II was known for the brutal conditions imposed on POWs in most countries. However, many German POWs in the United States had it relatively easy. In fact, some may have been treated better as prisoners in America than they would have been if they remained in the German Army.

❋ ❋ ❋ ❋

IF A GERMAN soldier were left with no option but to surrender, he would prefer it to be to the Americans. While more than three million of their comrades disappeared into the Soviet gulag system, nearly 400,000 German POWs were fortunate enough to be shipped to the United States. The first POWs arrived in May 1942, when 33 U-boat crewmen were rescued after the sinking of the U-352 off North Carolina's Outer Banks. Following the defeat of the Afrika Korps, a large number of German POWs arrived in the United States beginning in August 1943. Over the next two years, German prisoners were housed in 47 of the 48 states, though most went to rural areas in the South, Midwest, and West.

For Some, Waiting Out the War Wasn't So Bad

Thousands were sent to Camp Clinton outside Jackson, Mississippi. The camp was unique in that it housed the highest-ranking German officers, including 40 generals and 3 admirals. General Jurgen Von Arnim, who had been Rommel's replacement in North Africa, was among those held in Camp Clinton. But his sentence was hardly punishment—Arnim was given a car and driver. He frequently visited movie theatres in Jackson to enjoy the air-conditioning.

While officers were allowed to forgo labor as part of the Geneva Convention governing POW treatment, prisoners at 3 other Mississippi camps and 15 sub camps worked in cotton fields for much of the year. In Iowa, more than 10,000 German POWs were housed at Camp Algona and 34 sub camps in Iowa, Minnesota, and the Dakotas, providing millions of hours of manpower to farms in the region. Similar regional camp systems were organized across the country.

German POWs were generally astonished at how well they were treated in the U.S., often enjoying a quality of life better than they had in the German military. Though required to work, they earned 80 cents a day, which they used at camp canteens to purchase items such as candy bars and cigarettes. They were housed in tents with wood heaters or rough barracks covered with tarpaper roofs. Rations were plentiful, and most thought the food was good. Camps had their own dentists, doctors, libraries, and newspapers. Prisoners played sports, formed bands, and watched movies for recreation.

Defying Captors

While camps were segregated by branch of service and rank, German POWs maintained their own rigid chains of command. Conflicts sometimes arose between ardent Nazis and other prisoners. In a number of cases, "traitors" to the Fatherland were found murdered. Camp authorities began segregating Nazi fanatics and holding them at special camps in Oklahoma.

Some German prisoners staged incredible escape attempts. On December 23, 1944, 25 former U-boat crewmen at the Papago Park camp in Arizona escaped through a 178-foot tunnel, hoping to reach Axis sympathizers in Latin America. All were recaptured within six weeks. At Camp Clinton, a 100-foot-long tunnel was discovered only ten feet away from completion. Incredibly, in many cases, prisoners simply walked away—a frantic search for 30 POWs who disappeared from another camp in Mississippi found them strolling on the streets of nearby Belzoni. The prisoners told the FBI that they had been bored. A Luftwaffe pilot in Mississippi disappeared with the wife of a planter on whose plantation he had been working. The couple was soon found at a hotel in Nashville and told authorities they planned to make their way to New England, steal a plane, and fly to Greenland together.

Democracy in Prison?

In the last year of the war, an ambitious "reeducation" program sponsored by the War Department sent liberal arts professors into more than 500 camps. The professors spoke to the prisoners about the value of U.S.-style democracy and made anti-Communist presentations. Although the program was considered a somewhat laughable failure, most POWs did retain a positive view of America when they returned home.

Their experiences in the United States also apparently changed some attitudes. At Camp MacKall in Richmond County, North Carolina, prisoners organized political parties and voted on issues to understand the democratic process. At Camp Butner north of Durham, North Carolina, a local Jewish merchant of a particularly generous nature provided POWs with band equipment. After viewing newsreels on the liberation of Nazi concentration camps, the inmates reportedly burned their army uniforms in outrage.

Tens of thousands of German POWs remained in the United States for over a year after the war ended. President Truman decided to keep them in the United States to alleviate the labor shortage. The POWs often worked in agriculture, as America made the transition from wartime to peacetime farming.

At least one decided to stay after the war. Kurt Rossmeisl walked away from Camp Butner on August 4, 1945, and traveled to Chicago. He lived under the name Frank Ellis, got a job, and joined a local Moose lodge. He finally turned himself in on May 10, 1959, 14 years after the end of the war.

"It's Not Illegal If You Don't Get Caught"

Americans made many sacrifices during the war, but for some, forgoing sugar, meat, or gasoline was too much. Instead, they turned to the black market.

* * * *

SINCE THEIR COUNTRY'S founding, Americans have generally enjoyed the benefits of a free market economy. During the Second World War, however, the Roosevelt administration realized it would need to control the consumption of many basic goods in order to effectively fight a two-front war.

The World War II-era rationing and price-control system was the brainchild of Wall Street financier Bernard Baruch. He first suggested the scheme in early 1941, arguing that the government should apply rationing vertically (for example, rationing would affect not only automobiles, but also the steel, rubber, and cloth used to make them). To curb inflation, price controls would be needed. Americans would learn to sort through their ration books for items such as gasoline and meat. What Baruch did not take into account, however, was the rise of a vast black market fueled by many otherwise law-abiding citizens.

The OPA

The modern American black market is said to have been born on January 27, 1942. On that day, the Office of Price Administration (OPA) was given authority to enact civilian rationing and price control under Directive No. 1 of the War Production Board. Among the list of items classified as "scarce" were sugar, automobiles, tires, gasoline, and typewriters. Violators could face up to a year in jail and a $5,000 fine, but that was hardly a deterrent. Manufacturers, distributors, retailers, and consumers soon found ways to evade and sometimes profit from the price controls and rationing systems.

Consumers learned that with enough money, they could readily find what they wanted—regardless of government regulations. By some estimates consumer industries such as department stores, meat packers, and leather tanners realized profits as high as 1,000 percent during the war.

The subterfuge took many forms: trimming less fat from meat, counterfeiting gas vouchers, processing livestock through unregulated channels, and ignoring rent controls. Counterfeit vouchers, often sold through organized-crime syndicates, were the most common form of black-market exchange. One arrest in Detroit yielded 26,000 counterfeit vouchers that had been sewn into the lining of gang members' coats.

The black market could not have existed, however, if a large number of Americans had not been willing to engage in illegal trade. To most citizens, the transactions seemed so innocuous that they probably never thought twice about the corner gas station owner selling a few extra gallons for a bit more money or their friend the butcher providing them with a larger cut of meat for the same price.

Efforts to enlighten the public did little. In February 1944 Patricia Lochridge wrote an article for Woman's Home Companion titled "I Shopped the Black Market." In it, she detailed how homemakers, ministers, bankers, and other average Americans willingly engaged in illegal activity. Realizing the effect of this trade, the OPA launched campaigns that equated purchasing black market meat with doing "business with Hitler." For the most part, however, Americans ignored the pleas of the ineffective bureaucratic agency. In fact, many sympathized with those who were punished for transgressing the price controls.

Profiteering

Many of the items bought through black market channels had their origin in the military, which was where the goods had been funneled. While the penalty for selling goods within the armed services was severe, even rumors of executed transgressors did little to slow the brisk business. In the final months of the war, cigarettes were more valuable abroad than any country's currency. Robert F. Gallagher remembers that while serving as an MP in Belgium, he and his friends often used intermediaries to sell their cigarette rations to locals for a hefty profit.

Another common form of profiteering involved the illegal sale of currency. Some soldiers claimed to have made thousands of dollars buying and selling foreign currencies in the confusion and economic depression of postwar Europe. Gallagher: "It's not illegal if you don't get caught."

It is nearly impossible to quantify the amount of black market activity that occurred in the United States during the war. Some have claimed that at the height of the price controls, a majority of the citizens of New York engaged in black market exchanges, and 90 percent of the meat being shipped from San Antonio, Texas, came from black market sources.

The black market flourished in part because Americans mistrusted the goods' regulation. Equally important, the OPA was relatively powerless to enforce its controls: Popularly elected officials were reluctant to take measures of which the majority of their constituents would disapprove. Any society that has attempted to overregulate its market has had to increase its security and monitoring forces in kind. The police states engendered by Nazi Germany, Soviet Russia, and scores of Third World dictatorships did exactly that to secure the sanctioned exchange of goods and defend government property. Ironically, it was the war against Fascism and the police states of Nazi Germany and Imperial Japan that gave rise to regulation in America—and to the black market.

Last Stand of Germany's Volkssturm

In a last-ditch attempt to stem the tide of the advancing Allied armies, Adolf Hitler issued a call to arms. The creation of the Volkssturm, or "People's Storm," was a desperate act that needlessly cost thousands of Germans their lives.

* * * *

IN THE LAST weeks of the summer of 1944, Germany's Nazi regime found itself fighting a defensive war on two fronts. As American, British, and Canadian forces marched east from the beaches of Normandy, the Russian juggernaut rolled westward, catching the German Army in an ever-tightening vice.

Desperate Measures

German manpower losses were mounting at an alarming rate, especially on the eastern front where Hitler's "no retreat" policy cost the lives of tens of thousands of soldiers. So Hitler, on the advice of his personal secretary and Nazi Party Chancellery Chief Martin Bormann, ordered the formation of a nationwide militia. In October 1944, the Nazi dictator issued what can only be described as a desperate decree, drafting all men between the ages of 16 and 60 not already serving in the military into local militia units known as Volkssturm.

The order effectively conscripted thousands of men and boys into compulsory, quasi-military service. The units were composed of the elderly, members of the Hitler Youth, and those previously classified as unfit for military service.

Command of the Volkssturm fell under the Home Guard, which was controlled in the last two years of the war by SS leader Heinrich Himmler. The decision to place control of the men under Nazi party control, as opposed to the army, ensured the Nazification of the recruits. The Nazi leadership hoped fanatical resistance by self-sacrificing groups of Volkssturm fighters would inflict unacceptable casualties on Allied forces and create a stalemate, forcing the Allies to back away from their demands of unconditional surrender.

A Lack of Organization

The Volkssturm operated independently of the German Army until committed to battle. Because of weapon and ammunition shortages, Volkssturm members received only rudimentary training, usually at the hands of retired former World War I servicemen. Training focused on the art of close-quarters antitank warfare, as opposed to hit-and-run guerrilla tactics.

There was no standardization when supplying or arming units, and many were issued captured weapons. Some units were armed with the Mauser Kar-98k bolt-action 7.92mm rifle, and most Volkssturm members were shown how to use the Panzerfaust, or "armor fist," recoilless antitank grenade/rocket launcher, which was available in large quantities. This disposable, preloaded weapon fired an explosive charge to maximum range of roughly 100 yards and was capable of penetrating almost eight inches of armor. The Panzerfaust's simplistic design allowed the weapon to be mass-produced. During the Battle of Berlin, it was employed by Volkssturm units with effective results—the Russians lost an estimated 2,000 armored vehicles during the battle, due in large part to the antitank weapon.

The Volkssturm also had no standard uniform. Members were issued an armband marked with the words "Deutscher Volkssturm" and "Wehrmacht." Wearing such an armband meant captured Volkssturm would be treated like regular military combatants under the Geneva Convention and not as partisans, who would be summarily executed.

Group, section, company, and battalion leaders were issued collar tabs featuring from one to four pips to signify rank. Some men went into battle wearing their suit jackets, while others wore whatever military clothing they could cobble together.

Levy Breakdown

Conscripts were organized around geographic boundaries and were supposed to be called for duty only in the event enemy forces threatened their hometown or county. The February 1945 issue of Intelligence Bulletin, a soft-cover publication compiled by the Military Intelligence Service, indicated the physical area a Volkssturm unit was responsible for defending:

Squad: one city block

Platoon: equivalent to a U.S. precinct

Company: equivalent to a Congressional district

Battalion: equivalent to a U.S. county

Volkssturm personnel were divided into four categories, or levies. Levy IV contained men of limited physical ability. Levy III consisted of 16- to 20-year-olds and fell under the control of the local Hitler Youth organization. Teenagers in this levy were usually moved out of harm's way, although some Hitler Youth units did see action in Berlin. Most recruits fell under the auspices of either levies II or I. Placement in either of these levies was dependent on how valuable the conscript's civilian occupation was to the war effort—the more important the job, the less likely the person would have to fight.

The most extensive use of Volkssturm units was during the Battle of Berlin, where hundreds of civilians fought to their deaths rather than surrender and face an unknown fate at the hands of Russian captors. Berlin and the metropolitan area were ripped by more than 2 million artillery shells. From an assembled force of 2.5 million Russian infantry, a million took part in the all-out street assault on Berlin. Although close quarters briefly favored the defenders' portable antitank weapons, roughly 152,000 Berliners were killed during the battle. By some accounts, Volkssturm fighters comprised nearly half of German forces during the fight for the capital. Total German police and Volkssturm casualties during the war numbered 231,000.

The Daring Exploits of Hitler's Favorite Commando

An unremarkable engineering student who came to be considered the "most dangerous man in Europe" during the war went on to consult for other undercover militant groups.

⁂ ⁂ ⁂ ⁂

OTTO SKORZENY'S MILITARY career took him to more than a dozen countries on three continents. He had personal access to Adolf Hitler, who credited him with nothing less than "saving the Third Reich." President Eisenhower thought Skorzeny was the most dangerous man in Europe. Skorzeny even got away with dressing down the most feared Nazi of all, Heinrich Himmler, telling him to his face that he was useless.

Shadow Soldier

When war broke out in 1939, Skorzeny immediately volunteered for the Luftwaffe. However, at 31, he was deemed too old for flight training. He transferred to the Waffen-SS and served on the eastern front until he suffered a head wound from Russian artillery. Reassigned back to Berlin for desk duty, he was summoned to headquarters in April 1943. Hitler was reorganizing the feared Nazi intelligence machine, which included the creation of a special commando unit tasked with handling missions that the regular military was too "squeamish" to handle.

Skorzeny was offered the job based on a recommendation from Ernst Kaltenbrunner, Reinhard Heydrich's successor as head of RSHA intelligence. With no prior clandestine experience, he decided to borrow ideas from the best source of the time—captured British manuals. He soon put together a crack commando outfit and was the man directly responsible for some of the most fascinating operations of the war.

The Mussolini Rescue

On July 26, 1943, Skorzeny, along with a number of other crack officers, received a summons to attend a secret meeting with Hitler. Evaluating the men before him, the Führer dismissed everyone but Skorzeny. Hitler then informed him of the arrest of Benito Mussolini by King Vittorio Emanuele III. Hitler told Skorzeny he believed that the loss of Il Duce could lead to Italy deserting its alliance with Germany, and therefore, Skorzeny must free Mussolini from captivity. Skorzeny's commando units departed the next day for Italy and spent several weeks trying to determine where Mussolini was being kept, finally locating him in the isolated mountaintop resort of Gran Sasso.

The resort was accessible only by a single railroad line and was so heavily guarded that there was no question of using it for an approach. So Skorzeny settled on a bold plan, landing gliders under cover of darkness in a small boulder-strewn clearing on the mountaintop. Kidnapping an Italian general at gunpoint and forcing him to accompany the commandos in an attempt to keep the troops guarding Mussolini calm, the scar-faced Skorzeny rescued the dictator and took off again without firing a shot. The operation made headlines around the world and was considered so significant that the U.S. Army Historical Division asked Skorzeny himself to provide an account of the action after the war.

Hitler's Favorite Commando

The Mussolini episode was only one of Skorzeny's several encounters with wartime leaders. In September 1944, Hitler learned that Miklós Horthy, the Regent of Hungary, was attempting to negotiate a separate peace with the Soviets. The Führer dispatched Skorzeny to deal with the problem.

Skorzeny's solution was the perfect combination of finesse and violence. Disguising himself as a civilian, he located the regent's adult son, Milos Horthy. Leading his commandos in a raging gun battle with the younger Horthy's guards, Skorzeny took the man captive and personally ordered that Milos be rolled up inside a carpet and driven away. Stuffing the Hungarian into a plane bound for Vienna, the Nazis used his life as a bargaining chip in a telephone call with his father. The crying regent—his peace plan now in tatters—quickly agreed to abandon his overtures of peace.

The Battle of the Bulge

One of Skorzeny's last wartime exploits came during the opening stages of the Battle of the Bulge, Hitler's December 1944 last-ditch attempt to stave off ultimate defeat. In a personal meeting with Hitler, Skorzeny was given the task of capturing key bridges needed for the German offensive as well as sowing general discord among the Allied troops, a task which he carried out to perfection. Using captured American uniforms and jeeps, his commandos penetrated deep behind Allied lines, removing road signs and minefield markers, destroying supplies, and, on one occasion, giving false directions to an American regiment that promptly spent three days wandering around lost.

As news of German infiltrators began to run through Allied forces, paranoia struck all the way to the top of the U.S. command chain. There was speculation that Skorzeny intended to assassinate Eisenhower—a charge that Skorzeny denied to his death—but the possibility was taken seriously enough that Ike's guards kept him a virtual prisoner inside his own headquarters during the entire week of Christmas. Another rumor was that one of the German infiltrators was posing as British Field Marshall Bernard Montgomery. American troops were on such an edge that when the real Montgomery approached one checkpoint, they shot out his tires and forced him into a bunker, holding him at gunpoint until his identity was confirmed.

Postwar Legacy

After surrendering to the Allies and being acquitted of war-crime charges following the conflict, the former commando's services were successfully sought by both the Americans and Soviets seeking to expand their spy organizations. Skorzeny's first loyalty, however, remained to his wartime acquaintances. Using contacts and funds he managed to conceal during the collapse of the Reich, Skorzeny set up financial and travel networks to assist his former associates in escaping Nazi hunters by relocating them to Spain or South America. Skorzeny was reputedly even able to use secret letters between Mussolini and Winston Churchill that he had stolen during the Gran Sasso operation to blackmail the British leader into releasing war criminals.

However, Skorzeny was financially as well as ideologically motivated, and his talents were also sought by such groups as Juan Peron's Argentinean secret police, the Irish Republican Army, and Al Fatah and the PLO under Yasser Arafat—all of whom employed him as a consultant and adopted the shadow tactics of kidnapping and deception developed by Skorzeny during the war. Those methods continue to prove as difficult to combat in the world today as they did six decades ago.

Skorzeny died of cancer in 1975, still unrepentant. Confronted about his role in the war, Skorzeny forcefully stated, "I am proud to have served my country and my Führer."

✳ Chapter 6

Game-Changing Inventions

Bunsen's Burner: Scientific Error

A staple in chemistry classes for generations, this gas burner's hot blue flame has heated up the experiments of countless budding scientists. But the so-called "Bunsen burner" is actually a misnomer.

✳ ✳ ✳ ✳

AMONG THE ACHIEVEMENTS of 19th-century scientist Robert Wilhelm Bunsen are the co-discovery of chemical spectroscopy—the use of an electromagnetic light spectrum to analyze the chemical composition of materials—and the discovery of two new elements, cesium and rubidium. Bunsen is best known, however, for his invention of several pieces of laboratory equipment, including the grease-spot thermometer, the ice calorimeter, and a gas burner that became the standard for chemical laboratories the world over.

In 1852, Bunsen was hired as a lecturer at the University of Heidelberg and insisted on a new state-of-the-art laboratory with built-in gas piping. Although already in use, gas burners at the time were excessively smoky and produced flickering flames of low heat intensity. Bunsen had the idea of improving a burner invented by Scottish scientist Michael Faraday by pre-mixing gas with air before combustion, giving the device a hotter-burning and non-luminous flame. He took his concept to the university mechanic, Peter Desaga, who then designed and built the burner according to Bunsen's specifications, adding a control valve that regulated the amount of oxygen mixed with gas.

Bunsen gave Desaga the right to manufacture and sell the burner, and Desaga's son, Carl, started a company to fill the orders that began arriving from around the world. Bunsen and Desaga, however, did not apply for a patent, and soon other manufacturers were selling their versions. Competitors applied for their own patents, and Bunsen and Desaga spent decades refuting these claims. The court of history seems to have judged Bunsen the winner.

Chew, Chew, Pop!

A bad batch of chewing gum led to an iconic childhood treat.

* * * *

IF YOU'VE ALWAYS assumed that chewing gum and bubble gum were invented around the same time, well—sorry to burst your bubble! Chewing gum has been around for more than 800 years, but bubble gum is a relatively recent discovery—and an accidental one. In the 1920s, the Fleer Company, purveyors of fine candy and gum, bought gum base from a supplier and added their own sweeteners and flavorings to create a distinctive brand. With the company facing financial problems, Fleer President Gilbert Mustin began looking for creative ways to cut costs and hit upon the idea of having Fleer manufacture its own gum base rather than purchasing it.

Mustin set up a workroom near the accounting department, just down the hall from his office, and began tinkering with different formulas. In the end, he didn't have much of a knack for it and lost interest fairly quickly. Fortunately for Fleer and for gum lovers around the world, Mustin had enlisted the help of 23-year-old accountant Walter Diemer, and Diemer pressed on with the experiments, willingly chewing over the possibilities.

One day, Diemer made a batch of gum with an entirely wrong consistency. It was thinner, gooier, and much less sticky than traditional chewing gum, but it also had a tendency to form bubbles. Eager to try out the new product, Diemer mixed up a large batch and tossed in pink food coloring because it was the only color he could find. In December 1928, he took 100 pieces down to a nearby candy store; it sold out that same afternoon. And the interest stuck: Over the next year, the company earned more than $1.5 million from Diemer's Dubble Bubble. Although he never earned any royalties from the invention, Diemer went on to become a member of Fleer's board of directors and also took on the responsibility of training the company's sales staff in the art of blowing bubbles so they could demonstrate the product.

Inventing the Click

It's the input device we really couldn't live without. While you may be able to control some of your computer via keyboard "hot keys," the point-and-click mouse is still an essential component. In fact, it would be all but impossible to use most computers without one.

* * * *

How does a mouse work?

A computer mouse functions by detecting two-dimensional motion. This allows you to move the pointer around the screen to open documents, launch applications, and click on Web sites. It is even the primary means of control for many computer games.

Why is it called a mouse?

Care to take a guess? It's called a mouse because it resembles one. When the device was developed at the Stanford Research Institute (SRI), it had a cord attached to its rear—much like the tail on a mouse.

Who invented the mouse?

SRI inventor and engineer Douglas Engelbart is generally considered the father of the first PC mouse. But it was Bill English, an engineer who assisted with the construction of Engelbart's original designs, who first created the now-familiar inverted trackball system, which replaced the multiwheels used in Engelbart's device. English's innovation became the predominant model for more than three decades. It was also the basis of the first marketed integrated mouse, which was intended for personal computer navigation and was first sold in 1981 with the Xerox 8010 Star Information System. English can also be credited with adding the word mouse, as a computer term, to the English lexicon. He included it in his 1965 publication Computer-Aided Display Control.

While Engelbart's motion controls soon improved, his other contributions have remained a core part of the mouse. Most notable is his single-button design, a feature that remained part of Apple systems until 2005 (PC mice commonly have three or more buttons).

Best Inventors of the 19th Century

From the lightbulb to the internal combustion engine.

❋ ❋ ❋ ❋

THOMAS ALVA EDISON was self-taught. When he was seven years old, his mother had a disagreement with one of his teachers. She pulled him out of school and homeschooled him from then on. He went on to become one of the greatest inventors of all time. During his career, he invented the phonograph, the incandescent lightbulb, and the motion picture camera. The following were other prolific 19th-century inventors:

❋ Louis-Jacques-Mandé Daguerre (1789–1851) produced the first permanent photographic image.

❋ Barthelemy Thimonnier (1793–1857) invented the first functional sewing machine.

❋ Walter Hunt (1796–1859) invented the safety pin and the first eye-pointed needle sewing machine.

❋ Elias Howe (1819–67) invented the first sewing machine with a lockstitch mechanism.

❋ Isaac Singer (1811–75) invented the first sewing machine with an up-and-down motion mechanism.

❋ Gottlieb Daimler (1834–1900) invented the prototype of the modern gas engine.

❋ Sir Joseph Wilson Swan (1828–1914) invented an electric lightbulb that could burn for more than 13 hours.

❋ Alexander Graham Bell (1847–1922) invented the telephone.

❋ Karl Benz (1844–1929) invented the internal-combustion engine.

A Solution That Fits: The Origin of Jigsaw Puzzles

Puzzled over the origin of the jigsaw? We've put the pieces together!

* * * *

IN THE 1760S, engraver and cartographer John Spilsbury cut a wooden map of the British Empire into little pieces. Reassembling the map from the parts, he believed, would teach aristocratic schoolchildren the geographic location of imperial possessions and prepare them for their eventual role as governors. Spilsbury called his invention "Dissected Maps."

Spilsbury's "Dissected Maps" were soon popular among the wealthy. By the end of the 19th century, however, these "puzzles" functioned largely as amusements. The early decades of the 20th century were the heyday of the puzzle's popularity in the United States. The first American business to produce jigsaw puzzles was Parker Brothers; the company launched its hand-cut wooden "Pastime Puzzles" in 1908. Milton Bradley followed suit with "Premier Jig Saw Puzzles," named because the picture, attached to a thin wooden board, was cut into curved pieces with a jigsaw. As the Great Depression came to an end in the late 1930s, inexpensive cardboard puzzles were produced at prices nearly anyone could afford.

The Big Picture

For more than a century, jigsaw puzzles have delighted people of all ages; designers are constantly at work inventing new and more difficult challenges. There are three-dimensional picture puzzles, double-sided puzzles, and puzzles with curving and irregular edges. A puzzle advertised as "the world's largest jigsaw puzzle" has 24,000 pieces. Monochromatic puzzles include "Little Red Riding Hood's Hood" (all red) and "Snow White Without the Seven Dwarfs" (all white).

Hammond Organ: Tones of Endearment

Long before wizards such as Bob Moog, Keith Emerson, and Rick Wakeman made synthesizers and mellotrons essential entries in the lexicon of modern musical keyboard instruments, the Hammond organ was the straw that stirred the sound.

* * * *

As essential to the sounds that shaped the '60s as feedback and the fuzz box, the Hammond B-3 organ provided a whirring cascade of effects that stirred soul, jolted jazz, and revolutionized rock.

The Hammond organ as we know it was designed and developed by Laurens Hammond in 1933. A graduate of Cornell University, Hammond had invented a soundless electric clock in 1928. He used a similar technology to construct his revolutionary keyboard, adapting the electric motor used in the manufacture of his clocks into a tonewheel generator, which artificially re-created or synthesized the notes generated by a pipe organ. By using drawbars to adjust volume and tone, Hammond was able to electronically simulate instruments such as the flute, oboe, clarinet, and recorder.

Thaddeus Cahill, who invented the instrument dubbed the *telharmonium* in 1898, first formulated the concept of the tonewheel generator. While it was dynamic in design, it was cumbersome in concept, weighing seven tons and costing $200,000 to produce. Not exactly the potentially portable keyboard that Hammond was able to perfect 30 years later.

The effectiveness of Hammond's electric keyboard was greatly enhanced by the invention of the Leslie tone cabinet, a system that uses a rotating speaker to amplify, adjust, and enhance the intricacies of the Hammond sound. Invented by Donald Leslie in 1937, it proved to be a perfect partner to complement Hammond's keyboard, although there was bitterness between the two men, who thus never established a business partnership.

The Hammond B-3 was first manufactured in 1955. It remains a favorite with soul jazz musicians and rock and soul artists looking for a versatile, retro sound.

The Sands of Time

A seemingly simple bit of timekeeping technology has a surprisingly complex history.

* * * *

ACCURATE TIMEKEEPING WASN'T really possible until the Middle Ages, with the invention of the mechanical clock. But at least ancient people had the hourglass to give them an approximate measure of time passing, right? We've all seen these devices used in movies set in ancient Egypt and Rome. Why, even the Wicked Witch of the West had one in *The Wizard of Oz*.

A Grain of History

But actually, when sword-and-sandal flicks incorporated hourglasses into their story lines, they were indulging in a bit of cinematic license. Though no one knows for sure when the hourglass was invented, the first concrete evidence of anyone using one didn't appear until 1338, and even that evidence is indirect. The Italian artist Ambrosio Lorenzetti included an hourglass in a series of allegorical frescoes he created for the Palazzo Pubblico (Town Hall) in the city of Siena. The work focused on the theme of good government versus bad government, and a maiden holding an hourglass represented the virtue of Temperance. An earlier theory held that ancient Greeks used them, based on a carving of an hourglass on an ancient coffin. It was later discovered, however, that the carving had been added in the early 1600s.

Of course, the hourglass in Lorenzetti's fresco still doesn't tell us when these timekeepers first appeared, but it is clear that by the end of the 14th century they were common household items. The evidence this time comes from a "household treatise" dated to the 1390s. Belonging to the household of a Frenchman, the book provides instructions for performing various domestic chores, among them recipes for making preserves, glue, ink, and the filler for an hourglass. Wait a minute—a recipe for sand? Well, that's another common misconception about these simple devices. Because grains of sand vary in size, they would flow through an hourglass at an irregular rate and thus make it unreliable. It was more common to use marble dust or ground eggshells that had been laboriously processed to create fine particles of even size. The Frenchman's recipe indicates just how involved the task was: "Take the grease which comes from the sawdust of marble when those great tombs of black marble be sawn, then boil it well in wine like a piece of meat and skim it, and then set it out to dry in the sun; and boil, skim, and dry nine times; and thus it will be good."

Out to Sea

One additional piece of information raises speculation as to when the hourglass was actually invented. The device was often called a "sea clock." An inventory of the estate of King Charles V of France taken after his death in 1380 uses the term to describe the hourglass he kept in his study at a chateau. It's reasonable to infer from the term that the devices were commonly associated with ships. A bit of further digging reveals that by the 11th century, new methods of navigation were developed in the Mediterranean region that made use of sophisticated sea charts and the newly discovered magnetic compass. The thing about this method of keeping a ship on course, however, is that it requires an accurate way to gauge ship speed, which in turn requires an accurate way to measure the passage of time. It's possible that the hourglass was invented around this time in order to meet this need for maritime vessels.

Fake Forests

Take a walk in a polystyrene wonderland! Christmas wouldn't be Christmas without the sweet plastic smell of a fake tree. Or the slick shiny tinsel that will never melt in your hand like a real icicle. Or the white felt skirt that surrounds the towering faux pine instead of real snow.

❄ ❄ ❄ ❄

The Original

There are six general types of fake trees. The original is the German feather tree. These were made of goose or turkey feathers dyed green. The technique was invented in the 1800s when deforestation was a big problem in Germany. The feather trees were sparse to mimic the local white pines. Plastic versions are available if you're not only concerned about deforestation but also degoosification.

The Classic

This is your standard green plastic tree. It's a lot fuller than the feather tree and can even hold those heavy ceramic ornaments grandma gives you every year. You can't throw them out until she dies, but fortunately the tree can handle them on its wiry limbs.

The Midget

The midget is a small fake tree popular among single folks and urban couples. The former don't even take the lights off so that they can pull this baby out of the box and get their Christmas decorations up in under 16 seconds. The latter use the tree's small size to make their Amazon.com-wrapped gift boxes look larger by comparison.

The Snow White

You know how snow falls from the sky and lands on the tops of tree limbs? If you hate that, and prefer snow to cover every inch of the tree, then the Snow White is the way to go. Every inch of the tree is white. It's what Christmas will look like during a nuclear winter or after an asteroid hits earth covering everything with a fine pale ash. This bleak version is sometimes called the Cormac McCarthy.

Tinsel and Flock

Tinsel is a classic. Properly hanging tinsel on a tree involves the time-honored tradition of getting a strand stuck to every surface in the living room—except the tree—until one gets completely frustrated and just clumps the rest on one branch. People who spend time applying it strand by strand will have an epiphany on their deathbed about how they wasted their life on tinsel designs that nobody noticed anyway.

For a more realistic snow and ice effect, forget tinsel and forget white trees. Flocking is the answer. From homemade soap flakes to store bought sprays, this is how Martha Stewart would want you to spend your holiday season: not with friends, not with family, but on a 6-foot ladder spraying chemicals onto your tree.

The Abomination

Nothing says "I don't care about Christmas" more than the Abomination. This is for those parents too lazy to drag themselves down to the garden center so they have to pick up the last lonely tree at their local Wal-Mart. All the good trees are gone, as are all the halfway decent ones. All that's left is a skinny eyesore sitting naked and alone in what will next month become the Valentine's Day aisle. It's the pink aluminum tree.

For all their faults, fake trees have their good points. Every year that you lug it down from the attic represents another real, oxygen-giving tree saved. You don't have to water it. Most importantly, it doesn't shed needles that you continue to vacuum up six months later.

Taser: From Children's Book Concept to Riot Policing Tool

Like a .357 Magnum, the Taser makes troublesome suspects less troublesome. Unlike the .357, the tased suspect generally survives to stand trial, and the police save a bundle on coroner costs.

* * * *

Tom Swift

YOU'VE PROBABLY HEARD of the Hardy Boys and Nancy Drew. But unless you're a baby boomer or older, you may never have heard of Tom Swift books, which belonged to the same "teen adventure" genre. Tom, the precocious protagonist, is a young inventor who resolves crises and foils wickedness. One book in the series, called *Tom Swift and His Electric Rifle* (1911), has quite a stimulating legacy.

In 1967, NASA researcher Jack Cover, who grew up on Tom Swift, realized that he could actually make some of the gee-whiz gadgetry from the series. In 1974, he finished designing an electricity weapon he named the "Thomas A. Swift Electric Rifle," or TASER. (In so doing, he departed from canon. Tom Swift never had a middle initial, but Cover inserted the "A" to make the acronym easier on the tongue.)

How It Worked

Cover's first "electric rifle," the Taser TF-76, used a small gunpowder charge to fire two barbed darts up to 15 feet. Thin wires conducted electricity from the weapon's battery to the target, causing great pain and brief paralysis with little risk of death—except in the young, elderly, or frail. That was okay, since the police rarely felt compelled to take down children or senior citizens.

The police saw potential in the Taser. The TF-76 showed great promise as a nonlethal wingnut takedown tool.

Federal Shocker

Never underestimate the creativity-squelching power of government. The Bureau of Alcohol, Tobacco, and Firearms (BATF) wondered: How do we classify this thing? It's not really a pistol or a rifle. It uses gunpowder... *Aha!* The BATF grouped the TF-76 with sawed-off shotguns: illegal for most to acquire or possess. A .44 Magnum? Carry it on your hip if you like. An electric stunner that took neither blood nor life? A felony to possess, much less use. This BATF ruling zapped Taser Systems (Cover's new company) right out of business.

Second and Third Volleys

Taser Systems resurfaced as Tasertron, limping along on sales to police. In the 1990s, a creative idealist named Rick Smith wanted to popularize nonlethal weapons. He licensed the Taser technology from Cover, and they began changing the weapon. To deal with the BATF's gunpowder buzzkill, Smith and Cover designed a Taser dart propelled by compressed air. They also loaded each cartridge with paper and Mylar confetti bearing a serial number. If the bad guys misused a Taser, they wouldn't be able to eradicate the evidence.

Modern Tasers reflect the benefits of experience. In 1991, an LAPD Taser failed to subdue a violent, defiant motorist named Rodney King. The events that followed (including the cops beating King with billy clubs) put the Taser on the public's radar as something unreliable. This was not offset (on the contrary, it was compounded) by occasional deaths from tasing. The public might justly ask: "Does this thing really work? Does it work too darn well?"

One fact isn't in question. A nightstick blow to the head or a 9mm police bullet are both deadlier than a Taser. As a result, the debate revolves more around police officers' over-willingness to tase rather than whether or not police should carry Tasers in the first place. In 2007, the United Nations ruled that a Taser could be considered an instrument of torture.

I Scream, You Scream, We All Scream for Cones

Germs and sheer chance prompted an American classic.

⁕ ⁕ ⁕ ⁕

THE STORY OF the first ice-cream cone has become part of American mythology. A young ice-cream vendor at the 1904 St. Louis World's Fair runs out of dishes. Next to him is waffle-maker Ernest Hamwi, who gets the idea to roll his waffles into a cone.

But the ice-cream cone wasn't just born of convenience or a vendor's poor planning. Germs played a role, too. Italian immigrants introduced ice cream to the general public, first in Europe and then in the United States. Called *hokey pokey men* (a bastardization of an Italian phrase), these street vendors sold "penny licks," a small glass of ice cream that cost a single penny. The vendors wiped the glass with a rag after the customer was done and then served the next person. Forget the fact that people occasionally walked off with the glasses. Forget the fact that the glasses would sometimes break. This is about as sanitary as the space under your refrigerator. Unsurprisingly, people got sick. Ice-cream vendors needed a new serving method.

In the early 1900s, two people—Antonio Valvona and Italo Marchiony—independently invented an edible ice-cream cup. But this was just a cup; the cone did not appear until 1904.

Today, cones come in many varieties, but in 1904, Hamwi was making *zalabia*, a cross between a waffle and a wafer covered in sugar or syrup. He called his creation a *cornucopia* (a horn-of-plenty, a symbol of autumn harvest), and after the fair, he founded the Cornucopia Waffle Company with a business partner. A few years later, he started his own company, the Missouri Cone Co., and finally named his rolled-up waffles "ice-cream cones." How sweet it is!

The First Synthetic Fiber

Nylon, the first synthetic fiber, was invented by DuPont chemists in the 1930s.

* * * *

INITIALLY USED FOR stockings and toothbrush bristles, it was diverted for military uses during World War II and was used for everything from parachutes to bomber tires. Once the war ended, women were eager to buy nylons again.

Joe Labovsky was the last survivor of the team that invented nylon. Labovsky was born in Kiev, Ukraine, in 1912. His family emigrated to the United States in 1923. Labovsy's father worked as a tailor for the DuPont family, and he was able to find his son a job as a chemist helper to Wallace Carothers at DuPont in 1930.

Labovsky went on to study chemical engineering at the Pratt Institute in Brooklyn. He had trouble finding work after he graduated in 1934, but he did eventually find a job digging ditches—again at DuPont. One day Wallace Carothers walked by as Joe was working, and Carothers stopped to chat. He soon recruited Joe for the nylon project, and—because most of the others working on the project were research chemists—Labovsky's chemical engineering background was an asset to the team.

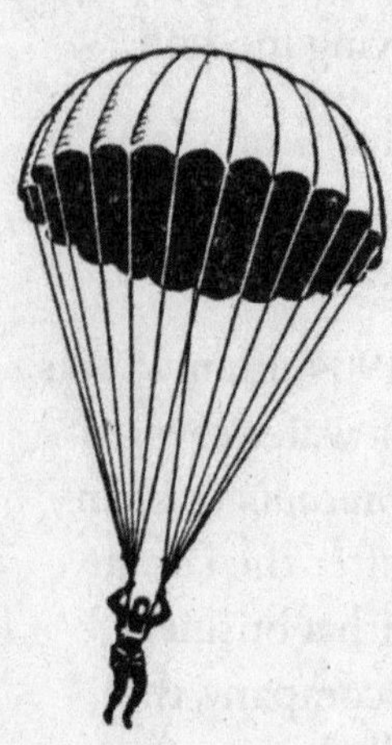

Everyone Loves a Slinky

The story behind the toy that "walked" into our hearts.

❋ ❋ ❋ ❋

IN THE POST–WORLD War II era, just about every kid owned one of these magical "walking" springs. It's no small wonder—folks of the day had every right to be mesmerized by a toy that actually performed feats of distraction. Hurray!

The Slinky is an extremely loose spring that will rest fully coiled in the palm of one's hand. If it stays in this dormant mode, you'll wonder what all the fuss is about. Thankfully, it rarely does. Curiosity compels people to stretch it like an accordion and commit it to its greatest trick: walking down stairs.

It all started in 1943, with naval engineer Richard James. Laboring to develop a stabilization method for shipboard monitoring systems, James took notice when a spring accidentally fell from a shelf to a stack of books. As it uncoiled then recoiled, the spring "stepped" down to the next level. This was followed by another step to an even lower perch. James committed this life-changing moment to memory. In his spare time, he and wife Betty worked to devise the jumpy toy of the future. In 1945, the Slinky finally bounced onto the toy scene.

It would be wonderful to report that the marital team team used this invention to advance their personal relationship, but such was not the case. In 1960, James recoiled from Betty and their six children, joined a missionary cult in Bolivia, and pumped a horde of Slinky profits into his new calling. Betty somehow managed to wrestle the suddenly floundering company back from her absentee husband, and the rest is history.

Slinky is now available in plastic and assorted colors, but the concept remains the same. According to the advertising jingle, "It's Slinky! It's Slinky! For fun it's a wonderful toy. It's Slinky! It's Slinky! It's fun for a girl and a boy." You betcha!

One Sharp Guy

King Gillette never ruled a country—he did better than that. He oversaw a disposable kingdom.

* * * *

THERE'S A QUIRKY billboard currently plastered along America's byways. It reads "Failed, failed, failed. And then . . ." Alongside the cryptic message is a photograph of Abraham Lincoln. It may take a few seconds, but the message ultimately becomes clear. The billboard is a variation of a popular maxim that we learned as children: "If at first you don't succeed, try, try again." Sounds like a page lifted from the life of King Camp Gillette (1855–1932).

As a boy, Gillette amused himself by watching his parents tinker. His dad, a patent agent, was forever inventing ways to do things better, and his mom, a homemaker, concocted wonderful recipes through experimentation. Their efforts planted a seed in the mind of young Gillette. He fantasized about inventing something that would make him rich and famous.

A Close Shave

By his mid-30s, Gillette had little to show for his efforts and was becoming increasingly bitter. Despite having poured his heart and soul into countless projects throughout the years—even earning several patents along the way—Gillette still hadn't found his breakthrough invention. Then, at age 40, while working as a salesman, he was given the advice to invent something disposable. Gillette was intrigued.

A Cutting-Edge Idea

It hit him while shaving one morning: Why bother to sharpen a razor if you could replace it with a disposable blade? *Voilà!* Gillette put his nose to the grindstone and began his quest. In 1903, at age 48, Gillette started selling his perfected disposable razor blades, and Gillette Safety Razor Company was born.

The King's Noble Vision

Over the years, King Gillette's disposable safety blade would become wildly successful. Always the innovator, Gillette started giving his razors away free of charge. This "get them hooked" strategy was ahead of its time and only added to the company's success. By 1999, the renamed "Global Gillette" had mushroomed into a $9.9 billion company. King Camp Gillette's magnum opus had finally been realized.

The safety razor is at least partly responsible for a personal hygiene trend among women. Underarm hair used to be no big deal. But the invention of the safety razor, combined with the acceptance of scantier clothing styles, changed all that. In fact, Gillette produced a razor designed especially for women, called the Milady Décolletée, in 1916.

Let's be grateful King Camp Gillette wasn't around to witness this cutting mistake: The brass at Gillette thought they'd come up with a great idea. For the cost of only one million dollars, they were able to supply Mach3 razors for all welcome bags given out to delegates at the 2004 Democratic National Convention. Unfortunately, security officers worried that the razors presented a threat to attendees and promptly confiscated any that were carried into the convention hall.

Flamethrower!

Controversial, terrifying, effective: Whether mounted on a tank or carried on an engineer's back, the flamethrower did the job—for those who dared use it.

❋ ❋ ❋ ❋

When were these hellish weapons invented?

Combatants have burned one another to death since antiquity, but the modern flamethrower was invented in Germany just before World War I.

Most Great War weapons improved during World War II. How does one improve a weapon of liquid fire?

The Germans were first to mount flamethrowers on armored vehicles, which could carry more fuel and propellant—and had an easier time getting the weapon into effective range than foot soldiers did. Americans invented napalm (jellied gasoline), which flew more accurately and stuck as it burned.

How exactly did World War II flamethrowers operate?

A flamethrower consisted of a tank (diesel, gasoline, or a mixture), a compressed propellant (such as butane or nitrogen), and a hose leading to an igniter. Pulling the trigger forced propellant into the fuel tanks, pushing a jet of fuel out of the nozzle. The igniter, of course, set it ablaze as it left the nozzle.

That sounds almost as scary for the wielder and his comrades as for his enemies.

No one was eager to stand too close to the flamethrower guy, and any flamethrower operator completely unafraid of his weapon was considered insane. Some German troops executed any Allied flamethrower operator they captured. A flamethrower burst is impossible to hide from, and defending troops took suicidal risks in hopes of destroying or escaping it.

Were flamethrowers useful in all combat situations?

The flamethrower's short range (15–50 yards for portable flamethrowers, about double that for tanks), inherent danger to the operator, and limited fuel always restricted its uses. It was best at neutralizing fortifications like bunkers and pillboxes, where a quick burst of flaming fuel through the firing ports meant a bad day for the occupants. Flamethrowers also worked against troops hidden in caves and other natural fortifications. If the flame reached an ammo pile, the end was quick.

In cases where starting a forest or house fire would give a tactical advantage, the flamethrower was quite efficient. It could even be used in desperation against heavy armor: The best bet was to try igniting the target's gas tank.

Most of these sound like special weapons situations, normally the province of combat engineers.

For the most part, only combat engineers and specialized troops carried flamethrowers. In the island-hopping Pacific campaign, however, U.S. Marines bristled with them. A 1944 U.S. Marine division at full strength had 17,465 Marines. Fully equipped, they had 153 mortars, 172 bazookas, 48 anti-tank guns, 48 howitzers, 46 tanks—and 267 flamethrowers. Considering the suicidal courage of Japanese defenders holed up in thousands of well-prepared bunkers, caves, and forts, Marine divisions carried more flamethrowers than artillery.

But no flamethrowing tanks for the Leathernecks?

The U.S. Army supported the Marines with flamethrowing tanks at Okinawa, for example. But the U.S. Marines were fundamentally riflemen, excelling at infantry weapons and tactics. A man could carry a flamethrower places no tank could go. Tanks had a hard time on islands with rough terrain or boggy ground, which were numerous in the Pacific.

Who made the most use of flamethrowing tanks?

Early on, the Germans; later the British. The Wehrmacht had thousands of obsolete prewar tanks, many of them captured, with undersize turrets that could not be fitted with larger guns-but could easily handle a flamethrower. None of the German conversions were resounding successes, but they all gave some useful service. The British Churchill Crocodile was a modified heavy Churchill tank that pulled a fuel trailer and could fire either flame (through a converted machine-gun port) or a standard tank cannon. Between its two main armaments, the Crocodile was very useful against any tough defensive position. Americans and Soviets each built modest numbers of different flamethrowing tank designs and conversions.

Inventing the Internet

Former vice president Al Gore received a Nobel Peace Prize, but did he also claim that he invented the Internet?

* * * *

On March 9, 1999, while being interviewed by Wolf Blitzer on CNN, Al Gore said, "During my service in the United States Congress, I took the initiative in creating the Internet. I took the initiative in moving forward a whole range of initiatives that have proven to be important to our country's economic growth and environmental protection, improvements in our educational system."

It was merely a case of unfortunate wording. Of course, no single person invented the Internet. Its forerunner, ARPAnet, evolved through the 1960s and '70s from defense researchers' information-sharing needs. In 1983, ARPAnet standardized the basic Internet information-transfer system still used today: TCP/IP, designed by Internet legends Vinton Cerf and Robert Kahn. With this, ARPAnet started to resemble the modern Internet.

When Gore joined Congress in 1977, his academic training was in journalism and law. Alone, he was no more qualified to create an Internet than the average Joe is to create the starship *Enterprise*. What Gore did was push funding to support and expand the Internet. He didn't invent the term "information superhighway," but he was using it back when the Net was still an information goat path.

In the greater context, Gore meant that he'd gotten off his congressional backside and done some work to help advance the country's fortunes. His clumsy way of saying that may have in some way cost him the presidency, but Cerf himself, considered to be the real "father of the Internet," has publicly acknowledged Gore's early legislative efforts on behalf of the technology.

Franklin Flies a Kite

As it turns out, Benjamin Franklin did not discover electricity. What's more, the kite he famously flew in 1752 while conducting an experiment was not struck by lightning. If it had been, Franklin would be remembered as a colonial publisher and assemblyman killed by his own curiosity.

❋ ❋ ❋ ❋

Before Ben

BLESSED WITH ONE of the keenest minds in history, Benjamin Franklin was a scientific genius who made groundbreaking discoveries in the basic nature and properties of electricity. Electrical science, however, dates to 1600, when Dr. William Gilbert, physician to Queen Elizabeth, published a treatise about his research on electricity and magnetism. European inventors who later expanded on Gilbert's knowledge included Otto von Guericke of Germany, Charles Francois Du Fay of France, and Stephen Gray of England.

The Science of Electricity

Franklin became fascinated with electricity after seeing a demonstration by an itinerant showman/doctor named Archibald Spencer in Boston in 1743. Two years later, he bought a Leyden jar—a contraption invented by a Dutch scientist that used a glass container wrapped in foil to create a crude battery. Other researchers had demonstrated the properties of the device, and Franklin set about to increase its capacity to generate electricity while testing his own scientific hypotheses. Among the principles he established was the conservation of charge, one of the most important laws of physics. In a paper published in 1750, he announced the discovery of the induced charge and broadly outlined the existence of the electron. His experiments led him to coin many of the terms currently used in the science of electricity, such as battery, conductor, condenser, charge, discharge, uncharged, negative, minus, plus, electric shock, and electrician.

As Franklin came to understand the nature of electricity, he began to theorize about the electrical nature of lightning. In 1751, he outlined in a British scientific journal his idea for an experiment that involved placing a long metal rod on a high tower or steeple to draw an electric charge from passing thunder clouds, which would throw off visible electric sparks. A year later, French scientist Georges-Louis Leclerc successfully conducted such an experiment.

The Kite Runner

Franklin had not heard of Leclerc's success when he undertook his own experiment in June 1752. Instead of a church spire, he affixed his kite to a sharp, pointed wire. To the end of his kite string he tied a key, and to the key a ribbon made of silk (for insulation). While flying his kite on a cloudy day as a thunderstorm approached, Franklin noticed that loose threads on the kite string stood erect, as if they had been suspended from a common conductor. The key sparked when he touched it, showing it was charged with electricity. But had the kite actually been struck by lightning, Franklin would likely have been killed, as was Professor Georg Wilhelm Richmann of St. Petersburg, Russia, when he attempted the same experiment a few months later.

The Lightning Rod

Although Franklin did not discover electricity, he did uncover many of its fundamental principles and proved that lightning is, in fact, electricity. He used his knowledge to create the lightning rod, an invention that today protects land structures and ships at sea. He never patented the lightning rod but instead generously promoted it as a boon to humankind. In 21st-century classrooms, the lightning rod is still cited as a classic example of the way fundamental science can produce practical inventions.

On a Roll

Toilet paper is one invention that has been flushed with success since its 14th-century origins.

❋ ❋ ❋ ❋

LIKE PASTA AND gunpowder, toilet paper was invented in China. Paper—made from pulped bamboo and cotton rags—was also invented by the Chinese, although Egyptians had already been using papyrus plants for thousands of years to make writing surfaces. Still, it wasn't until 1391, almost 1,600 years after the invention of paper, that the Ming Dynasty Emperor first used toilet paper. The government made 2×3 foot sheets, which either says something about the manufacturing limitations of the day or the Emperor's diet!

Toilet paper didn't reach the United States until 1857 when the Gayetty Firm introduced "Medicated Paper." Prior to the industrial revolution, many amenities were available only to the wealthy. But in 1890, Scott Paper Company brought toilet paper to the masses. The company employed new manufacturing techniques to introduce perforated sheets. In 1942, Britain's St. Andrew's Paper Mill invented two-ply sheets (the civilized world owes a great debt to the Royal Air Force for protecting this London factory during The Blitz!). Two-ply sheets are not just two single-ply sheets stuck together; each ply in a two-ply sheet is thinner than a single-ply sheet. The first "moist" toilet paper—Cottonelle Fresh Rollwipes—appeared in 2001.

What was the rest of the world doing? Some pretty creative stuff! Romans soaked sponges in saltwater and attached them to the end of sticks. There is little information about what happened when the stick poked through the sponge, but the Romans were a hearty, expansionist people and probably conquered another country for spite. Medieval farmers used balls of hay. American pioneers used corncobs. Leaves have always been a popular alternative to toilet paper but are rare in certain climates, so Inuit people favor Tundra moss. Of all people, the Vikings seemed the most sensible, using wool. It's not easy being a sheep.

Breaking Morse Code

In these high-tech times, most people are glued to their cell phones 24/7. We have Antonio Meucci, an Italian American inventor, to thank for this. If it weren't for him, we'd have to listen to dashes and dots—the basis for Morse code—when we call our significant other to find out if we're running low on coffee.

* * * *

THE TRANSITION FROM Pony Express to transcontinental telegraph lines took years. The first telegraph was invented by Charles Wheatstone and William Fothergill Cooke in 1831. That same year, Joseph Henry would develop the basic principles of enabling an electric current to travel long distances.

American inventor and painter Samuel Finley Breese Morse found a practical use for Henry's principles. With technical assistance from chemistry professor Leonard Gale and financial support from Alfred Vail, Morse invented the Morse Code. In 1837, he unveiled his new way of interpreting communications over telegraph wires to the public in New York, Philadelphia, and Washington, D.C. He received a patent for his code in 1840, and by 1843, the U.S. Congress had approved an experimental line.

In a watershed moment for telecommunications, Morse sent his first message via Morse Code on May 24, 1844: "What hath God wrought?" The communication traveled from Washington to Baltimore.

By 1861, most major cities were connected by lines that transmitted electrical signals, and each dash and dot was translated by operators at telegraph stations. This was a boon to communications (and often the bane when the enemy cut lines) during the Civil War. The short signals were called *dits* (seen as dots), and the long signals were called *dahs* (seen as dashes); in combination, the signals represented the letters of the alphabet and ten numerals. For example, the most well-known use of the Morse code is the SOS signal—or dit, dit, dit, dah, dah, dah, dit, dit, dit.

Neon Signs: Bright 'n' Gassy

The neon sign serves both art and function.

* * * *

THE NEON SIGN is one of advertising's most effective tools. Its colorful, often soothing light is pleasing to the eye and effortlessly draws our attention to products ranging from automobiles to beer.

The neon sign has been an American advertising icon for decades, but the science behind it dates back to the turn of the century. Neon, derived from the Greek word *neos* (meaning "new") is a relatively rare gaseous element first identified in 1898 by British researchers William Ramsay and M. W. Travers. However, it was French engineer Georges Claude who, around 1902, discovered that a glass tube filled with neon gas glowed brightly when electrically charged. Claude realized the glow would make an effective light source, and he debuted the first neon lamp in Paris in December 1910.

Neon glows fire-red when hit with electricity, but Claude learned that different colors could be produced by mixing neon with other gases such as argon and mercury. He also found that the glass tubes could easily be shaped into letters and designs. This discovery led to the development of the neon sign.

In 1923, Claude introduced his innovative invention to the United States. The first two neon signs, reading "Packard," were sold to Earle C. Anthony, who owned a Packard dealership in Los Angeles.

Neon signs took America by storm and quickly became an integral part of indoor and outdoor advertising. The signs were such a novelty at first that people would literally stop and stare at them.

Like all great inventions, there have been some amazing spin-offs from the neon sign. In the 1930s, for example, the concept led to the development of the fluorescent lightbulb. And the very first experimental color television receivers used neon to produce the color red, complemented by mercury-vapor and helium tubes for green and blue.

Chill Out: The Popsicle Story

As luck would have it, some of the best inventions actually happened by accident. The world would be a sadder place without penicillin, microwave ovens, ice cream cones, Post-it Notes, potato chips, superglue, Slinkies, or heaven forbid—Popsicles.

* * * *

THE POPSICLE WAS "invented" by an industrious 11-year-old boy named Frank Epperson on an unseasonably cold San Francisco evening in 1905. After accidentally leaving his drink in a cup on the front porch overnight, Epperson discovered that the drink had frozen around the wooden stir stick. The next morning, he pulled the frozen drink out of the cup by the stick and voilà . . . the Popsicle was born.

Actually, Epperson's frozen invention originally took the neighborhood by storm as the "Epsicle." It wasn't until 1923 while running a lemonade stand at the Neptune Beach amusement park in Oakland, California, that he realized the money-making potential of his discovery. His own children loved the cool treat, begging him for one of "Pop's 'sicles." In 1924, Epperson applied for the first patent of the "Popsicle," the first "drink on a stick."

The Popsicle Goes Big Time

A year later, Epperson sold the patent and rights to the brand name "Popsicle" to the Joe Lowe Company in New York. As it turned out, he made a wise business decision—during the first three years, he earned royalties on the sale of more than 60 million Popsicle ice pops. Popsicles soon began to appear all over the world—they were affectionately known as "Ice-lollies" in Great Britain and "Icy Poles" in Australia.

Popsicles grew in popularity with kids and adults alike. As soldiers returned home from World War II and began building families, the average breadwinner could afford the convenience of having their own refrigerators and freezers. That meant that busy homemakers could buy large quantities of Popsicles in "multipacks" and store them indefinitely in the freezer, dispensing them to the kids whenever they deserved a treat (or needed a bribe). In the mid-'40s, cartoonist and adman Woody Gelman created the "Popsicle Pete" mascot to help market the product in magazines, comic books, and television commercials. Eventually, cardboard advertisements were distributed to vendors touting the new marketing slogan, "If it's Popsicle, it's possible."

Popsicle Spin-Offs

The treats continued to sell well and, in 1965, they became part of the Consolidated Foods Corporation lineup. At that time, 34 different flavors were offered by the company. Several years later, "Creamsicles" (a sherbet pop on a stick with a vanilla ice cream center) were sold in orange and raspberry flavors. The new item became so popular that "National Creamsicle Day" is now celebrated every August 14.

The Popsicle continued to make the corporate rounds when the Gold Bond Ice Cream Company purchased the U.S. operations of Popsicle Industries; it was purchased three years later by Unilever. In 1993, the Popsicle underwent another change when the Unilever company name was changed to the

Good Humor–Breyers Ice Cream Company, where the brand remains today.

* Popsicles also are used in craft projects, as children, adults, and just about anyone who went to camp create bridges and houses made from Popsicle sticks. But the "stick bomb" is one of the more notorious creations made from used Popsicle sticks. After weaving five sticks together in a specific pattern, the stick bomb is thrown to the floor, where it "explodes" with a loud pop.

* Woody Gelman was also the writer and co-creator of the Bazooka Joe comics found in Bazooka gum, as well as the sci-fi trading card series Mars Attacks.

Some Toys Are Sillier Than Others

Is Silly Putty a toy or an industrial compound? Both, actually.

* * * *

WHAT'S NOT TO like about Silly Putty? It bounces, it stretches, it breaks. It's good, clean fun. But surprisingly enough, Silly Putty didn't begin its career as a toy. It's actually a by-product of war.

During World War II, the Japanese military began a series of conquests in southeast Asia with the intention of cutting off the Allies' supply of rubber. In response, the U.S. began to look for ways to develop synthetic rubber products. Because, after all, it's hard to fight an effective war without rubber.

Enter James Wright. While working at General Electric in 1943, Wright combined boric acid with silicone oil. The result was a polymerized substance that could bounce and stretch and had a high melting temperature. Unfortunately, it couldn't replace rubber and thus served no purpose for the military.

General Electric was determined to find a use for Wright's invention. The company bounced the putty across the country, hoping to drum up interest. In 1949, the putty landed in the hands of a New Haven, Connecticut, toy-store owner named Ruth Fallgatter. Fallgatter teamed up with marketing guru Peter Hodgson to sell the product through her toy catalog as a bouncing putty called Nutty Putty. It was an instant hit.

In 1950, Hodgson christened the new toy "Silly Putty" and went on to sell the product in the now-famous egg-shape containers. Originally marketed to adults, the product ultimately found success with kids ages 6 to 12—especially after 1957 when a commercial for Silly Putty debuted on the *Howdy Doody* show.

Silly Putty has done a brisk business over the years, with more than 300 million eggs sold since 1950. Silly Putty also has practical uses—it is useful in stress reduction, physical therapy, and in medical and scientific simulations. It was even used by the crew of *Apollo 8* to secure tools in zero gravity. Because there's nothing silly about being hit in the head with a wrench when you're in space.

Tremendous Tinkertoys

If you have kids, know kids, or ever were a kid, you know that even when surrounded by expensive, designer toys, children are often perfectly happy banging pots and pans together or playing with a box. This was the principle behind Tinkertoys, a Christmas gift favorite for over a century.

* * * *

Hey, Kid: Good Idea!

ONCE UPON A time in the early 1900s, two guys were on their way to work. Their names were Charles Pajeau and Robert Pettit, and they were on a train headed from the Chicago suburb of Evanston into the city. While they rode, they talked about how bored they were with their jobs: Pajeau was a stonemason, Pettit worked at the Board of Trade.

Neither man felt satisfied. The talk turned to inventions, and Pajeau shared an idea with Pettit.

Prior to their conversation, Pajeau had seen some kids playing with pencils and spools of thread. They invented moving parts, made up stories to go along with what they engineered, and generally seemed to have a blast with their improvised toy. Pajeau figured he could design a toy on the wooden spool/pencil concept; Pettit knew he could market his new friend's brilliant idea.

Ladies and Gentlemen . . . Tinkertoys!

In his line of work, Pajeau was used to dealing with angles, math, and basic engineering principles. The first Tinkertoy set was entirely handcrafted, and it consisted of 2-inch-diameter wooden spools with holes drilled every 45 degrees around the perimeter, plus one through the center. The perimeter holes in a Tinkertoy spool don't go all the way through, but the center hole does. This allows for the spool to move freely around the "axle," which is made with any of the various lengths of sticks in the set. Ever the math whiz, Pajeau based the toy on the Pythagorean progressive right triangle, and he made sure everything was exact. Pettit helped get the toy to the American Toy Fair, but the set didn't take off until the pair tried a unique marketing scheme: the Toy Tinkers (which is what they called their company) hired a group of dwarves dressed in elf costumes to play with Tinkertoys in a Chicago department store window. The idea paid off: Customers came in droves to see the elves and bought Tinkertoy sets by the hundreds. Within a year, over a million sets had been sold and the two working stiffs from Chicago were on the toymaker map for good.

Tinker Adapts

By the late 1940s, more than 2.5 million Tinkertoy construction sets had been produced. The Tinkertoy continued to sell, but the inventors were both in their seventies. When Pettit passed away, Pajeau decided to sell the rights to his creation to A. G. Spalding Bros., Inc., a company that grew the line by adding color and additional pieces to play with. The Tinkertoy continued to find its way under Christmas trees around the world, even garnering press as a tool for professional engineers. Students of computer science and robotics at Cornell and MIT used the toy as a model to develop complex machines. Then, in 1985, game giant Playskool purchased the rights. In 1992, they released a revamped Tinkertoy set to commemorate its 80th anniversary. The new sets were all plastic, with easy-to-assemble parts including "flags," "pulleys," "elbows," and of course, "rods" and "spools." Available in both junior and jumbo versions, kids (and adults) can now build bigger structures than ever before. Every set includes instructions on how to create machines that move and building structures with intricate moving parts. Until the 1960s, the company averaged sales of over 2.5 million sets per year. While that number has decreased significantly, Tinkertoys, that classic toy of motion and construction, continue to be a hit for Christmas, birthdays, and ordinary trips to the toy store.

It Takes All Types

The long, hard road to the common typewriter

* * * *

CHRISTOPHER LATHAM SHOLES didn't set out to invent a typewriter. Inside Milwaukee's Kleinsteuber Machine Shop in the winter of 1867, Sholes tried to invent a device to print consecutive numbers on paper—for creating book pages, tickets, or bank notes.

Another tinkerer in the shop suggested to Sholes that a machine printing letters and words might be better. After reading a *Scientific American* article about such a device, Sholes, an ex-newspaperman, was intrigued. With the help of two other inventors, Sholes began building his own word machine.

A Rough Start

By fall, Sholes' machine was complete, but not perfect. Resting on a kitchen table, the device was the size of a small piano and styled like one, with two rows of ebony and ivory keys moved by wires. It could only print capital letters on tissue-thin paper. Also, the paper faced downward, meaning the typist could not see them as they were being typed. Undaunted, the men began a letter-writing campaign to potential investors.

One letter recipient was an old newspaper colleague of Sholes' named James Densmore. He was itching to invest, and without ever seeing the machine, offered the three inventors $200 each, plus funding for manufacturing in exchange for interest in the enterprise. The next spring, Densmore first saw the machine. Sholes also had a simpler, smaller second version. Pleased, Densmore secured a patent for each. He made 15 of the new version and tried selling them. However, the machines were not a hit. Densmore prodded the inventors to keep tinkering.

Initially, they attacked the challenge with vigor. Together the trio toiled to simplify the machine so that it could be manufactured quickly and cheaply. Densmore also begged them to devise a design that allowed for paper of regular thickness rather than tissue.

Enough, Already!

By 1869, Sholes' companions tired of the toiling. They sold their rights to the original patent to Densmore and left Sholes to work alone. That year, Sholes, who was tiring of the invention himself, typed a note to Densmore. "I am satisfied," he said. "The machine is now done." On heavy cardstock he added a smug note scrawled in pencil: "Is this paper thick enough?"

However, in response to criticisms of the new version, Densmore demanded that Sholes make further improvements. Sholes protested at each request, but Densmore pressed Sholes to toil on. Over the next few years he made several versions of his machine, but none caught on.

Giving Up

Densmore wouldn't quit. In a last-ditch effort in 1872, Densmore built his own manufacturing outfit on land wedged between the Milwaukee River and the Rock River Canal. He warned Sholes not to give up his shares, but Sholes sold them anyway. Initially, it looked as if Sholes' pessimism was justified. The factory was failing and Densmore was spending more to manufacture the machines than he was earning as profit.

An old friend suggested Densmore show the machine to E. Remington and Sons, a maker of firearms, sewing machines, and farm equipment in Ilion, New York. In 1873, Densmore took the machine to Remington's headquarters and made the only deal he could: After working out a way to be paid, he would allow Remington to mass-produce the machines.

Remington began producing his typewriters, yet it was a decade before Densmore earned returns on his investment. As for Sholes, he considered the Remington Typewriter an insult. He disowned his part in its creation and declared it too complicated, even the portions of its design he himself had created.

It wasn't until the end of his life that Sholes was ready to take his rightful place in history. In one of his last letters, he wrote, "Whatever I may have felt in the early days of the value of the typewriter, it is obviously a blessing to mankind, and especially to womankind. I am glad I had something to do with it. I builded wiser than I knew, and the world has the benefit of it."

Making His Mark

Indoor cats and their owners should give thanks to Ed Lowe, the inventor of Kitty Litter.

* * * *

Stumbling Upon Paydirt

BORN IN MINNESOTA in 1920, Ed Lowe grew up in Cassopolis, Michigan. After a stint in the U.S. Navy, he returned to Cassopolis to work in his family's business selling industrial-strength absorbent materials, including sawdust, sand, and a powdered clay called fuller's earth. Due to its high concentration of magnesium oxide, fuller's earth has an extraordinary ability to rapidly and completely absorb any liquid.

Back in those days, domestic kitties did their business in litter boxes filled with sand, wood shavings, or ashes. One fateful morning in 1947, a neighbor of Lowe's, Kaye Draper, complained to him about her cat tracking ashes all over the house. She asked if she could have a bag of sand from his warehouse.

Instead, Lowe gave her a sack of fuller's earth. Draper was so pleased with the results that she asked for more. After a while, her cat used only fuller's earth—it was the first Kitty Litter-using critter in the world.

Sensing that he was on to a good thing, Lowe filled ten brown bags with five pounds of fuller's earth each and wrote "Kitty Litter" on them. He never explained exactly how he came up with the name, but it was certainly an inspired choice.

The Idea Catches On in Catdom

Initially, convincing pet shop owners to carry Kitty Litter proved to be a challenge. Lowe's suggested price of 65 cents per bag was a lot of money at that time—the equivalent of about $5 today. Why would people pay so much for cat litter, the shop owners asked, when they could get sand for a few pennies? Lowe was so sure Kitty Litter would be a success that he told the merchants they could give it away for free until they built up a demand. Soon, satisfied customers insisted on nothing but Kitty Litter for their feline friends, and they were willing to pay for it.

Lowe piled bags of Kitty Litter into the back of his 1943 Chevy and spent the next few years traveling the country, visiting pet shops and peddling his product at cat shows. "Kitty Litter" became a byword among fastidious cat owners. The *Oxford English Dictionary* cites this advertisement from the February 9, 1949, issue of the Mansfield, Ohio, *Journal News* as the phrase's first appearance in print: "Kitty Litter 10 lbs $1.50. Your kitty will like it. Takes the place of sand or sawdust."

The Kitty Litter Kingdom

By 1990, Lowe's company was raking in almost $200 million annually from the sale of Kitty Litter and related products. He owned more than 20 homes, a stable of racehorses, a yacht, and a private railroad. He even bought up 2,500 acres of land outside of Cassopolis, where he established the Edward Lowe Foundation— a think-tank dedicated to helping small businesses. Lowe sold his business in 1990 and died in 1995. As far as anyone knows, he never owned a cat himself.

Is Jell-O Made from Horses?

Could this fun, wiggly dessert be the final resting place for the likes of Black Beauty and Mister Ed? Sure. But let's not be too picky—any creature with bones can become Jell-O. It's an equal opportunity dessert.

* * * *

JELL-O IS MADE from gelatin, which is processed collagen. Collagen makes your bones strong and your skin elastic and stretchy (there's that jiggly wiggle). To make gelatin, you take bones, skin, tendons, and whatnot from animals (primarily cows or pigs), grind everything up, wash and soak it in acid (and also lime, if cow parts are used), and throw it in a vat to boil. The acid or lime breaks down the components of the ground animal pieces, and the result is gelatin, among other things. The gelatin conveniently rises to the top of this mixture of acid and animal parts, creating an easy-to-remove film.

In the Victorian era, when gelatin was really catching on, it was sold in the film state. People had to clarify the gelatin by boiling it with egg whites and eggshells, which took a lot of time. In 1845, a crafty inventor patented a powdered gelatin, which was to be extracted from the bones of geese. In 1897, this powdered gelatin was named Jell-O and went on to become the line of dessert products that, to this day, we always have room for.

Why does the list of ingredients in Jell-O include gelatin and not cow and pig pieces? Because the U.S. federal government does not consider gelatin an animal product, since it is extensively processed. Gelatin is also found in gummy bears candy, cream cheese, marshmallows, and other foods.

What if you like Jell-O, cream cheese, marshmallows, and such, but would rather not eat the boiled bones and skin of animals? There are alternatives. Agar and carrageenan are made from seaweed and can be used to create delicious gelatin-like goodies. So while it's unlikely your Jell-O contains traces of Mister Ed or Black Beauty, it could test positive for Elsie.

A Mover of People

The Segway was born to revolutionize urban transportation.

⁕ ⁕ ⁕ ⁕

INVENTOR DEAN KAMEN lives to solve problems. His fertile mind has produced a pocket-size infusion pump to deliver insulin and other medications, a wheelchair that can climb stairs, and the Segway Human Transporter, a two-wheel standing scooter that changes the way people get around in the city.

The key to the Segway's function is balancing technology, which is comprised of microprocessors and gyroscopes that prevent the user from falling over. To go forward, the user leans slightly forward; to go backward, he or she leans slightly back. A "LeanSteer" handlebar is used to turn left or right. The vehicle has a maximum set speed of 12.5 miles per hour—though police vehicles can go twice as fast—and can travel about 24 miles on a single battery charge.

Kamen and his team began developing the Segway in the mid-1990s at a cost of $100 million. The device was finally unveiled, to tremendous publicity, in December 2001. Kamen conceived the Segway as an innovative way to relieve traffic congestion and began lobbying governments in the United States and abroad to allow the unique vehicle to be driven on sidewalks. Many municipalities have given their approval, though others have expressed concern that the vehicles could potentially be dangerous on crowded sidewalks. The Segway is especially popular among people who like both its futuristic design and its environmental friendliness. It has also gained favor among some police departments, which use it to patrol city streets. It has even been evaluated by the United States Postal Service as an assistive device for delivering the mail. In the end, it seems Kamen's invention offers a unique solution to a common problem associated with urban life. What more could an inventor hope for?

Brilliant and Loopy

A few inches of wire twisted into loops brought order to the office.

* * * *

YOU'RE SITTING AT your desk shuffling through a 12-page report; you've spent four hours preparing it, and it's due in five minutes. One last look reveals that everything is in order—but wait. It's not ready to go yet. You can't just hand over a messy sheaf of papers. You have to put it all together, neat and organized; without even thinking, you reach for a paper clip.

Few things are as unappreciated yet so widely used as the lowly paper clip. A few inches of wire twisted to form a loop within a loop—what could be simpler? Believe it or not, this familiar office product wasn't widely available until around 1900, when the Cushman and Denison manufacturing company began selling the Gem paper clip that is so familiar today.

Why hadn't anyone come up with such a simple device before? The first reason is that it wasn't until the mid-1800s that people could produce malleable strands of thin wire in mass quantity. Why did it take 50 years to come up with the paper clip? Well, it didn't. The first wire clip for holding papers was patented by Samuel B. Fay as early as 1867, and numerous other inventors around the world very quickly produced their own versions in various shapes and designs. In fact, at least 50 other paper clip patents were issued in the latter half of the 19th century. But none of these inventors was able to devise a machine that could cheaply manufacture their products. It wasn't until William Middlebrook of Connecticut patented a machine for rapidly and economically producing the Gem design that the paper clip could be made cheaply and in bulk. The only real competition the Gem has seen in the last hundred years is from the Gothic paper clip, which offers pointed rather than rounded loops and extends the two ends of the wire all the way to the end, making it less likely to tear paper.

Shop 'Til You Drop

An Oklahoma entrepreneur used a simple folding chair to change the way the world shopped.

❋ ❋ ❋ ❋

IN THE LATE 1930s, Sylvan Goldman, like any good businessman, was trying to find a way to increase sales. At his two grocery store chains in the Oklahoma City area, Standard and Humpty Dumpty, he noticed that when the wire hand baskets his stores provided became full or heavy, most of his customers headed for the checkout line. He imagined this problem could be remedied if shoppers had a way to conveniently carry more items as they wandered the aisles. Puzzling over the problem in his office one evening, he was struck by inspiration when a simple wooden folding chair caught his eye. What if that chair had wheels on the bottom and a basket attached to the seat? Or better, why not *two* baskets?

Goldman explained his idea to Fred Young, a carpenter and handyman who worked at the store, and Young began tinkering. After many months and many prototypes, the two men hit on a design they thought would work. Goldman's first carts used metal frames that each held two enormous baskets—19 inches long, 13 inches wide, and 9 inches deep. When not in use, the baskets could be removed and stacked together, and the frames folded up to a depth of only five inches, thus preserving the most precious commodity of any retail store: floor space.

Nowadays, most of us couldn't imagine a grocery store without shopping carts, but Goldman's customers were reluctant to use the strange new contraptions at first. Ever the salesman, he hired models of various ages to troll his stores and shop with his "folding carrier baskets." Eventually, the innovation caught on, not only at Goldman's stores but also at retail outlets across the country. In 1937, Goldman founded the Folding Carrier

Basket Company to manufacture his carts for other stores. They became so popular that by 1940 he was faced with a seven-year backlog of new orders.

Brassieres: A *Bust*-ling Business

A simple strap of linen led to the padded, wired contraption that is the modern bra.

* * * *

IN ROMAN TIMES, women who had active jobs often wore straps of fabric around their busts to keep things stable. Thereafter, women vacillated wildly between incredibly restricting corsets and less restrictive support. In 1889, French *couturier* Herminie Cadolle created a two-piece undergarment that began to topple the reign of the corset. Cadolle's *soutien-gorge*, or "breast supporter" (the top half of the two-piecer), was an instant hit at the Great Exposition of 1900. Alas, it was still expensive to purchase, since it was made primarily of the same materials as the traditional corset.

In 1913, socialite Mary Phelps Jacob was dressing for her New York debut when she realized that the gauzy dress she'd selected would never go with her heavy corset. She enlisted the help of her maid, and the two of them stitched together two handkerchiefs and ribbon. Jacob called her invention the "backless brassiere," and the name stuck, both with her friends and the unknown person who sent Jacob a dollar to create one for her. (So *that's* what girls gossip about at those debutante balls!)

Jacob was a society girl, not a businesswoman, and though she had the wherewithal to apply for a patent in 1914 for her new undergarment, she either didn't enjoy running the business or couldn't keep up with the demand. She eventually sold her patent to the Warner Brothers Corset Company for $1,500.

Perhaps if she'd been as creative with her product name as she had with the name she'd created to run the business—Caresse

Crosby—she might have had better luck with the item. As it was, Jacob ended up becoming a fairly major literary influence, establishing two publishing imprints, and also founded the organization Women Against War.

You'll Shoot Your Eye Out!

An unlikely classic becomes a must-watch at Christmastime.

⁕ ⁕ ⁕ ⁕

NINE-YEAR-OLD RALPHIE PARKER wants a Red Ryder BB gun for Christmas. That, in a nutshell, is the entire plot of *A Christmas Story*. Wrapped around that is a tapestry of oddball characters and themes that draws viewers in to pre-World War II America. Rosecolored nostalgia is wryly adjusted by the ever-present narrator: Ralph as an adult looking back both fondly and not so fondly. The oddball characters in the movie aren't even so strange; they're more a commentary that normalcy is just too rare.

Oh, Fudge

Ralphie's dream for the Red Ryder BB gun is scuttled by his mother, teacher, and Santa, who all scold, "You'll shoot your eye out." So consumed is Ralphie with the gun that he daydreams about submitting an A essay about the rifle and defending his home against intruders. Ralphie's quest is sidetracked by other humorous moments of suburban childhood. School kids "triple dog dare" a classmate to stick his tongue on the frozen flagpole. His father (an artist with obscenity and affectionately named the "Old Man") hears Ralphie accidentally repeat one of his own favorite words ("the queen mother of dirty words, the f-dash-dash-dash"). Ralphie launches into his own screed of foul language when he stands up to the bully with yellow eyes. Younger brother Randy is layered so tightly in sweaters and a snowsuit by his mother that he can't move his arms. Ralphie's parents wage war over a grotesque lamp in the shape of a woman's leg in netted stockings. Neighborhood dogs trash the family's holiday turkey, sending the Parker tribe to a Chinese restaurant where waiters horribly mispronounce seasonal songs. From start to finish, *A Christmas Story* mashes the familiar with the bizarre.

"My Old Man was one of the most feared furnace fighters in Northern Indiana"

Several scenes reveal a more cynical view of Christmastime. When Ralphie finally gets his Little Orphan Annie secret decoder badge, he is dismayed to learn the encrypted message from the radio show is little more than an ad for Ovaltine. But there are warm moments, too. Ralphie's mother changes the subject after telling the Old Man about the fight in order to keep Ralphie out of trouble. Later, Ralphie's disappointment over not getting the BB gun is erased when the Old Man brings out one last present. The climax—when Ralphie fires his gun and really does almost shoot his eye out (thanks to a ricochet)—launches into the turkey/ Chinese restaurant disaster. The family bursts into laughter, and despite all the lies, swearing, and back-stabbing, we're still treated to a happy ending.

Triple Dog Dare

Ralphie is the creation of Jean Shepherd, a writer and radio host. He largely invented the talk radio format with his long, discursive monologues. Shepherd's semiautobiographical stories and monologues would become the foundation for the movie. Before he brought Shepherd's stories to screen, director Bob Clark cut his teeth on horror films, including *Black Christmas,* in which a killer stalks a sorority house. He later scored a hit writing and directing the sex farce *Porky's* (and its sequel). Perhaps it was *because* and not *in spite* of his previous work that *A Christmas Story,* retained an edge uncharacteristic of holiday fare. Clark brought Shepherd's stories to the big screen, collaborating with the author on the script and enlist ing him in narration duties as well. The idea of an adult narrating his childhood would later inspire *The Wonder Years* television series.

The Norden Bombsight

Civilian populations in some areas may have been spared during the war due to the tinkering of a humble engineer.

* * * *

In 1920, Dutch immigrant Carl Norden began working on an advanced bombsight for the U.S. Navy. Two years later, he partnered with Theodore Barth, and during the next four years the team designed and built a bombsight from locations in both the United States and Switzerland. The firm incorporated in 1928 and finished a revolutionary bombsight the same year. Its key feature was a timing mechanism that indicated when the bombardier should release his payload.

Norden demonstrated his bombsight's capabilities in 1931 to senior Navy personnel, who were duly impressed and placed a contract for 40 sights. The Army Air Corps also placed an order. In 1935, the Air Corps tested the new bombsights, which had been installed on B-10 bombers: By the end of testing, bombardiers were able to drop 50 percent of their bombs within just 75 feet of their targets.

How Did It Work?

The Norden bombsight was a complex piece of equipment consisting of an analog computer, gyros, levels, mirrors, electric motors, and a small telescope. The final version, the Mark XV, comprised more than 2,000 individual components.

Approximately 30 to 45 minutes before reaching the target, the bombardier would run through a series of precise, carefully predetermined functions to prepare the bombsight for use. He would input altitude, airspeed, wind speed, and the angle of drift into the bombsight's computer, which then calculated both the trajectory of the bombs and the exact time the payload should be released.

Antiaircraft fire and weather conditions severely decreased the accuracy of the bombsight. To help counter these adverse conditions, Norden had a hand in the invention of an automatic pilot, which worked in conjunction with the sight. The autopilot helped reduce turbulence and overcontrol by anxious pilots, which increased the accuracy of the sight. The actual bombing run began when the airplane reached a point specified at the preflight briefing. The bombardier then ran through a series of final steps that began with aligning the vertical and horizontal cross hairs on the target and ended with the bomber's autopilot engaging and the bombs dropping.

The Bombsight Becomes a Must-Have

Norden's factory had prewar production capabilities of about 800 bombsights per month. After the Japanese attack on Pearl Harbor, Norden factories began producing some 2,000 units per month. By the fall of 1945, more than 43,000 sights had been manufactured.

Norden was paid just $250 for his bombsight. In the spirit of patriotism typical of the times, he sold the rights to his invention to the U.S. government for just one dollar more. Norden was satisfied with the fact that his invention enabled the military to more accurately strike targets, thus minimizing the collateral damage done to civilian populations.

Getting Carded

Modern holiday well-wishers aren't the first to be crunched for time. Back in 1843, Sir Henry Cole invented the Christmas card when the demand of writing personal holiday greetings to friends and family on his list threatened to squelch his holiday spirit.

* * * *

COLE, A RENAISSANCE man who wrote and published books on art collections and architecture, ran in social circles where gifts and tokens were synonymous with friendship.

Knowing that more than a few of his uppity buddies would be offended if they didn't receive personal greetings from him, he commissioned well-known painter of the time John Callcott Horlsey to design a "one-size-fits-all" card that could be sent to everyone on his list. The lithographed and hand-colored sepia-toned sketch was printed on cardboard and featured a classic Victorian Christmas scene of a family eating and drinking merrily, with the caption, "A Merry Christmas and a Happy New Year to You." On the other side, Cole ingeniously asked for a picture to encourage the care and feeding of the poor, his own personal cause.

Christmas Card Craze

The idea slowly gained momentum in the Christmases to come, with even Queen Victoria and the British royal family sending specially designed cards focusing on the events of the past year. Americans joined in the tradition early on by importing cards from England. But in 1875, Louis Prang, a German immigrant who wrote and published architectural books, printed images in color with a series of lithographic zinc plates. The process allowed up to 32 colors to be printed in a single picture, with the finished product resembling an oil painting. His first cards featured flowers and birds, then he moved to snow scenes, fir trees, and glowing fireplaces. So in demand were these cards that Prang couldn't fulfill all of the orders. Every year following, demand increased, and at one point Prang was printing five million cards a year. His efforts earned him the moniker "The Father of the American Christmas Card."

Glamorous Greetings

Well-known illustrators such as children's book author Kate Greenaway and Ellen Clapsaddle designed cards for the public. Early cards could be elaborate. Some were cut into fancy shapes such as bells, candles, and birds. Others made noises. The more elaborate fitted together like puzzles or were embellished with fancy trims and buttons. The first pop-up card in the United States was designed in the mid-1800s, when

New York engraver Richard Pease created a card with a small Santa Claus with his sleigh and reindeer, and drawings of holiday celebrations were placed in each corner. Other popular images included skating rinks with children skating and family scenes around the fireplace. In 2009, more than 1.8 billion people sent Christmas cards, and anything goes in regard to the style. Clever verses, sports figures, comic strips, celebrities, and animals grace the fronts of cards, and high-tech options with intricate paper cuttings, embedded sounds and songs, and heavily embossed illustrations have become all the rage. Cards have also evolved into other mediums. In 1992, Vodafone sent the first holiday message using a cell phone; today, many people celebrate the holiday by emailing cards to friends and family.

Pass the General, Please!

For all of their military importance, generals have made many other notable contributions to civilization. Civil War general Ambrose Burnside, for example, had such an impressive pair of muttonchops that he inspired a new term: "sideburns." Then there's General Tso and his scrumptious chicken dish.

⁕ ⁕ ⁕ ⁕

General Tso's Incongruous Legacy

WHO WAS GENERAL Tso, anyway? Was he as good a general as he was a chef? As it turns out, Tso was a brilliant Chinese general who gained fame during the Taiping Rebellion of 1850–64. By the time Tso helped crush the rebellion, his was a household name. Thus, it would seem likely that General Tso's chicken is a recipe that was invented for or named after him during his lifetime, in the same way that beef Wellington was named after the Duke of Wellington—right? Wrong.

Unfortunately, it appears that General Tso has about as much to do with the classic Chinese recipe that bears his name as General Eisenhower has to do with Mike and Ike's. According to most food historians, General Tso's chicken as we know it today wasn't even invented until the 1970s, when it was devised in New York City, far away from the southern provinces of China where Tso earned his glory.

Most accounts claim that General Tso's chicken—a sweet, spicy, crispy chicken dish—was introduced in 1973 by Peng Jia, a onetime chef to Chinese military and political leader Chiang Kai-shek and the proprietor of a Manhattan Chinese restaurant named Peng's. Peng claimed that he actually invented the dish while working for Chiang sometime in the 1950s, but that the original recipe was far different from the dish that so many Americans love today.

A Taste Sensation

It may be hard to believe now—every town has a "Best Hunan" languishing in a strip mall, after all—but when Peng opened his New York restaurant in 1973, Hunan cuisine was virtually unheard of in the United States. Instead, most Chinese restaurants in the U.S. featured Cantonese cuisine, which is far blander and sweeter than its Hunan counterpart.

Peng—concerned the American palette was unprepared for the fiery, sour taste of his original dish—sweetened the recipe. The dish was an instant hit and gained massive exposure when Henry Kissinger—whose every move was covered in the social columns and gossip rags of the day—made Peng's restaurant a regular hangout and General Tso's chicken his usual meal. Soon, General Tso's chicken could be found on Chinese menus from coast to coast.

Why the name General Tso's chicken? Peng never said. But considering that he invented the dish for Chiang Kai-shek, the name was probably one of several that the chef used to honor the military greats who had come before the Nationalist leader.

Rejected in the Homeland

Interestingly, Peng had less success with his dish in his native China. In 1990, he returned to Hunan province, where he opened a restaurant that featured the dish that had made Hunan ground zero for international Chinese cuisine. The establishment closed quickly.

The reason? General Tso's chicken, the symbol of Hunan cooking throughout the world, was too sweet for the Hunan people.

Why Don't Grape-Nuts Contain Grapes or Nuts?

In a 1992 Saturday Night Live sketch, Jerry Seinfeld played the host of a quiz show for comedians. Seinfeld poked fun at his own penchant for riffing on the banalities of daily life by posing some questions, including, "What's the deal with airplane food?" and "What is the deal with Count Chocula?" and "Grape-Nuts—you open it up, no grapes, no nuts! What's the deal?"

⁕ ⁕ ⁕ ⁕

THE "CONTESTANTS" WERE stumped; apparently, so were *SNL*'s writers, because no good answer to the Grape-Nuts query was presented.

To be fair, this is a question that surely has been asked often since 1897, when C. W. Post invented the grain-heavy breakfast food. Essentially a shredded brick of baked wheat and malted barley, Grape-Nuts cereal has nothing remotely resembling grapes or nuts on its ingredient list—and it never has.

For most of us, breakfast cereal is about as inseparable from an American childhood as Saturday morning cartoons—but it wasn't always that way. Until the late nineteenth century, a typical American breakfast consisted of eggs, bacon, and sausage. Heart disease was rampant, though its causes were poorly understood and treatments were virtually unknown. Those who escaped cardiovascular disease ran the risk of developing gastrointestinal disorders because of the near-absence of fiber in the typical American diet.

This began to change in 1863, when Dr. James Caleb Jackson concocted a fiber-rich bran nugget that he hoped would bring relief to the bowels of his patients. Unfortunately, these nuggets required overnight soaking just to be chewable. And Jackson opted to call his breakfast item Granula, a name that evokes a blood-sucking grandmother, not a healthy meal.

Americans were not impressed. But the idea stuck around, and decades later, another doctor, John Harvey Kellogg—if the last name sounds familiar, it should—created his own version of a fiber-rich breakfast item. Never much of a wordsmith, Kellogg also named his creation Granula; he ultimately changed it to Granola after a trademark dispute with Jackson.

One of Dr. Kellogg's patients was C. W. Post. Confident that he, too, could make an edible breakfast food, Post set out to create his own cereal made of baked wheat and barley. Not being much of a scientist, Post believed that the sucrose that formed during his cereal's baking process was grape sugar. Nor did Post have much of a palate: He thought that his creation tasted nutty. Hence the head-scratching name.

Though Post might have been somewhat disconnected from reality, he was a clever marketer. By the turn of the twentieth century the nation had developed a taste for breakfast cereal, and Post positioned Grape-Nuts as a healthy option. Americans bought both the marketing claims and the cereal—Grape-Nuts has become one of history's best-selling breakfast foods even though it has no grapes, no nuts, and no flavor.

Clothes Encounters

This was the way we washed our clothes...

* * * *

ONCE UPON A time, people washed clothes by hand, pounding them against stones or using such high-tech gadgets as washboards. It was an arduous process that involved hauling or pumping water, heating it in large kettles, and scrubbing with caustic substances such as lye soap. No wonder the *Lady's Book* of 1854 said, "Our spirits fall with the first rising of steam from the kitchen, and only reach a natural temperature when the clothes are neatly folded in the ironing basket." It's also no wonder that lots of folks tried to come up with gadgets that would make washing clothes easier.

In 1797, a man named Nathaniel Briggs was granted the first U.S. patent for "an improvement in washing clothes." Unfortunately, the records of the patent were destroyed when a fire broke out at the patent office in 1836, so we have no idea what that device looked like.

Early washing machines were rotating drums with attached cranks—you put in the soap, water, and clothes, closed the door, and started cranking. C. H. Farnham received a patent for a hand-cranked washing machine in 1835. From there, it was an obvious leap to attach a motor to the crank and power it with steam, electricity, or gasoline.

Who gets credit for first attaching an electric motor to a washing machine and thereby "inventing" the electric washer? Hard to tell. Many sources credit Alva J. Fisher, who received a patent for such a washer in 1910. But other people had received patents for motor-driven washers before Fisher, and Fisher did not claim to have invented the electric washer in his patent. Oddly, the Hurley Machine Company had been selling the electric Thor washer, designed by Fisher, since 1907. While Hurley is no longer in business, two companies that began selling electric washers about the same time are Maytag and Whirlpool.

"Everything that can be invented has been invented."

—U.S. Office of Patents Commissioner Charles H. Duell, 1899

✳ Chapter 7

Amazing 19th-Century Americans

The Legendary John Muir

The man who kept America's wildernesses wild began as just a boy from the woods of Wisconsin.

✳ ✳ ✳ ✳

Son of the Midwest

MENTION THE NAME John Muir, and people instinctively think of the West: The vast, snow-capped mountains of the Sierra Nevada, Mount Rainier, Arizona's petrified forest, the Grand Canyon, and, of course, the wild glacier-carved wilderness of Yosemite, where Muir took President Theodore Roosevelt camping in 1903. Muir's thousand-mile walks and multiyear explorations of these wild places are legendary, as is the role he played in protecting and preserving them for the future. But as much as Muir influenced the shaping of the American West, it was Wisconsin that shaped him first.

Muir wasn't supposed to have grown up in Wisconsin. He was born in Dunbar, Scotland, in 1838. His father yanked Muir and his brother out of school and uprooted the family for a move to the New World, aiming to settle in the backwoods of Canada when Muir was just 11 years old. But on the rough six-week sail across the Atlantic, the elder Muir heard other emigrants talk of Wisconsin and Michigan. The prospects were better, they said, and the soil richer. There were woods, they had heard—but not as dense as Canada's, where a man could spend his whole life clearing a patch of ground to farm.

By the time the ship arrived on American shores, Muir's father was losing his lust for Canada. What was the ultimate nail in the family's Canada trip? While docked in Buffalo, New York, the elder Muir met a grain dealer. He asked the man where his grain originated. Most, the man replied, came from Wisconsin. Muir's father made his decision on the spot. The family sailed up the Hudson, through the Erie Canal, and across the Great Lakes, finally arriving at the gateway to Wisconsin's storied oak savannahs, Milwaukee.

Love at First Sight

The Muir clan headed inland, settling beside the Fox River, just northeast of Portage. The day John Muir set eyes on the landscape that would be his new home, he and his brother discovered in a tree a blue jay's nest, full of green eggs and just-hatched fledglings.

"This sudden splash into pure wilderness—baptism in Nature's warm heart—how utterly happy it made us," he recalled in his 1913 book, *The Story of My Boyhood and Youth.* "Nature streaming into us, wooingly teaching her wonderful, glowing lessons, so unlike the dismal grammar ashes and cinders so long thrashed into us. Here without knowing it, we were still at school; every wild lesson a love lesson, not whipped but charmed into us. Oh, that glorious Wisconsin wilderness!"

Under his father's stern command, Muir did not attend school in Wisconsin, but rather, toiled on the family's farm. He managed to escape now and again to savor Wisconsin's woods and waters. One of his favorite haunts was along the Fox River and Pucaway Lake, where he marveled at the "millions of ducks" who congregated in the surrounding wild rice marshes. It was there he began his long love affair with bird-watching.

By 1856, the farm's soil had been depleted, and Muir and his family moved six miles southwest, to a place that would come to be known as Hickory Hill Farm. Muir's first task? Digging the family a well. Dynamite proved ineffective in blasting through the thick layers of sandstone below the soil, so his father put the young Muir to work. Day after day, from sunup to sundown, the teenager crouched in the ever-deepening hole, cutting away at the sandstone with a hammer and chisel. The task almost killed him. One day, roughly 80 feet down, the earthen hole filled with carbonic acid gas, knocking Muir nearly unconscious. Luckily, Muir's father, suspicious of the sudden quiet, peered down the cavernous hole, realized his son was in danger, and hauled young Muir out in a bucket. Instead of sympathy, however, his father tossed piles of straw into the hole to stir up the poison and carry down some pure air. Within the next day or two, Muir was sent back in, toiling on until, 90 feet down, he finally hit water.

To secure more time to himself—and away from the endless demands of his father—Muir struck a deal with his dad: he could have all the hours to himself that he wished, so long as they were gotten before his workday began. Thrilled, Muir took to waking up at 1:00 A.M. each day, relishing the quiet hours before the rest of the family awoke and satiating his curious mind by reading books and building wild contraptions he dreamt up. Among them was a rotary saw powered by the water of a dammed brook, unique door locks and latches, an automatic horse feeder, a clock-set firelighter, and, perhaps most famous of all, an alarm clock that not only told the time but connected to his bedstead to launch him onto his feet at a designated hour.

Encouraged by neighbors who thought his mechanical inventions were the work of a genius, Muir decided to take his clocks to the State Fair in Madison in hopes of securing a job in a machine shop. They were a hit, and Muir was hired to work in a machine shop in Prairie du Chien. Within a few months of starting his new job, however, he had set his sights on another goal—college.

Higher Education

At age 22, Muir enrolled at the University of Wisconsin in Madison, where he quickly gained notoriety for his inventions, including a desk he had whittled from wood, which accelerated his studies by pushing a new book in front of him after several minutes had elapsed. Though Muir scrimped and saved to pay for his tuition, room and board, and other school supplies—often surviving for weeks at a time on bread, potatoes, and molasses—he didn't work toward a degree. He simply took whatever classes interested him: Greek, Latin, chemistry, math, and physics were among his favorites. However scattered his curriculum at the University of Wisconsin, it was there that a seminal moment occurred that would forever alter the course of his life.

One June day a fellow student who loved "imparting instruction" (read: know-it-all) called Muir over to see a flower he had plucked from a nearby locust tree. He showed Muir the subtle characteristics that revealed, to the trained eye, the giant, thorny hardwood's relation to a weak, spindly pea vine. The revelation that nature's designs were not singular or similar by coincidence, but rather, unified with boundless variety, astounded Muir. He rushed into the woods and meadows, wild with enthusiasm for botany, a line of study he believed revealed traces of God's thoughts.

Without regard for a diploma or future employment, Muir decided to leave school to explore the world's wildest corners. It was not without some sadness that he left the place that had ignited his life's passion. As he recalls in *The Story of My Boyhood and Youth,* "From the top of a hill on the north side of Lake Mendota, I gained a last wistful, lingering view of the beautiful University grounds and buildings where I had spent so many hungry and happy and hopeful days. There with streaming eyes I bade my blessed Alma Mater farewell. But I was only leaving one University for another, the Wisconsin University for the University of the Wilderness."

Abraham Lincoln

Although more than two centuries have passed since Abraham Lincoln was born on February 12, 1809, this farm-boy-cum-president is still the object of worldwide fascination.

⁕ ⁕ ⁕ ⁕

Lincoln's Looks

HONESTLY, ABE WAS no babe. At a gangly 6 feet 4 inches, Lincoln stood nearly a foot taller than the average man and wore a size 14 shoe. With a wingspan measuring seven feet, Lincoln's long arms gave him a decided advantage when engaged in his favorite sport of wrestling.

A wart on his cheek and a scar over his eye marked the right side of Lincoln's face, a souvenir from an 1828 robbery attack. Lincoln's thin face was so homely that an 11-year-old girl bribed him to "let [his] whiskers grow" in exchange for her brothers' votes.

Abe was well aware of his shortcomings in the looks department. When charged with being "two-faced" by political rival Stephen Douglas, Lincoln retorted: "If I had another face, do you think I would wear this one?"

Funny Fellow

With his bawdy wit and knack for storytelling, Honest Abe had the makings of a great stand-up comedian. After observing a well-dressed woman in a plumed hat fall while navigating the muddy streets of Springfield, Illinois, Lincoln commented, "Reminds me of a duck." When someone asked why, Lincoln quipped, "Feathers on her head and down on her behind." Lincoln had a story for every occasion and often peppered his legal pleadings and stump speeches with relevant anecdotes.

This gift of gab probably aided Lincoln's political career. Many voters related to Honest Abe's folksy charm and hardworking "rail splitter" background.

However, not everyone was amused by Lincoln's offbeat sense of humor. Some cabinet members questioned the president's "buffoonery," especially the time when he delayed the start of a crucial meeting to discuss the Emancipation Proclamation to read aloud a chapter from an amusing book by Artemus Ward.

At the conclusion of the reading, Lincoln laughed heartily while the stone-faced cabinet members refused to crack a smile. So he read them a second chapter. Upon receiving the same cold response, Lincoln threw down the book in exasperation, saying "Gentlemen, why don't you laugh? With the fearful strain that is upon me night and day, if I did not laugh I should die."

Ironies and Oddities

Lincoln was shot on April 14, 1865. It was also a Good Friday, which helped boost the president (who died the next day) into the role of martyr/messiah for many supporters. Ironically, Edwin Booth, the brother of Lincoln's assassin John Wilkes Booth, had once saved the life of Lincoln's son Robert. The oldest of Lincoln's four sons, Robert was the only one to survive into adulthood.

Lincoln's wife, Mary Todd, was fascinated with psychic phenomena and publicly frequented spiritualists. After their son Willie's death, she was said to have held a séance at the White House to make contact with her son's spirit. Lincoln, desolate with grief over the loss of his favorite son, is rumored to have attended.

Lincoln himself reportedly had psychic abilities. Numerous accounts of his visions exist, but perhaps the most chilling of these is the premonition he had prior to his assassination. Lincoln dreamt he heard crying. He went on to dream that upon investigating its source, he discovered a body laid out in the East Room, stationed with guards. When he asked the mourners who had died, they replied, "The president. He was killed by an assassin."

Some people claim that Lincoln must have had a sense of what was to happen. When he left for Ford's Theatre the evening of his assassination, he paused to tell his trusted bodyguard, William Crook, "Goodbye." Crook reports that Lincoln habitually bid him "Good night." It was the first time the president had ever told him otherwise.

The Haunted Mind of Edgar Allan Poe

Although he died in 1849, Edgar Allan Poe remains one of the world's best-known writers of horror fiction. His stories are continually reprinted and have inspired numerous motion pictures, television shows, and literary works.

* * * *

Madness Unmasked

HORROR WASN'T ALL that Poe wrote. Over the course of his career, the mustachioed wordsmith also invented the modern detective story and penned several well-received works of humor and satire. But it was his work in the horror genre that made him a household name.

A review of Poe's deliciously demented short fiction reveals him to be a master of what we now call the Gothic style of horror, in which characters slowly descend into madness that often leads to horrific consequences. Indeed, his short stories and poems are filled with individuals driven to the brink of insanity—some by their own devices and others by ghosts and spirits in various manifestations.

"The Fall of the House of Usher" is a prime example of this style. First published in 1839, it's the tale of a man named

Roderick Usher who suffers from an unknown malady. His sister Madeline is also ill and has a propensity to fall into death-like trances.

After Madeline dies suddenly, she is interred in the family tomb for two weeks before her final burial. During that time, Roderick experiences all manner of unexplained phenomena, including horrific sounds that echo throughout the house. The manifestations eventually become so unbearable that Roderick falls into hysteria, convinced that his sister—whom he admits he entombed while she was still alive—is responsible. When Madeline arrives at Roderick's bedroom door, she reaches for him and they both fall to the floor dead.

Morella's Curse

Another fine example of Gothic horror from Poe's supernatural canon is "Morella," the story of a woman who dabbles in the black arts.

As Morella's physical body deteriorates, her husband wishes for her to die so that she may find peace. His wish is granted when Morella passes away during childbirth. Unfortunately, because of her experimentation with the black arts, her soul cannot cross over; instead, it is transferred to the baby, who grows up bearing a terrifying resemblance to her mother.

Stricken, the father refuses to give the girl a name and restricts her to the house. Eventually, however, he decides to have her baptized. During the ceremony, the priest asks for the child's name, and when the father whispers "Morella," the child turns her head and declares, "I am here!" before dying on the spot. Heartbroken, the father brings the little girl to the family tomb, where he finds no trace at all of his wife's body.

Similar themes can also be found in the short story "Ligeia," in which the title character—a beautiful and intelligent woman—falls ill and eventually dies. Her husband, the story's narrator, marries Lady Rowena, who also becomes ill and dies.

Distraught, the grieving husband spends the night with the corpse of Lady Rowena, which comes back to life—as Ligeia.

Genius Revealed

Edgar Allan Poe was an extraordinarily talented writer who had a remarkable gift for establishing a mood of Gothic foreboding. His characters are often insane or drug-addicted (or both), which leaves analysis of their actions and revelations open to interpretation by the reader. Through it all, however, Poe never fails to deliver haunting descriptions of the mind-bendingly grotesque and macabre. As a result, his stories and poems excel at sending shivers down the spines of even the most jaded readers.

Ghosts and bizarre images of the undead in various forms play integral roles in many of Poe's literary works, though readers are occasionally left wondering what is real and what is merely a figment of the protagonist's fevered imagination. In the end, however, it doesn't really matter: A thrill is a thrill, regardless of the cause—a fact that Edgar Allan Poe knew only too well.

Mathew Brady, Dean of the Photographers

Acting more like a director than a photographer, Mathew Brady was the first to provide visual documentation of war.

* * * *

THE CONNECTION BETWEEN Mathew Brady and Civil War photography isn't completely accurate—his assistants captured the immortal images attached to his name. He had a drive to propound his own myth, seeking personal glory and fortune as chronicler of the war. But Brady was an innovator—prior to his time, no one else had even done as much as consider how to take pictures of a war.

An Entrepreneur Is Born

Even before the war, Brady was no ordinary entrepreneur. Born in upstate New York, he arrived in the big city in 1840 as a teenager and trained in basic photography with Samuel Morse, the inventor of the telegraph. Just five years later, Brady launched his own empire. Never without ambition, Brady styled himself as photographer to the illustrious, from chronicling leaders such as Andrew Jackson and the Prince of Wales to entertainers and writers such as singer Jenny Lind and Edgar Allan Poe. His Broadway gallery became a compulsory stop for both those who were seeking immortality and those who merely wished to observe it.

Documenting History

With the country poised to erupt in 1861, Brady began preparing to document history. Brushing aside concerns for his personal safety, he parlayed his connections into a ringside seat at the country's darkest hour: President Lincoln signed a note reading "Pass Brady" to smooth any obstacles.

The resourceful Brady put together rolling, horse-drawn darkrooms staffed by his best photographers. Possibly due to failing eyesight, he worked more as a director than a hands-on camera operator. Two of his associates, Alexander Gardner and Timothy O'Sullivan, were responsible for the lion's share of what we know today of the war's horrific images.

Seeing the war's carnage up close awakened Brady to the commercial possibilities of putting it on public display. Antietam provided his first opportunity for an exhibition of the photographic documentation. Recording an endless array of images of corpses on the battlefield, Brady's team caused a media sensation when the series "The Dead at Antietam" played to packed houses at his New York City gallery shortly after the battle. Viewers were horrified by the compelling photographs, but they kept coming. As The New York Times wrote, "Mr. Brady has done something to bring home to us the terrible reality and earnestness of war." The net effect of Brady's work galvanized public opinion in the North against the war.

Brady's penchant for claiming all the credit galled his associates, and soon Gardner and O'Sullivan struck out on their own. Gardner in particular prided himself on his technical innovations. Undeterred, Brady hired others who in time made their own mark in photography. By war's end, he had an inventory of some 10,000 images.

Interest Wanes

When the fighting ended, however, interest in Brady's work quickly dissipated. The public was weary of the war in all its aspects. Brady had spent himself into oblivion, and competitors soon wrested his share of the market away. It took him ten years and much quibbling to convince Congress to buy out his negatives for a paltry $25,000. Pushed by Representative James Garfield, a Union vet, the government eventually ponied up, but not before Brady had sold off most of his assets, including his New York City gallery.

Brady's wife died in 1887, and he never recovered. He became an alcoholic and died penniless in 1896. Not until the next century would historians take stock of his achievement and restore his name to its place in photographic history.

The First Texan

Sam Houston, whose name is almost synonymous with Texas, lived a life full of action, controversy, and risk.

* * * *

A STATUE OF SAM Houston towers over the East Texas city of Huntsville, easily visible to motorists traveling along Interstate 45. At 67 feet tall, it is the world's largest freestanding statue of an American figure. Samuel Houston was such an important character in the history of Texas, though, that many Texans argue that the statue isn't nearly large enough.

Born in Virginia in 1793, Houston successfully led the Texas rebels in their battle for independence from Mexico, famously defeating Santa Anna's army in the decisive battle of San Jacinto in 1836. He was then elected as the new Lone Star Republic's first president and had the city of Houston founded in his name. When Texas joined the Union in 1846, Houston became one of the new state's U.S. senators and later served two terms as governor.

Houston's renowned fondness of alcohol, women, and brawling, however, meant that his life was never far from controversy. He spent years living among the Cherokee, who referred to him by his Indian name "the Raven," or more simply as "Big Drunk." Houston took a Native American as the second of his three wives, and he was once arrested and convicted for beating a U.S. congressman.

Houston came close to being nominated to run for U.S. president in 1860, but his ardent opposition to secession and his refusal to take the oath of loyalty to the newly formed Confederate States of America saw him removed from the governor's mansion in 1861.

The Land of Promise

Sam Houston's public life didn't start with his arrival in Texas in 1832. By that point, he had already served under Andrew Jackson in the War of 1812, suffering three near-fatal wounds at the Battle of Horseshoe Bend; had served two terms in Congress; and had been elected governor of Tennessee. He resigned as governor in 1829 at the age of 36, after his 11-week marriage to 19-year-old Eliza Allen ended amid mysterious circumstances. Since both Houston and his young bride maintained a lifelong silence about their brief marriage, the circumstances of its demise remain vague to this day.

Houston left Tennessee and went to live among the Cherokee in modern-day Oklahoma, with whom he had spent two years as a teenager. He was granted Cherokee citizenship and became a tribal emissary, a role in which he took great pride. When William Stanbery, a U.S. representative from Ohio, delivered a speech on the House floor in 1832 that Houston believed insulted him over an Indian rations contract, the Raven confronted him on Pennsylvania Avenue in Washington, D.C., and thrashed Stanbery with a hickory cane. In the subsequent criminal trial, Houston was found guilty but instead of paying the fine, he simply left the country. His second wife remained among the Cherokee while Houston moved to the Mexican state of Texas, a place he described as a "land of promise."

The Texas War of Independence

Houston quickly gained prominence within the rebellion movement that was building against the Mexican government. He was a member of the convention that met at Washington-on-the-Brazos in 1836 to declare independence from Mexico, and after the fall of the Alamo and the Goliad Massacre, he took charge of the ragtag Texan army.

Despite being outnumbered, Houston led his forces in the Battle of San Jacinto on April 21, 1836, and defeated the Mexican army led by that country's president, Santa Anna. The battle lasted less than 20 minutes, but along with the capture of Santa Anna the following day, it proved decisive and paved the way for the Republic of Texas to become an independent country.

Sam Houston was officially a Texas hero and became the first regularly elected president of the republic, serving until 1838. As the constitution of the fledgling nation barred a president from succeeding himself, Houston became a Texas congressman until 1841, when he was again eligible to be elected for another term as president. After Texas joined the United States in 1845, Houston served as first a senator and then governor.

The Problem of Slavery

Although Houston was a slaveowner and opposed abolition, he consistently voted against the expansion of slavery into other states beyond the South. As the issue became more and more heated, he also used his position as governor to vehemently oppose the growing support for Texas to secede from the Union. This proved a hugely unpopular move to his constituents, as did his refusal to pledge allegiance to the newly formed Confederacy in 1861. President Abraham Lincoln repeatedly offered Houston the use of federal troops to keep him in office, but Houston flatly declined, wishing to avoid bloodshed and civil unrest in his beloved Texas. Instead, he peacefully left the governor's mansion in March 1861, with only a prophetic warning for his opponents. An American Civil War, Houston predicted, would result in a Northern victory and the destruction of the South. The following month, hostilities broke out between the North and South when Confederate forces attacked a U.S. military base in South Carolina.

Houston retired to Huntsville with his third wife, Margaret Lea, who had provided him with eight children. A year later, Houston died from pneumonia at age 70. Fittingly, his final recorded words were: "Texas, Texas, Margaret!"

In addition to the city of Houston and the statue in Huntsville, many other places in Texas have been named in Sam Houston's honor, including a state university, a national forest, a regional library and research center, and a U.S. Army installation.

* Spanish Texas Governor Manuel de Salcedo (1808–1813) respected his Anglo citizens quite a bit. His words: "The Anglo-Americans are naturally industrious. If this were not true, they would not love to live in deserts, where their sustenance depends on their industry."

* What was the currency of the Republic of Texas? The Texas dollar banknotes' reddish reverses led to the colloquial name Texas redbacks. Unfortunately, most of the time the redback wasn't worth much. In some times and places, TX$1.00 was worth about US$0.02.

Laura Ingalls Wilder: From Little House to Big Success

It seems that everyone wants to claim a piece of author Laura Ingalls Wilder. Anyone who watched the Little House on the Prairie TV series knows that Walnut Grove is in Minnesota and there's a bust of Laura on display in Missouri where she settled in her later years. Laura also lived in Kansas, Iowa, South Dakota, Florida, and New York. Near the tiny village of Pepin, Wisconsin, is where it all began.

❋ ❋ ❋ ❋

Little House in the Big Woods

CAROLINE AND CHARLES Ingalls were living near Pepin in a small log cabin when their second daughter was born on February 7, 1867. They named her Laura Elizabeth. Little did they know that 100 years later Laura Ingalls Wilder, their little "Half Pint," as she was known on the TV show, would become a household name around the world.

Charles Ingalls was an explorer at heart and dreamed of settling in unknown territories out west. (Wanderlust may have come naturally to Laura's pa—he was a descendant of Richard Warren, who traveled to the New World aboard the *Mayflower.*) The Ingalls family lived in Pepin for just two years before they moved to Kansas. But times were hard on the prairie, and in another two years, they were back in the big woods of Wisconsin where Laura and her sister Mary attended the Barry Corner School.

Book It

It was those early years that became the inspiration for *Little House in the Big Woods,* the first of eight books written by Laura Ingalls Wilder. Although she didn't keep a diary as a young girl, Laura often wrote down her thoughts on scraps of paper and saved them. When Laura's sister Mary lost her sight, her father told Laura that she would need to be Mary's eyes—describing the things that Mary could not see for herself. That attention to detail served Laura well when she finally sat down to write about her life in 1932, at the age of 65.

Little House in the Big Woods tells the story of Laura's childhood and describes pioneer life in Wisconsin's vast woods in the 19th century. She tells of hunting and cooking over a fire and describes how they used to smoke venison, weave rugs, tap trees to get real maple syrup, and even make bullets. Laura's original intention was to write a single book called *When Grandma Was a Little Girl*, but her editors at Harper Brothers suggested she turn it into a series. Her books were hugely successful and remain so today.

Although it's based on her life, certain creative embellishments made their way into the text. For instance, fans may be surprised to find out that the character of Nellie Oleson, Laura's nemesis, is not actually based on one individual, but rather a combination of three people.

The Confederacy's President

Jefferson Davis had a dashing and distinguished career as a leader in both the United States and the Confederate States.

* * * *

IN THE YEARS leading up to the Civil War, Jefferson Davis had one of the best resumes in Washington, D.C., with a long and illustrious career in and out of government. But the decisions he made as war approached made him one of the most controversial figures in U.S. history. Was he a hero or a villain? That depends on the observer's point of view.

Early Life

Born in Kentucky in 1808, Davis trained at West Point and served in the Army for seven years after he graduated. He distinguished himself in the military—one early assignment was to escort the captured Native American chief Black Hawk to prison. The chief reportedly appreciated the kind treatment Davis showed him. The Army also brought Davis more than he might have expected. While in the service, he met his first wife, Sarah Knox Taylor, the daughter of his commanding officer, Zachary Taylor, future president of the United States. He retired from duty to wed her, but shortly after the wedding both Davises came down with malaria, and Sarah died only three months into the marriage. Davis spent most of the next decade growing cotton on his Mississippi plantation, but when he reentered public life, he seemed to have renewed energy.

A Political Career

In 1844, he won a seat in Congress, and the next year he married 18-year-old Varina Howell. Davis didn't stay in the House of Representatives long, however, resigning his seat to rejoin the Army for the Mexican War. After being wounded in 1847, he was appointed as a senator, and he served in that post until 1851, when he resigned to run for Mississippi governor. He lost that race, but President Franklin Pierce made the Mississippian his secretary of war, and Davis remained in that post throughout Pierce's term. Washington life agreed with Davis, and after the Pierce administration, he returned to the Senate.

Though an ardent supporter of states' rights, Davis opposed secession. On July 4, 1858, Senator Davis delivered an antisecessionist speech aboard a ship near Boston. He reiterated his support for preserving the Union again that autumn, but he was preparing for how the South might have to react if an "abolition president" were to be elected in 1860.

Leaving the Union

After South Carolina split its ties with Washington, Davis finally succumbed to pressure from his constituents and colleagues, and in January 1861 he announced Mississippi's secession, subsequently resigning from the Senate himself. The following month in Alabama, delegates from the states that had so far seceded met to write a provisional Constitution and organize a government. They named Davis provisional president of the Confederate States of America—he would be popularly elected later that year. Yet still Davis tried in vain to appoint a peace commission to travel to Washington to negotiate a way of avoiding war. That effort came to nothing when Confederates fired on Fort Sumter. Jefferson Davis and his family moved to the new Confederate capital of Richmond, Virginia, on May 29. These early days of the war with their success in battle looked as if they might lead to victory, or at least a victorious stalemate and international recognition for the South.

A Micromanager

That optimism didn't last, however. Although Davis had a strong military background, he faced a torrent of criticism for his handling of the war. Defying what he'd learned as a cadet at West Point and during his seven years of military service, Davis insisted on defending all Southern territory with nearly equal strength. This failed strategy diluted the South's already limited resources and made it vulnerable to strategic thrusts by the Union army, which dominated the Western Theater, capturing New Orleans and, later, Atlanta. Davis twice gave Lee permission to invade the North even though the Confederate military faced heavy pressure in the West. Yet, despite his confidence in Lee, he refused to appoint him General-in-Chief of the army until almost the very last minute, on January 31, 1865.

As the war turned sour for the South, those closest to Davis saw him gradually change from a confident, poised leader to a stubborn chief executive, increasingly withdrawn when leadership and guidance were desperately needed. "He did not know the arts of the politician and would not practice them if understood," said Varina Davis.

The End Is Nigh

As the South was falling, Davis and his government escaped Richmond. He was caught in May 1865 at Irwinville, Georgia, and subsequently jailed at Fort Monroe, Virginia, to wait for trial on charges of treason against the United States. In 1867, however, in the spirit of reconciliation, prominent citizens of Northern and Southern states, led by abolitionist and New York Tribune editor Horace Greeley, bailed him out of jail.

In 1869, federal prosecutors dropped the charge of treason against him. He lived the rest of his days in freedom, traveling to Canada, Cuba, and Europe, though his American citizenship, stripped when he was indicted for treason, was never restored in his lifetime. Still popular in the South, Davis was even elected to the U.S. Senate in 1875, but the new 14th Amendment of the Constitution barred him from federal office for having served in the Confederacy.

At Jefferson Davis's 1889 funeral—one of the largest ever held in the South—supporters ran a continuous march, day and night, from New Orleans, where he died, to Richmond, Virginia, where he lies buried. Davis once again became an American citizen 89 years later, when Congress restored his citizenship (on the heels of similar action to posthumously forgive Robert E. Lee) in legislation signed by President Jimmy Carter, a Southerner himself.

Sarah Emma Edmonds (AKA Frank Thompson)

This patriotic and devoted woman was determined to fight for her country—even if it meant posing as a man.

✻ ✻ ✻ ✻

DURING THE CIVIL War, women, their hearts full of worry and sorrow, watched their husbands, brothers, and sons march away to the battlefield. While such emotional partings were difficult, some women also felt deep regret that they couldn't suit up to defend their country as well. The life of the average woman in the 1860s was one of restrictions and clearly defined gender roles: They were to maintain the home and raise the children. Joining the military certainly was not an option. Still, as many as 400 women snuck into brigades from both North and South by posing as male soldiers. Some successfully maintained their disguise, while others were discovered and discharged for "sexual incompatibility." These female soldiers were trailblazers who put their lives on the line for their beliefs.

Sarah Emma Edmonds, born in 1841, believed that everyone should have the chance to fight for freedom and liberty—no matter what their gender. By age 17, she had already proven herself bold and willing to buck convention, fleeing her home in New Brunswick, Canada, to escape her overbearing father. She stole away in the night to create a new, unencumbered life for herself in the United States, settling in Flint, Michigan.

Starting Over

Edmonds knew that if she was going to experience the world in the way she wished, she'd have to reinvent herself entirely. As a woman, she could never fulfill her dreams of adventure—too many doors were closed to her on the basis of gender. But she believed those doors would open if she could become a man. Discarding the identity of Sarah Emma Edmonds, she became Franklin Thompson, a book salesman. Dressed as a man and acting with assertiveness and confidence, she was soon able to support herself independently. Edmonds saw America as a land of unlimited potential, and she was determined to make the most of it.

When war broke out, Edmonds saw another opportunity to prove her mettle and joined up with a Michigan infantry as a male nurse and courier. "I am naturally fond of adventure," she later explained, "a little ambitious and a good deal romantic and this together with my devotion to the Federal cause and determination to assist to the utmost of my ability in crushing the rebellion, made me forget the unpleasant items."

Upping the Ante

Posing as Franklin Thompson, Edmonds blended in with the men of her unit, served admirably, and aroused no suspicions during her tour of duty. Since her disguise seemed to be working so well, she volunteered to spy for General George McClellan at the start of the Peninsula Campaign. Edmonds continued to be effective in her use of disguises. She infiltrated the Confederates at Yorktown as a black slave by darkening her skin with silver nitrate and wearing a wig. After several days there, she returned to McClellan and shared the information she had gained. Edmonds's next assignment found her portraying a heavyset Irish woman named Bridget O'Shea. As O'Shea, she crossed enemy lines, peddled her wares, and returned with an earful of Confederate secrets. In August 1862, Edmonds assumed the guise of a black laundress in a Confederate camp. One day, while washing an officer's jacket, she found a large packet of official papers. After giving the jacket a thorough "dry cleaning," Edmonds returned to Union camp with the packet.

The Game's Up

All this time, army officials continued to believe Edmonds was Franklin Thompson. In the spring of 1863, she contracted malaria. She knew that she couldn't visit an army hospital for fear of being found out as a woman. She reluctantly slipped away to Cairo, Illinois, and checked into a private hospital. Although Edmonds had planned to return to her previous duty after her recovery, she discovered that during her sickness her alias, Private Thompson, had been pegged as a deserter. Edmonds couldn't reassume that identity without facing the consequences, and so she remained dressed as a woman and served as a nurse to soldiers in Washington, D.C.

After the Fighting Stopped

Two years after the Civil War ended, she married Linus Seelye, a fellow Canadian expatriate. The couple eventually settled down in Cleveland, Ohio. Determined that the world know her story and see that a woman could fight just as well as a man, she wrote the best-selling Nurse and Spy in the Union Army, which exposed her gender-bending ways. She also fought hard for her alter ego, petitioning the War Department to expunge Frank Thompson's listing as a deserter. Following a War Department review of the case, Congress granted her service credit and a veteran's pension of $12 a month in 1884. She died five years later and was buried in the military section of a Houston cemetery.

Lincoln's Neptune

As secretary of the Navy and a devoted diarist, Gideon Welles had a reputation as the most effective member of Lincoln's Cabinet.

❋ ❋ ❋ ❋

TO GET AN accurate picture of what life was like in the 19th century, historians often look to the media of the time, examining newspapers and speeches. But to truly get an insider's view on how a person felt about the state of the world—or the state of his or her own personal life—historians turn to people's diaries. One faithful diarist was Gideon Welles, U.S. secretary of the Navy from 1861 to 1869. His diary was published in 1911, and it provided readers with an important view of Lincoln and his Cabinet throughout the Civil War.

On the Rise

Before the Civil War, Gideon Welles had worked as a lawyer, later becoming the founder and editor of two Connecticut newspapers, the Hartford Times and the Hartford Evening Press. It was there that he expressed his support for Lincoln.

As the war started, the U.S. Navy was in disarray. Many of its officers had resigned to join the Southern cause, and it had fewer than 100 ships—only 12 of which were available, as the rest were on missions. Assembling an effective fighting force would be a major task for anyone. Although Welles had no experience in naval affairs, he made up for that fact by focusing on administration, working closely with his aides and officers to increase the Navy's power exponentially.

A Resemblance to Neptune

Journalist Noah Brooks described Welles as a "kind-hearted, affable and accessible man. He is tall, shapely, precise, sensitive to ridicule, and accommodating to the members of the press, from which stand-point I am making all of these sketches." Not everyone found him genial, however; Welles often gave the impression of being a curmudgeonly old man. Lincoln called him his "Neptune," after the surly Roman god of the sea—if nothing else, his long white beard made him look the part. The New Englander found it difficult to befriend his fellow Cabinet members. He was extremely outspoken in his dislike of everything British and in his conservative beliefs, which placed him at odds with fellow department heads William Seward, Salmon P. Chase, and Edwin M. Stanton.

Expanding the Navy

What was important to Welles was the work that he did. By 1865, the Navy had increased tenfold. Under Welles's watch, the United States had pioneered ironclad ships and had begun to gain a significant presence around the world. He was very consistent in his faithfulness to Lincoln, and he was quite sympathetic in the portrayal of the President in his diary. This attribute was never more evident than in his description of Lincoln's final hours.

On the evening of April 14, 1865, Welles was awakened and informed the President had been shot. He arrived at the house across from Ford's Theatre where the fallen President had been taken. "The giant sufferer lay extended diagonally across the bed, which was not long enough for him," Welles later wrote. "He had been stripped of his clothes. His large arms, which were occasionally exposed, were of a size which one would scarce have expected from his spare appearance. His slow, full respiration lifted the clothes with each breath that he took."

Welles remained at his post through the administration of Andrew Johnson. He wrote several books prior to his death in 1878. At three volumes, *The Diary of Gideon Welles* was published posthumously in 1911, an invaluable insight into the personalities of the Civil War.

James Strang, Island King

Few people would believe that a separate empire with its own full-fledged king once existed within the borders of the United States of America. But James Jesse Strang was indeed crowned ruler of a Lake Michigan island kingdom in the mid-1800s.

* * * *

Growing the Garden

STRANG WAS BORN in 1813 in Scipio, New York. He moved to Wisconsin in 1843 with his wife, Mary, to a large parcel of land just west of what would become the city of Burlington. He set up a law practice and met the Mormon prophet Joseph Smith on a trip to Nauvoo, Illinois. Strang's rise to fame began, as Smith immediately appointed him an elder in the faith and authorized him to start a Mormon "stake" in Wisconsin named Voree, which meant "Garden of Peace."

Mormons from around the country flocked to Voree to build homes on the rolling, forested tract along the White River. A few months after Strang became a Mormon, Joseph Smith was killed. To everyone's amazement, Strang produced a letter that appeared to have been written and signed by Joseph Smith, which named Strang as the church's next Prophet.

Another leader named Brigham Young, whom you may have heard of, also claimed that title, and Young eventually won. As a result, Strang broke away to form his own branch of Mormonism.

Secrets from the Soil

In September 1845, Strang made a stunning announcement. He said that a divine revelation had told him to dig under an oak tree in Voree located on a low rise called the "Hill of Promise." Four followers armed with shovels dug under the tree and unearthed a box containing some small brass plates, each only a few inches tall.

The plates were covered with hieroglyphics, crude drawings of the White River settlement area, and a vaguely Native American human figure holding a scepter. Strang said he was able to translate them using special stones, like the ones Joseph Smith had used to translate similar buried plates in New York. The writing, Strang said, was from a lost tribe of Israel that had somehow made it to North America. He managed to show the plates to hundreds of people before they disappeared.

As his number of followers grew, he created sub-groups among them. There was the commune-style Order of Enoch, and the secretive Illuminati, who pledged their allegiance to Strang as "sovereign Lord and King on earth." Infighting developed within the ranks, and area non-Mormons also raised objections to the community. Some Burlington residents even went so far as to try to persuade wagonloads of Voree-bound emigrants not to join Strang.

In 1849, Strang received a second set of messages, called the Plates of Laban. He said they had originally been carried in the Ark of the Covenant. They contained instructions called *The Book of the Law of the Lord,* which Strang again translated with his stones. The plates were not shown to the group at that time, but they did eventually yield support, some would say rather conveniently, for the controversial practice of polygamy.

Polygamy Problems

Strang had personal reasons for getting divine approval to have multiple wives. In July 1849, a 19-year-old woman named Elvira Field secretly became Strang's second wife. Just one problem—he hadn't divorced his first one. Soon, Field traveled with him posing as a young man named Charlie Douglas, with her hair cut short and wearing a man's black suit. Yet, the "clever" disguise did little to hide Field's ample figure.

At about that time, Strang claimed another angel visited to tell him it was time to get out of Voree. Strang was to lead his people to a land surrounded by water and covered in timber. This land, according to Strang, was Beaver Island, the largest of a group of islands in the Beaver Archipelago north of Charlevoix, Michigan. It had recently been opened to settlement, and the Strangites moved there in the late 1840s.

The Promised Island

On July 8, 1950, Strang donned a crown and red cape as his followers officially dubbed him King of Beaver Island. Falling short of becoming King of the United States, he was later elected to Michigan's state legislature, thanks to strong voter turnout among his followers. Perhaps reveling in his new power, eventually, he took three more young wives, for a total of five.

On the island, Strang's divine revelations dictated every aspect of daily life. He mandated that women wear bloomers and that their skirts measure a certain length, required severe lashings for adultery, and forbade cigarettes and alcohol. Under this strict rule, some of the followers began to rebel. In addition, relations with the local fishermen soured as the colony's businesses prospered.

On June 16, 1856, a colony member named Thomas Bedford, who had previously been publicly whipped, recruited an accomplice and then shot Strang. The king survived for several weeks and was taken back to Voree by his young wives. He died in his parents' stone house—which still stands near Mormon Road on State Highway 11. At the time, all four of his young wives were pregnant. And back in Michigan, it wasn't long before local enemies and mobs of vigilantes from the mainland forcibly removed his followers from Beaver Island.

James Jesse Strang was buried in Voree, but his remains were later moved to a cemetery in Burlington. A marker, which has a map of the old community, stands just south of Highway 11 where it crosses the White River, and several of the old cobblestone houses used by group members are preserved and bear historical markers. Strang's memory also lives on in a religious group formed by several of his followers, the Reorganized Church of Jesus Christ of Latter-Day Saints.

Billy the Kid

Henry McCarty, aka William Bonney or Billy the Kid, was a Western icon romanticized as a fast-shooting outlaw who killed 21 men—one for each year of his life. However, the body count proves to have been exaggerated.

* * * *

IN 1877, AT Fort Grant in Arizona Territory, Billy the Kid was repeatedly slapped and then thrown to the ground by a burly blacksmith named F. P. Cahill. Because he'd been bullied by Cahill for months, the slightly built 17-year-old pulled out a revolver and fatally wounded him. Then, in a Fort Sumner saloon early in 1880, the Kid was challenged by gunman Joe Grant. When Grant's six-gun misfired, the Kid pumped a slug into the man's head. The following year, the Kid shot his way out of jail in Lincoln, New Mexico, and gunned down guards J. W. Bell and Bob Olinger in the process.

Cahill, Grant, Bell, and Olinger are the four men who are known to have fallen under the revolver fire of Billy the Kid. But the Kid was also a central figure in New Mexico's bloody

Lincoln County War. Early in the conflict, the Kid and several comrades, lusting for vengeance, blasted two prisoners, Frank Baker and William Morton. Then the Kid led an ambush on Lincoln's main (and only) street, and the bushwhackers killed Sheriff William Brady and Deputy George Hindman. During the five-day Battle of Lincoln in July 1878, the Kid was blamed for killing Bob Beckwith, though there is a strong case that Beckwith fell to friendly fire.

Emily Dickinson: Amherst's Reclusive Belle

What caused the one-time seminary student to choose a life of isolation? Was it a failed love affair—or something more sinister? Will the world ever really know?

* * * *

EMILY DICKINSON'S STRANGE, brooding, reclusive lifestyle had tongues wagging and rumors flying in her time, and her private life remains an enigma today. Though she penned 1,789 poems, only 9 were published during her lifetime; her literary fame as a founder of neo-modern American poetry came posthumously. The answer for why she chose to shut herself away from the world may lie in her poetry.

A Secret from the World

Dickinson died at her family home in Amherst, Massachusetts, on May 15, 1886, at the tender age of 56, after spending her entire adult life in almost total isolation. The cause of her death was Bright's disease, a form of kidney trouble untreatable in the 19th century. With the exception of her family and an inner circle of close correspondents and visitors, her passing went unheralded—indeed, almost unnoticed.

Seventy years passed before her immense literary contributions received the recognition they rightfully deserved. Her unconventional staggered meter, her lack of conventional punctuation, an idiosyncratic style, and the immersion of her deeply personal life into many of her poems were condemned by early literary critics—some pronounced her work "grotesque." It was not until 1955 that "The Belle of Amherst," as she was familiarly known, became recognized as an American original. Her personal life, however, continued to be a mystery.

A Solitary Life

Dickinson was born in 1830 into a prominent and highly political New England family; she was the second of three children. Her father was a Massachusetts state senator who was later elected to the U.S. House of Representatives. Her mother was chronically ill for most of her life. Ironically, both Emily's older brother and younger sister exhibited some of the same tendencies that led Emily to be labeled a recluse. William Austin Dickinson married Emily's most intimate friend, Susan Gilbert, and moved next door to the family home, where Emily and her sister Lavinia lived. "Vinnie," like her older sister, also lived at home and remained a spinster. Their puritanical and religious upbringing created very strong family ties—almost an unwillingness to completely cut the umbilical cord. It was Vinnie who discovered and helped assemble Emily's poetry for publication around the turn of the century.

Emily's withdrawal from society was not innate; in her early years she was described as sociable, friendly, even gregarious. She attended seminary but could not bring herself to sign an oath dedicating her life to Jesus. She left the seminary after a year and never returned; whether it was embarrassment over the incident or simply that she was homesick is unknown. Following her voluntary withdrawal from the school, Emily began to display the self-denying tendencies that would lead to her lifelong austerity. With the exception of brief visits to relatives, she began the reclusive lifestyle that some labeled agoraphobia. Her detractors offered darker explanations, from failed love affairs and bisexual relationships to lesbianism and an illicit relationship with her brother's wife. Only circumstantial evidence exists to give any credence to Dickinson's supposed indiscretions.

Writing from the Inside

After the seminary, Emily began the poetic foray that would eventually lead to the establishment of the modern American style of poetry. At the time, her balladlike lyrics and halting meter were highly unconventional, yet they would eventually influence decades of modern poets. Her lyrical poetry has also become the basis for songs by such musical stalwarts as Aaron Copland, Michael Tilson Thomas, and Nick Peros. The use of common meter makes her poetry adaptable to almost any type of music, from hymns to popular ballads.

Early in her poetry-writing career, Dickinson sought the advice and approbation of literary critic and Atlantic Monthly writer Thomas Wentworth Higginson. He recognized the potential talent in the budding poet but attempted to tutor her in employing the more orthodox style of the romantic poetry popular at the time. Fortunately for the world, Emily rebelled at the idea and abandoned her plans to publish her poetry. Yet, the initial efforts of Higginson and editor Mabel Loomis Todd first brought Dickinson's works to the attention of the American public by publishing three heavily edited editions of her poetry between 1890 and 1896. Her poetry attained some popularity, but it did not become well known for another 60 years. In 1955, Thomas H. Johnson, America's preeminent Dickinson scholar, published a complete edition of her poetry in her original, unorthodox style.

Still Unknown Even Today

Though Dickinson's poetry grows more popular with time, the reasons behind her voluntary exile from society remain enigmatic. Her introspective personality manifested itself early in life, and as time passed Emily developed several eccentricities: She began dressing exclusively in white and never left her family's home, though she did entertain a few visitors.

Dickinson's sexual orientation has been questioned as well. There is evidence that she had an unrequited love affair with at least one man, known only as "Master," which is how Emily addressed her correspondence. Although scholars have speculated on his identity, only Emily knew the truth. It has never been established whether or not the letters she wrote were ever sent. Emily and her sister-in-law, Susan, did exchange intimate letters, which gave rise to the rumors that they were having a lesbian relationship. According to some experts, it was not unusual behavior for women in the puritanical environment of the late 19th century to exchange intimate correspondences. But there is simply no proof that Emily ever consummated a physical relationship with anyone—man or woman.

The Personal in the Poetical

Many scholars are convinced that the answers to the paradoxical and hermitic lifestyle of Dickinson lie in the immersion of her personal life into her poetry—poetry in which she bares her innermost thoughts, almost as if revealing the answers to the many questions scholars have sought to solve. Though no definitive answers to the mystery have emerged, it may keep historians guessing forever.

Calamity Jane

You probably know her from the HBO series Deadwood. But what's the real story behind Wild Bill Hickok's gal pal? And how did she get that nickname?

* * * *

SHE SWORE AND she drank. She was an accomplished rider and was handy with a gun. She scouted for the army. She mined for gold. She wore pants. She did just about everything that 19th-century American women weren't supposed to do. But Calamity Jane got away with it all—and became a Wild West legend.

As with most of the legendary figures of the Wild West, it's hard to separate fact from fiction when it comes to Calamity Jane. We don't even know her real birthdate. What we do know is that she was born Martha Jane Canary (or Cannary) in Missouri, probably in 1852. Orphaned not long after her family left Missouri for the Montana Territory, teenage Martha turned to prostitution—a trade she'd ply occasionally throughout her life. Over the next decade or so, Martha drifted through the mining camps and army posts of the West, where she earned the respect of her masculine peers by showing that she could do anything a man could do—and in some cases, do it better. A heavy drinker, Calamity was thrown into various frontier jails for disturbing the peace.

She first started wearing men's clothes while scouting and carrying messages for the army during the Indian campaigns of the early 1870s. She won the name "Calamity Jane" when she rode to the rescue of an officer caught in an ambush—he declared that she'd saved him from "a calamity." (Though some say her moniker came from her threat that if any man messed with her, "it would be a calamity" for him.)

Welcome to Deadwood

Around 1876, Calamity landed in Deadwood, South Dakota, where she developed a serious crush on the handsome gunslinger Wild Bill Hickok. Later in life, she claimed they'd been married and had a child together, but they were probably just "friends." She stayed in Deadwood after Hickok caught a bullet during a poker game. A couple of years later, tough Calamity showed her tender side by nursing victims of a smallpox epidemic that swept the town.

By then, word of this crossdressing hell-raiser had reached the publishers back East who were busy churning out "dime novels" to satisfy the public's taste for tales of Western derring-do. Calamity Jane quickly became a recurring character in these outlandish fictions. She was usually described as a beautiful woman; in illustrations, her baggy trousers mysteriously turned into skin-tight leggings. Photographs of Calamity don't exactly bear out this image.

It was all downhill from there for Calamity. She resumed drifting around the West, married briefly, and in the 1890s she toured the country with Buffalo Bill's famous Wild West Show. She got fired for drinking and carousing and ultimately died in poverty in 1903. But she gets to spend eternity with Wild Bill; as per her deathbed request, she's buried next to him in the Deadwood cemetery.

Geronimo's Surrender

On September 4, 1886, Geronimo, the last Apache chief still fighting the U.S. Army, surrendered to General Nelson Miles in Skeleton Canyon near the Arizona–New Mexico border. Most of Geronimo's people went into captivity, first in Florida and then in the Southwest. The last surviving witness to the bitter surrender was Jasper Kanseah, Geronimo's nephew and the band's youngest warrior.

❋ ❋ ❋ ❋

KANSEAH WAS BORN on the Chiricahua reservation; his father died before his birth. Just a small child when the Chiricahua reservation was closed and the Chiricahua were forced to move to the San Carlos reservation in 1876, Kanseah was left with only his grandmother after his mother died during the journey. Kanseah had trouble keeping up along the way, and his grandmother carried him at times.

Kanseah was about 11 years old when Geronimo came to San Carlos. Geronimo took Kanseah with him when he left and trained him as a warrior. After Geronimo's surrender, American authorities sent Kanseah to Carlisle Indian School, which he was lucky to survive—many Apaches died there of disease.

After nine years of schooling, Kanseah returned to captivity. Finally freed in 1913, Jasper Kanseah did several stints as a cavalry scout and was honorably discharged each time. He died in 1959, approximately 86 years old.

Women's Suffrage Leaders

The following are a few prominent women's suffragists.

* * * *

* Lucretia Mott (died 1880) was an abolitionist and women's rights activist. She helped organize the Seneca Falls Woman's Rights Convention. She refused to use cotton cloth, cane sugar, and other goods produced through slave labor, and she often sheltered runaway slaves in her home.
* Sojourner Truth (died 1883) was a former slave and prominent abolitionist and women's rights activist. She was famed for her simple yet powerful oratory, especially her 1851 "Ain't I a Woman?" speech.
* Lucy Stone (died 1893) recruited many (Susan B. Anthony and Julia Ward Howe, for instance) to the cause of women's suffrage. After her marriage, she and her husband put their house in her name. She refused to pay property taxes on it at one point, claiming she should not have to pay taxes because she did not have the right to vote.
* Amelia Bloomer (died 1894) was a suffragist, editor, temperance leader, and tireless volunteer. She also became known for wearing "bloomers" (trousers underneath a skirt).
* Frances E. Willard (died 1898) served as president of the Woman's Christian Temperance Union from 1879 to 1898 and lectured across the country on prison, education, and labor reform.
* Elizabeth Cady Stanton (died 1902) helped organize the Seneca Falls convention and worked to liberalize divorce

laws and other laws that made it difficult for women to leave abusive relationships.

* Susan B. Anthony (died 1906) helped found the Equal Rights Association. In 1872, she voted in a presidential election in an attempt to show that under the Constitution, women should already have the right to vote. She was arrested, convicted, and ordered to pay a fine. She never paid the fine, and the authorities did not pursue the case.
* Emmeline Pankhurst (died 1928) was a British suffragist who founded the Women's Franchise League and the Women's Social and Political Union. She went to jail 12 times in 1912.
* Carrie Chapman Catt (died 1947) founded the League of Women Voters.

Jersey Jumper

With enough ingenuity, a person can get famous in all sorts of ways. Just ask Sam Patch, who made a name for himself by going over the falls.

* * * *

Factory Boy

SAM PATCH WAS probably born in 1807 in Rhode Island. As a young boy, Sam worked long shifts in the Pawtucket mills, often for 12 hours a day. It was dull, hard work, but he found something to keep him occupied and his friends entertained: waterfall jumping.

Waterfall jumping was an art. Jumpers dove in feet first, and they sucked in a deep breath just before they hit the water. Once underwater, they stayed under long enough to start frightened spectators buzzing. Finally, when all appeared lost, they burst out of the water like a fish on a line, much to the delight of a relieved audience.

In his mid-20s, Patch left Pawtucket for Paterson, New Jersey. Beside being a factory town, Paterson was also home of Passaic Falls, an idyllic waterfall with a 70-foot drop that was second only to Niagara on the East Coast.

Taking the Plunge

On September 30, 1827, Patch dove off a cliff and into the swirling waters of the falls. He emerged from the waters with a gasp; the "Yankee Leaper" had been born. Patch realized that he could either work for pennies in a hot, sweaty factory or parlay his jumping skills into fame and possibly fortune.

Patch developed a routine. He would lay out his coat, vest, and shoes on the ground, as if he may not need them again. He gave a speech containing one, or both, of his favorite sayings: "There's no mistake in Sam Patch," and "Some things can be done as well as others."

Soon, word of Patch's jumps got around. He attained celebrity status, and with fame's siren song ringing in his ears, Patch set out to jump whenever and wherever. On August 11, 1828, he jumped 100 feet from a ship's mast into the Hudson River at Hoboken, New Jersey. Patch's antics earned him much publicity, as hundreds of onlookers would gather to watch him jump from increasingly tall heights. In addition to being the "Yankee Leaper," he also became known as "Patch the Jersey Jumper."

On October 7, 1829, Patch leaped into immortality by jumping 85 feet into Niagara Falls from a platform off of Goat Island. This time, he was billed as the main entertainment. As an onlooker recorded: "Sam walked out clad in white, and with great deliberation put his hands close to his sides and jumped." Ten days later, he jumped Niagara again from 120 feet, netting a cool $75 for his work.

In November, Patch, now accompanied by a pet black bear, showed up in Rochester, New York, to jump the thundering Genesee Falls. He performed the jump on November 6,

1829; some reports say he threw the bear over the side as well. Showing that there was "no mistake in Sam Patch," he decided to repeat the plunge on Friday, November 13.

He may have chosen an ominous date to jump the Genesee, and he may or may not have been drunk, but he was certainly tempting fate. Sure enough, a third of the way down, he lost his form and hit the water like a sack of potatoes. One report later said that he had ruptured a blood vessel on the way down. Either way, thousands of spectators anxiously watched the water, but Patch never resurfaced.

Post-Patch

Even in death Patch was controversial. Some say his body wasn't found for months, while others say that it was found two days later. After his body was found, he was buried in a grave with a marker reading: "Here Lies Sam Patch. Such Is Fame."

But death did not stop Patch. He lived on for years in novels, comics, and popular plays such as *Sam Patch* and *Sam Patch in France*. There was a Sam Patch cigar, and President Andrew Jackson even named his favorite riding horse Sam Patch.

Clearly, Sam Patch was not all wet.

The Brain of the Confederacy

Mystery surrounds the life of Judah P. Benjamin, a Confederate leader subjected to anti-Semitism and suspicion despite his loyalty to the Southern cause.

✻ ✻ ✻ ✻

EVEN BEFORE BECOMING Jefferson Davis's most trusted confidant, Senator Judah P. Benjamin had made an exceptional career for himself. The Sephardic Jew who devoted his powers of oratory to the cause of slavery was a novelty in the prewar Senate. Admirers claimed that intelligence blazed from his eyes. He enchanted allies with his disarming wit, his old-world sophistication, and, as one lawmaker recalled, a voice "as musical as the chimes of silver bells."

Representing Louisiana in the Senate, he withdrew from that body when Louisiana left the Union. The newly formed Confederate government put his smiling face on its two-dollar bill and his quick mind to work as secretary of war. The wife of Confederate President Jefferson Davis reported that Benjamin was at Davis's side 12 hours a day. One biography asserts that his workdays began at 8:00 A.M. and lasted until the crack of dawn the next morning. But even as a powerful Confederate, he remained an outsider in his own South. Plantation society repaid Benjamin's devotion to the Confederacy with anti-Semitism and suspicion. One way he coped with this predicament was through humor. In one instance when Davis was headed for a service at an Episcopal church and Benjamin realized he was obliged to accompany his president, Benjamin joked, "May I not have the pleasure of escorting you?"

Humble Beginnings

Judah Philip Benjamin was born in the present-day Virgin Islands to a dry-goods seller who founded one of America's first reform synagogues. After growing up in the Carolinas, the immigrant son attended Yale at only 14 years old, but he was expelled under cloudy circumstances. A newspaper piece published just before the Civil War claimed that his expulsion had been for playing cards and pickpocketing, a charge that Benjamin dismissed as libel by a man who falsely claimed to be his classmate.

Fortunately for the young Benjamin, one could become a self-trained lawyer in the 19th century. Harvard was the only law school at that time, so most lawyers—including Abraham Lincoln—were self-taught. Benjamin's facile legal mind readily cut to the heart of a case, and clients paid him handsomely. In this way, he achieved one of his aims, that of owning a plantation. His climb up the social ladder accelerated when he married Natalie St. Martin, a stunning Creole beauty from a well-to-do Catholic Louisiana family. The marriage was somewhat mysterious, characterized by long periods of separation and rumors that Mrs. Benjamin sometimes looked elsewhere for love. She spent most of her married life in Paris.

A Future in Politics

Benjamin sold off his plantation and about 140 slaves in 1850. His rising political career took him to the Senate two years later. President Millard Fillmore considered appointing Benjamin to the Supreme Court, but the senator chose to stay in Congress instead. He also wrote books and was a walking encyclopedia on everything from techniques for planting sugar to the French language, his wife's native tongue. For relaxation, he regaled friends with suspenseful ghost stories and verses by his favorite poet, Lord Alfred Tennyson.

Benjamin had a knack for enraging abolitionist enemies such as Senator Ben Wade of Ohio, who disdainfully called him "an Israelite in Egyptian clothing" for defending the right to own slaves. Benjamin took what might be called a libertarian view of slavery. To him, the pro-abolitionist positions of the Republican party were violating the sacred American right of property. In one speech, he asked how the North would like it if animal-rights fanatics, believers in "the sinfulness of subjecting the animal creation to the domination and service of man," were to steal cattle and descend on farms with torches, making "the night lurid with the flames of their barns and granaries."

There is no record of Benjamin being religious, but he was quick to confront any attack on his heritage with devastating oratorical parries. "It is true that I am a Jew," he snapped back at Wade, "and when my ancestors were receiving their Ten Commandments from the immediate Deity, amidst the thunderings and lightnings of Mt. Sinai, the ancestors of my opponent were herding swine in the forests of Great Britain." Known for his attacks, he was just as well-known for the ease with which he'd patch up the injury, using courtesy and tricks such as allowing his adversary to win a round of ten-pin bowling.

The election of Abraham Lincoln and the secession of Southern states that followed it found Benjamin declaring his enemies beyond reconciliation. In an angry farewell address to the Senate on December 31, 1860, he told Northerners to wild applause, "You may carry desolation into our peaceful land, and with torch and fire you may set our cities in flames ... but you never can subjugate us; you never can convert the free sons of the soil into vassals, paying tribute to your power ... Never! Never!"

The Davis Connection

When the Confederacy was formed, Jefferson Davis tapped Benjamin to be attorney general. To anyone who saw how they bickered in the Senate, their friendship seemed unlikely. Benjamin had been speaking on an army appropriations bill when Davis snidely interrupted to complain that he'd had no idea he'd have to listen to "the arguments of a paid attorney in the Senate chamber." Benjamin demanded an apology. To avert the duel that was expected in those days, Davis admitted his error and apologized on the Senate floor. From that point, the two gradually came to be close allies in the Southern cause. From attorney general, Davis moved Benjamin over to Confederate secretary of war, but he soon ran into trouble in that post. In February 1862, Roanoke Island, off the coast of North Carolina, was lost to General Ambrose Burnside's forces after the rebel government failed to send reinforcements. Benjamin scored loyalty points by shouldering responsibility. Had he been less faithful, he might have embarrassed the President by revealing the real reason for the defeat: He had no troops to send.

Benjamin couldn't continue as secretary of war under the circumstances, but he remained in the Confederate Cabinet by sliding to a new role as secretary of state. He worked hard to lure the British into the war, an effort that ultimately failed. The Confederates hoped to entice European allies with promises of cotton, but that scheme proved no more successful than another idea he promoted. In late 1864 and early 1865, Benjamin devised a desperate plan to impress the United Kingdom and shore up Southern forces by freeing the slaves—if they joined the gray army. "Let us say to every Negro who wishes to go into the ranks, 'Go and fight—you are free!'" Benjamin said, and General Robert E. Lee agreed. It was, however, a plan that came too late in the war to come to fruition.

New Country, New Life

When the South lost the war, Benjamin disguised himself and escaped to England, where he again took up his career as an attorney. Back at home, on both sides of the Mason-Dixon line, anti-Semitic conspiracy theories had sprouted to blame "Judas" Benjamin for scuttling the Confederate war effort, running off with the Southern treasury, or plotting the death of Lincoln in his capacity as Southern spy chief. For that last accusation, Benjamin and Davis both feared that Benjamin would be arrested and hanged. Later investigation and scholarship has determined that Benjamin was not involved in the Lincoln assassination.

The popular memory Benjamin left behind was captured in poet Stephen Vincent Benet's 1928 depiction of a lonely but malevolent outsider—"Seal-sleek, black-eyed, lawyer and epicure/Able, well-hated, face alive with life"—who is haunted by the question: "I am a Jew. What am I doing here?"

Benjamin died in 1884 in Paris and was buried under a headstone that read, "Phillipe Benjamin." His personal letters had been burned, leaving historians with the mystery of what it was really like to be a Jewish leader of the Confederacy.

A Woman in the White House?

When Victoria Woodhull ran for president in 1872, some called her a witch, others said she was a prostitute. In fact, the very idea of a woman casting a vote for president was considered scandalous—which may explain why Woodhull spent election night in jail.

* * * *

KNOWN FOR HER passionate speeches and fearless attitude, Victoria Woodhull became a trailblazer for women's rights. But some say she was about 100 years before her time. Woodhull advocated revolutionary ideas, including gender equality and women's right to vote. "Women are the equals of men before the law and are equal in all their rights," she said. America, however, wasn't ready to accept her radical ideas.

Woodhull was born in 1838 in Homer, Ohio, the seventh child of Annie and Buck Claflin. Her deeply spiritual mother often took little Victoria along to revival camps where people would speak in tongues. Her mother also dabbled in clairvoyance, and Victoria and her younger sister Tennessee believed they had a gift for it as well. With so many chores to do at home (washing, ironing, chipping wood, and cooking), Victoria only attended school sporadically and was primarily self-educated.

Soon after the family left Homer, a 28-year-old doctor named Canning Woodhull asked the 15-year-old Victoria for her hand in marriage. But the marriage was no paradise for Victoria—she soon realized her husband was an alcoholic. She experienced more heartbreak when her son, Byron, was born with a mental disability. While she remained married to Canning, Victoria spent the next few years touring as a clairvoyant with her sister Tennessee. At that time, it was difficult for a woman to pursue divorce, but Victoria finally succeeded in divorcing her husband in 1864. Two years later she married Colonel James Blood, a Civil War veteran who believed in free love.

In 1866, Victoria and James moved to New York City. Spiritualism was then in vogue, and Victoria and Tennessee established a salon where they acted as clairvoyants and discussed social and political hypocrisies with their clientele. Among their first customers was Cornelius Vanderbilt, the wealthiest man in America.

A close relationship sprang up between Vanderbilt and the two attractive and intelligent young women. He advised them on business matters and gave them stock tips. When the stock market crashed in September 1869, Woodhull made a bundle buying instead of selling during the ensuing panic. That winter, she and Tennessee opened their own brokerage business. They were the first female stockbrokers in American history, and they did so well that, two years after arriving in New York, Woodhull told a newspaper she had made $700,000.

Woodhull had more far-reaching ambitions, however. On April 2, 1870, she announced that she was running for president. In conjunction with her presidential bid, Woodhull and her sister started a newspaper, Woodhull & Claflin's Weekly, which highlighted women's issues including voting and labor rights. It was another breakthrough for the two since they were the first women to ever publish a weekly newspaper.

That was followed by another milestone: On January 11, 1871, Woodhull became the first woman ever to speak before a congressional committee. As she spoke before the House Judiciary Committee, she asked that Congress change its stance on whether women could vote. Woodhull's reasoning was elegant in its simplicity. She was not advocating a new constitutional amendment granting women the right to vote. Instead, she reasoned, women already had that right. The Fourteenth Amendment says that, "All persons born or naturalized in the United States . . . are citizens of the Unites States." Since voting is part of the definition of being a citizen, Woodhull argued, women, in fact, already possessed the right to vote. Woodhull, a persuasive speaker, actually swayed some congressmen to her point of view, but the committee chairman remained hostile to the idea of women's rights and made sure the issue never came to a floor vote.

Woodhull had better luck with the suffragists. In May 1872, before 668 delegates from 22 states, Woodhull was chosen as the presidential candidate of the Equal Rights Party; she was the first woman ever chosen by a political party to run for president in the United States. But her presidential bid soon foundered. Woodhull was on record as an advocate of free love, which her opponents argued was an unacceptable attack on the institution of marriage (though for Woodhull it had more to do with the right to have a relationship with anyone she wanted). Rather than debate her publicly, her opponents made personal attacks.

That year, Woodhull caused an uproar when her newspaper ran an exposé about the infidelities of Reverend Henry Ward Beecher. Woodhull and her sister were thrown in jail and accused of publishing libel and promoting obscenity. They would spend election night of 1872 behind bars as Ulysses Grant defeated Horace Greeley for the presidency.

Woodhull was eventually cleared of the charges against her (the claims against Beecher were proven true), but hefty legal bills and a downturn in the stock market left her embittered and impoverished. She moved to England in 1877, shortly after divorcing Colonel Blood. By the turn of the century she had become wealthy once more, this time by marriage to a British banker. Fascinated by technology, she joined the Ladies Automobile Club, where her passion for automobiles led Woodhull to one last milestone: In her sixties, she and her daughter Zula became the first women to drive through the English countryside.

Hearst at His Worst

Media mogul William Randolph Hearst built an empire out of sensationalism and dishonesty. Owning a large percentage of America's newspapers in the first half of the 20th century, he exemplified the notion of yellow journalism.

❋ ❋ ❋ ❋

In the 1890s, long before the advent of broadcast television, cable, and the Internet, even before radio, America relied on the daily newspaper to deliver the news. Propped up at the breakfast table, opened and folded on the trolley car, or perused from a park bench, newspapers became the trusted source for "what's what" across the country. Publishers such as Horace Greeley and James Gordon Bennett, taking advantage of their influential powers, forged a path of biased, politically and monetarily swayed reporting that would soon be undertaken by other newspaper publishers such as Joseph Pulitzer and, especially, famed magnate William Randolph Hearst.

Wee Willie

Hearst was born in 1863 in San Francisco, the son of a successful mining entrepreneur. Even as a young lad, he knew how to stir up trouble. For example, he set off some fireworks in his bedroom one night, yelled, "Fire!" down the hallway, locked the door, and waited as the fire department came crashing through it. While attending Harvard University, he was expelled—possibly for a combination of pranks and poor grades. Inheriting his father's Comstock mining fortune in 1887, Hearst decided that publishing would be a lucrative, secure venture and took over a number of faltering newspapers, including the San Francisco Examiner.

All the News That's (Un)Fit to Print

In efforts to up his numbers, Hearst became something of a champion of the people. He focused on revealing corruption in the San Francisco Police Department, city hall, and at Folsom

Prison. He relished reporting on lurid crimes and blazing fires, and he filled the Examiner with political cartoons of all sorts. One writer for the paper masqueraded as a homeless person and was admitted to the city's hospital. When her story revealed the rampant cruelty and neglect she found there, the entire hospital staff was fired.

The number of publications under the Hearst banner grew, and he eventually built an unprecedented empire. At its height, he owned a whopping 28 newspapers and a variety of magazines; by the close of his career, he'd owned a total of 36 newspapers and multiple magazines. Among these were some of the most influential and widely read publications in the country, including the Washington Herald, the Seattle Post-Intelligencer, and the Baltimore Post.

Hearst's chief rival, Joseph Pulitzer, ran the New York World, another newspaper deeply rooted in aggressive reporting. In 1895, a strange cartoon character began appearing in Pulitzer's paper. A bald boy with a toothy smile, he was called the "Yellow Kid" (the Kid appeared in a color cartoon and sported a nightshirt of this shade). After a fierce bidding war, the popular "Yellow Kid" was moved to Hearst's New York Journal in 1896 and, with both papers' reputation for using their pages to stir up trouble, the concept behind their nature became known as "yellow journalism." As the years went on, so did the stories peppered with increasingly exaggerated and biased claims. Hearst often published articles colored with shades of his own prejudices and agendas—the opposite of fair news reporting.

A War, Made to Order

Hearst and Pulitzer received a lot of credit for getting America into the Spanish-American War in 1898. The accusations may be exaggerated, although their publications did keep Spain's occupation of Cuba on the front pages. An oft-told story (with questionable accuracy) has Hearst rebuffing a photographer's claim that there was no war in Cuba, stating, "You furnish the pictures—I'll furnish the war." For example, one week after America declared war on Spain, Hearst ran a front-page headline asking, "How Do You Like the Journal's War?" The devastating explosion that destroyed the U.S. battleship Maine while it was docked in the Havana harbor was immediately blamed on Spanish terrorists, yet it is more likely that an internal fire ignited the ammunition stored onboard. The World and the Journal led the way in exacerbating patriotic sentiment before and during the five-month conflict, with Hearst actually traveling to Cuba as a reporter and filing regular stories of the carnage.

Fanning the Flames

With the war behind them, the journalistic holdings of Pulitzer and Hearst took separate paths. Pulitzer, feeling somewhat guilty about the direction his paper had taken, righted his New York World. It became a respected and prestigious daily, before closing in 1931. But Hearst hitched the Journal to the 1900 political campaign of William Jennings Bryan, a progressive Democrat who was running for president against Republican incumbent William McKinley. Hearst painted McKinley as a puppet of wealthy industrialists, with Bryan as a hero who would come to the country's rescue. Hearst papers also took shots at McKinley's running mate, Theodore Roosevelt. He was portrayed in political cartoons as an ugly, young boy in a Rough Rider hat who bullied McKinley.

Despite Hearst's efforts to defeat McKinley, the president returned to office. Prior to the inauguration in March 1900, Hearst editorials stooped to shockingly suggest that McKinley be assassinated. When fiction became fact in September of the next year, Hearst was accused of being somewhat responsible for McKinley's death through his papers' editorials. The backlash was so bad that he changed the name of the Journal to the more patriotic-sounding American.

Hearst did not limit his sensationalist provocations to American affairs. When the Japanese were victorious in the Russo-Japanese War in 1905, Hearst's papers warned their readers to beware of "the yellow menace." During the early years of World War I, Hearst's editorials had a pro-German slant, painting an anti-British picture. Between 1928 and the mid-1930s, he featured editorials by Benito Mussolini, dictator of fascist Italy. In 1930, Hearst asked a newly elected leader in Germany to write articles for him. The new columnist—Adolf Hitler—took advantage of the forum and wrote of his plans to bring Germany out of its economic woes. He also wrote of Germany's innocence in World War I and the unfair conditions of the Versailles Treaty. Hitler was ultimately fired from Hearst's stable for, among other reasons, missing too many deadlines.

What About Citizen Kane?

In 1941, a 26-year-old first-time filmmaker named Orson Welles directed a film called Citizen Kane. The similarities between the title character, Charles Foster Kane, and William Randolph Hearst were hardly coincidental. The movie told of the life of a newspaper publisher famous for printing sensational stories, rampant artwork collecting, having an affair with a famous performer, and failed political aspirations—all factual occurrences from Hearst's life. Hearst was so outraged at first mention of the film's impending release, he offered RKO Pictures $800,000 to burn all prints and the master negative. When the studio refused, he barred every member of his media empire from even mentioning the film. Despite all Hearst's efforts to quell Kane, the film came to be regarded as possibly the greatest of all time by hordes of critics and filmgoers.

Opinions differ on whether or not Hearst ever saw Citizen Kane, but he wasn't without opportunity. In a chance meeting in a San Francisco elevator, Welles came face-to-face with Hearst. He offered the media magnate an invitation to see the film. Hearst silently declined and exited the elevator.

✳ Chapter 8

That's Incredibly Wrong!

The Wrong Side of the Bed: Left vs. Right

This is truly a matter of critical importance, lest you spend every minute of every day in a funk. Those who get up on the wrong side of the bed, as we all know, are a blight on society, casting a pall on everyone who comes into their presence.

✳ ✳ ✳ ✳

The Right Side Is the Right Side

To help you avoid being the one under the rain clouds, we'll let you in on the secret: It's the left side. Do whatever you have to do in order to get up on the right side—pushing your bed against the wall works nicely—and you should be fine.

But what's wrong with the left side? If you were a Roman living around the time of Julius Caesar, you might ask, "What *isn't* wrong with the left side?"

The ancient Romans distrusted everything left: left-handedness, entering a house with the left foot forward, even setting the left foot down first when getting out of bed. If it could be done on the right side, with the right hand, or with the right foot, then that was the way it should be done.

Are Left-Handers Sinister?

There is an entire history of myths that associate evil with left-handedness. For instance, the devil is reputed to be left-handed, and it is with the left hand that he baptizes his followers. Or so they say. Also, according to the Satanic Bible, Satanists follow the "Left-Hand Path," which is the path opposite that of the Christian faith.

From the very beginning of recorded time, the left side has been distrusted. In fact, the Latin word for "left" (*sinistro*) is the root of the word "sinister." In almost all cultures, this belief has been held.

Left-handed people are no longer treated as devil worshippers, as they once were by practitioners of the Catholic faith. Sure, biases continue to exist—scissors, notebooks, and guitars all spring to mind—but lefties are treated as average human beings in most respects.

Still, just to be safe, remember to get up on the right side of the bed. That is to say, the correct side, which is also the right side. And just for good measure, you should probably let your right foot hit the floor first. Your day might be better for it.

On Top of the World

Think you're at the top of the world when you climb Mt. Everest? Think again. You'll need to scale Ecuador's Mt. Chimborazo to make that claim.

* * * *

YOU'VE SCALED MT. Everest and are marveling at the fact that you're getting your picture taken at the highest point on the planet. You're as close to the moon as any human being can be while standing on the surface of the earth. Actually, you're not. To achieve that, you'd have to climb back down and travel to the other side of the world—Ecuador, to be exact—so you can schlep to the summit of Mt. Chimborazo.

It's true that Everest, at 29,035 feet, is the world's tallest mountain—when measured from sea level. But thanks to the earth's quirky shape, Chimborazo, which rises only 20,702 feet from sea level, is about a mile and a half closer to the moon than Everest's peak.

Earth is not a perfect sphere. Rather, it's an oblate spheroid. Centrifugal force from billions of years of rotating has caused the planet to flatten at the poles and bulge out at the equator.

In effect, this pushes the equator farther from the earth's center than the poles—about 13 miles closer to the moon. The farther you move from the equator, the farther you move from the moon. Chimborazo sits near the equator, while Everest lies about 2,000 miles north—enough to make it farther from the moon than Chimborazo, despite being about 8,000 feet taller.

Murphy's Law

Perhaps you've heard of "Murphy's Law," which basically means, "If anything can go wrong it will." It's a simple adage that seems to perfectly sum up so many situations in our topsy-turvy lives.

⁕ ⁕ ⁕ ⁕

AH, MURPHY, DEAR Murphy. Your failures have served as such a global inspiration. Murphy likely had no idea his name would become synonymous with screw-ups when he uttered that now infamous phrase. Although, to be fair, the phrase we know isn't *exactly* what he said. Here's the story.

Meet Murphy

As the tale goes, there once was a U.S. Air Force captain named Edward A. Murphy. Back in 1949, Captain Murphy was working on a military air test at Edwards Air Force Base, trying to figure out how much force a human body could tolerate during a plane crash.

It's probably no surprise that things didn't go as planned. After the test, Murphy and his crew realized all of the sensors they'd installed on the test subjects—the devices that, after the experiment ended, should have allowed them to get the data they needed—failed to work.

Days later, the captain held a news conference at which he inadvertently cemented his place in history. "If there are two or more ways to do something and one of those results is a catastrophe," he was quoted as saying, "then someone will do it that way." Thus, Murphy's Law was born.

Murphy's Law in Action

Whether it's the traffic lane next to yours always moving faster, the dropped piece of bread always landing butter-side down, or a single sock always disappearing during laundry, the law is still used as the all-purpose scapegoat. But not everyone agrees with the origins of Murphy's Law—some folks swear Murphy mistakenly received credit for the creed. If so, the fact that its true creator was cheated out of the glory is, one might say, a perfect example of Murphy's Law in action.

Abner Strikes Out

Generations of baseball fans have been led to believe that the game was invented by Civil War hero Abner Doubleday in Cooperstown, New York, in 1839. Historians tell a different story.

* * * *

THE DOUBLEDAY MYTH can be traced to the Mills Commission, which was appointed in 1905 by baseball promoter Albert Spalding to determine the true origins of the game. Henry Chadwick, one of Spalding's contemporaries, contended that the sport had its beginnings in a British game called *rounders,* in which a batter hits a ball and runs around the bases. Spalding, on the other hand, insisted that baseball was as American as apple pie.

Can We Get a Witness?

The seven-member Mills Commission placed ads in several newspapers soliciting testimony from anyone who had knowledge of the beginnings of the game. A 71-year-old gent named Abner Graves of Denver, Colorado, saw the ad and wrote a detailed response, saying that he'd been present when Doubleday outlined the basics of modern baseball in bucolic Cooperstown, New York, where the two had gone to school together. In his account, which was published by the *Beacon Journal* in Akron, Ohio, under the headline "Abner Doubleday Invented Baseball," Graves alleged to have seen crude drawings of a baseball diamond produced by Doubleday both in the dirt and on paper.

The members of the Mills Commission took Graves at his word and closed the investigation, confident that they had solved the mystery of how baseball was invented. The commission released its final report in December 1907, never mentioning Graves by name, and an American legend was born.

Sadly, Graves's story was more whimsy than fact. For one thing, he was only five years old in 1839, when he claimed to have seen Doubleday's drawings. But even more important, Doubleday wasn't even in Cooperstown in 1839—he was a cadet attending the military academy at West Point. In addition, Doubleday, a renowned diarist, never once mentioned baseball in any of his writings, nor did he ever claim to have invented the game.

Nonetheless, the Doubleday myth received a boost in 1934 when a moldy old baseball was discovered in an attic in Fly Creek, New York, just outside Cooperstown. It was believed to have been owned by Graves and as such was also believed to have been used by Abner Doubleday. The "historic" ball was purchased for five dollars by a wealthy Cooperstown businessman named Stephen Clark, who intended to display it with a variety of other baseball memorabilia. Five years later,

Cooperstown became the official home of the Baseball Hall of Fame.

Who Was Baseball's Daddy?

If anyone can lay claim to being the father of American baseball, it's Alexander Cartwright. He organized the first official baseball club in New York in 1845, called the Knickerbocker Base Ball Club, and published a set of 20 rules for the game. These rules, which included the designation of a nine-player team and a playing field with a home plate and three additional bases at specific distances, formed the basis for baseball as we know the sport today. Cartwright's Hall of Fame plaque in Cooperstown honors him as the "Father of Modern Base Ball," and in 1953, Congress officially credited him with inventing the game.

Baby Ruth's Truth?

Many people believe that the Baby Ruth candy bar was named for baseball great Babe Ruth. Others contend that the honor belongs to President Grover Cleveland's daughter Ruth.

* * * *

GERMAN IMMIGRANT OTTO Schnering founded the Curtiss Candy Company in Chicago in 1916. With World War I raging in Europe, Schnering decided to avoid using his Germanic surname and chose his mother's maiden name for the business. His first product was a snack called Kandy Kake, a pastry center covered with peanuts and chocolate. But the candy bar was only a marginal success, and it was renamed Baby Ruth in 1921 in an effort to boost sales. Whenever pressed for details on the confection's name, Curtiss explained that the appellation honored Ruth Cleveland, the late and beloved daughter of President Grover Cleveland. But the company may have been trying to sneak a fastball past everybody.

Cleveland's daughter had died of diphtheria in 1904—a dozen years before the candy company was even started. One questions the logic in naming a candy bar after someone who had passed away so many years earlier. The gesture may have been appropriate for a president, but a president's relatively unknown daughter?

The more plausible origin of the name might be tied to the biggest sports star in the world at the time—George Herman "Babe" Ruth. Originally a star pitcher for the Boston Red Sox, Ruth became a fearsome hitter for the New York Yankees, slamming 59 home runs in the same year the candy bar was renamed Baby Ruth. Curtiss may have found a way to cash in on the slugger's fame—and name—without paying a dime in royalties. In fact, when Ruth gave the okay to use his name on a competitor's candy—the Babe Ruth Home Run Bar—Curtiss successfully blocked it, claiming infringement on its own "baby."

The Not-So-Windy City

Called the "Windy City" for well over a century, Chicago is famed for its swirling wind currents. But all may not be as it seems. Both the city's nickname and the assumptions behind it are "airy" affairs.

⁕ ⁕ ⁕ ⁕

PERHAPS THE MOST interesting aspect of Chicago's "Windy City" nickname is the fact that there's no certainty about its origin. But there are theories. One possible explanation appeared in an article in the September 11, 1886, edition of the *Chicago Tribune*. It claimed that the nickname referred to the "refreshing lake breezes" blowing off Lake Michigan. Another explanation ignores climate and credits 19th-century Chicago promoters William Bross and John Stephen Wright with inspiring the phrase. In this version, witnesses to the duo's loud boasts eventually branded them "windbags." The backhanded term "Windy City" grew from this, and the rest is history.

A Lot of Hot Air

The origin of Chicago's nickname may be up for debate, but the veracity of its claim is not. One need only consult weather records that display average annual wind speeds for U.S. cities. Certainly, if Chicago's "Windy City" tag derives from its much-celebrated wind currents, this would be the place to confirm it.

According to National Climatic Data Center findings from 2003, Blue Hill, Massachusetts, blows harder than all other U.S. cities, with a 15.4-mile-per-hour average annual wind speed. Dodge City, Kansas, and Amarillo, Texas, nab second- and third-place honors, with wind speeds of 14.0 mph and 13.5 mph, respectively. Lubbock, Texas, comes in tenth on the center's top-ten list with a 12.4-mph clocking. How fast do air currents move in the notorious Windy City? With average annual winds pushing the needle to just 10.3 mph, it appears Chicago's blustery status is full of hot air.

Gastric Geography

The adage "You are what you eat" may be accurate in terms of describing the relationship between consumption and health, but it would be unwise to confuse the foods we eat with the countries they are named for.

* * * *

One of the most popular fast-food staples in the junk-food jungle is the French fry. Poll the average carbohydrate consumer on the country of origin of this particular potato product, and the most likely answer would be France. And that answer would be wrong. French fries were likely first cooked up in Belgium. The verb "to french" refers to the technique of cutting something into long, thin strips.

On a similar note, the popular salad enhancer known as French dressing doesn't come from France, either. In fact, the sugary-sweet substance isn't widely available in that country. Popular folklore tells us that the wife of Lucius French, the man who founded Hazleton, Indiana, created the recipe to "dress up" the vegetables she prepared for her husband.

Another example of a product that does not share its domicile with its domain is the popular pastry known as the Danish. The idea for the fruit-filled doughy delight was dreamed up in Austria, which is why the sweet supplement is known as *Vienerbrod* (or "Vienna bread") in Denmark. It was, however, introduced to American appetites by a Danish baker named L. C. Kiltteng, who claims to have first baked the buttery biscuit for the wedding of President Woodrow Wilson in December 1915. After introducing the pastry to people in communities as diverse as Galveston, Texas, and Oakland, California, Kiltteng established the Danish Culinary Studio on Fifth Avenue in New York City, where bakers from far and wide were schooled in the fine art of creating the dainty delicacy.

'Til Port Do Us Part

The captain of a ship holds wide-ranging legal powers when that vessel is at sea, but it's a nautical myth that any ship's captain can perform a legally binding marriage.

✻ ✻ ✻ ✻

✻ Unless the captain of a vessel happens to also be an ordained minister, judge, or recognized official such as a notary public, he or she generally doesn't have the authority to perform a legally binding marriage at sea. In fact, a suitably licensed captain is no more qualified to perform marriages than a similarly licensed head chef, deck hand, or galley worker. There are a few specific exceptions: Captains of Japanese vessels can perform marriages at sea, as long as both the bride and groom hold valid Japanese passports. And thanks to a quirk in Bermuda law, captains with Bermuda licenses are also legally authorized to officiate weddings aboard ship.

* The myth that any ship's captain has the power to marry at sea has been propagated by countless romantic movies and is so widely believed, even among sailors, that the U.S. Navy specifically forbids it. Section 700.716 of the U.S. Navy Regulations reads: "The commanding officer shall not perform a marriage ceremony on board his ship or aircraft. He shall not permit a marriage ceremony to be performed on board when the ship or aircraft is outside the territory of the United States, except: (a) In accordance with local laws... and (b) In the presence of a diplomatic official of the United States..."
* As an alternative to a wedding at sea, couples may to want to consider exchanging vows aboard a ship that is docked in a port. Ultimately, though, if you want to avoid a legal battle to validate your cruise-line marriage, you may want to heed the adage displayed on many vessels: "Any marriages performed by the captain of this ship are valid for the duration of the voyage only."

It's Iron-IC

Popeye credited his trusty can of spinach for his bulging biceps. But his assumptions about spinach were based on a widespread misconception about its iron content.

* * * *

POPEYE DIDN'T START the rumor about the nutritional value of spinach. He simply popularized the widely held belief, based on a single scientific study, that spinach is a superior source of iron.

Thanks to a Typo

But the leafy greens' reputation wasn't so ironclad. An 1870 German study of spinach claimed it had ten times the iron content of other green leafy vegetables. This claim, uncontested for 70 years, turned out to be based on a misprint—a misplaced decimal point. The iron content of spinach was overestimated by a factor of ten! By the time the error was discovered in 1937, Popeye had already helped spread the myth far and wide—and encouraged several generations of children to tolerate the unappealing vegetable in hopes of developing their hero's brawn.

The hype about spinach didn't end then, however, because the error wasn't publicized. It wasn't until an article on the mistake was published in a 1981 issue of the *British Medical Journal* that the public learned of the true iron content of spinach.

In the 1990s, spinach received another blow when it was discovered that its oxalic acid content prevents the body from absorbing more than 90 percent of the vegetable's iron. Oxalic acid binds with iron and renders most of it unavailable.

Still Packs a Wallop

Luckily for Popeye's legacy, however, the spirited sailor was not wrong to think that spinach has abundant nutritional merit. It's a terrific source of vitamins A, B1, B2, B3, B6, C, E, and K, as well as magnesium, calcium, and potassium.

If it's muscles you're after, you'll need to pump iron rather than consume it.

Castro Strikes Out!

Though an avid fan of the game, Fidel Castro never came close to playing professional baseball. While we're at it, the bearded Cuban was never an aspiring movie star, either.

* * * *

FORMER CUBAN PRESIDENT Fidel Castro is one of the most controversial and divisive national leaders of the 20th century. He is also one of the longest lasting, ending his almost-50-year political reign when he handed power to his brother, Raul, in February 2008 due to poor health.

Not surprisingly, Castro is the subject of numerous myths, and the most popular is that he was given a pitching tryout in the 1940s by the New York Yankees (or the Washington Senators, depending on which version is told).

A Love of Baseball

Castro was a longtime fan of the great American pastime, which remains equally popular in Cuba, but there is no evidence that he was ever scouted by an American major league team. He never played baseball professionally in his native country, though it's possible he may have played a bit of extracurricular ball during his college years. But even if that's true, the revolutionary leader certainly didn't possess the athletic skills that would have drawn the eye of a major league scout.

It's likely that this myth started as a way to humiliate Castro by portraying his revolutionary ambitions as payback for the fact that he was found lacking by an American baseball team—especially one called the Yankees. Such myths carry great power in trivializing the motivations of one's enemies.

Castro Takes the Mound

That said, there is strong evidence that Castro did have one brief, shining moment on the mound—when he faced third baseman Don Hoak, who would later go on to play for the Brooklyn Dodgers, Pittsburgh Pirates, and other teams.

According to his wife, singer/actress Jill Corey, Hoak played in the Cuban leagues during the winter of 1950–51, just prior to joining the Brooklyn Dodgers. In a game between Cienfuegos and Marianao, a large group of rowdy fans ran onto the playing field in the middle of an inning. Before being chased out of the park by security guards, one of the fans took to the mound and threw several pitches to Hoak, who was at bat at the time. According to Corey, the "pitcher" was none other than a young Fidel Castro. Hoak himself confirmed the bizarre incident in an article he wrote with journalist Myron Cope titled *The Day I Batted Against Castro*.

The Mythical Movie Extra

A shattered dream of baseball glory isn't the only urban legend involving Fidel Castro. It is also commonly reported that he was an extra in several Hollywood movies in the 1940s, including *Two Girls and a Sailor* (1944), *Holiday in Mexico* (1946), and *Easy to Wed* (1946). But like the spurious baseball tryout, this rumor is also untrue. In fact, Castro wasn't even in the United States during the time he allegedly was making Hollywood movie history. Rather, he was a student in Havana.

The rumor apparently started after bandleader Xavier Cugat, who appeared in several of the movies often linked to Castro, made mysterious references in a magazine interview to a dancer he had hired. Cugat refused to give the dancer's name because the man was "a South American general" at the time.

Fidel Castro was one of the modern era's longest-serving political figures, outlasting nine U.S. presidents. But he was never rejected by the New York Yankees, the Washington Senators, or any other major league team, nor was he a minor league Hollywood hopeful.

Voltaire's Defense

Voltaire, the infamous 18th-century French Enlightenment writer, is supposed to have said, "I disapprove of what you say, but I will defend to the death your right to say it." Noble as this concept may be, it was actually one of Voltaire's many biographers who penned the words.

❋ ❋ ❋ ❋

VOLTAIRE WAS AN outspoken advocate of free speech. It is difficult to believe that the most powerful words ever written in support of this freedom cannot be attributed to the master himself, but the fact remains that the famous quote comes from Evelyn Beatrice Hall's 1907 book *The Friends of Voltaire*, published 129 years after Voltaire's death.

Hall wrote under the pseudonym Stephen G. Tallentyre at a time when it was difficult for women to publish nonfiction. At one point in the book, Hall discusses Voltaire's support of a fellow writer, Helvetius, who had been censored by the French government. The direct quote from Hall's book is: "The men who had hated [Helvetius' book] flocked round him now. Voltaire forgave him all injuries, intentional or unintentional... 'I disapprove of what you say, but I will defend to the death your right to say it,' was his attitude now."

Thus, through her indulgent dramatization of Voltaire's life, Hall inadvertently succeeded in summarizing his views on censorship in terms that were more eloquent than anything uttered by Voltaire himself (Hall later explained that the line was meant as a paraphrasing of his views). Voltaire did, however, write a similar line in a 1770 letter, which translates as, "Monsieur l'abbe, I detest what you write, but I would give my life to make it possible for you to continue to write."

A Real Stretch

During the Cold War, everyone was acutely aware of the Red Scare. But was the Marxist menace really so bad that highways in the United States were built as emergency runways?

* * * *

THE INFORMATION HIGHWAY has perpetrated more than its share of misconceptions, including one about highways in the United States. According to rumor mongers, the Federal-Aid Highway Act of 1956, which launched the Interstate Highway System and created a 42,800-mile ribbon of roads across the country, contained a clause that stated that one out of every five miles of newly paved blacktop had to be completely straight. The purpose of this rigid regiment of roads was to supply the U.S. military with a set of highway landing strips in case its aircraft came under attack and needed an emergency runway. These straight stretches were designed to be easily visible from the air, allowing a perplexed or panicked pilot to guide an aircraft to a safe stop.

However, from an aerial standpoint, this proposition makes no sense. Given the rate at which modern aircraft travel, a landing strip every five miles would be both unfeasible and unnecessary.

Although President Dwight D. Eisenhower fully supported the Interstate Highway System as a vital and viable link to secure the country's economy, safety, and defense, he never proposed any kind of one-out-of-five-mile rule, and Congress certainly didn't include such a requirement in the fine print of the Federal-Aid Highway Act. In other words, this myth has no basis in law or fact.

* Airplanes have landed on interstate highways, but that course of action has been taken only in cases of an emergency when no other alternative landing space was available.

An Earth-Shaking Moment

Although it's widely regarded as the preferred instrument for measuring earthquakes, the Richter scale has long played second fiddle to the moment magnitude scale.

* * * *

WHILE LISTENING TO your car radio, you hear an earthquake bulletin. The excited newscaster says, "Looks like this is the big one we've dreaded. You seldom see anything this high on the moment magnitude scale!" The *what*?

The Richter Scale

You were expecting the guy to reference that other scale, the one developed by seismologist Charles Richter in 1935, considered by many to be the chief method for measuring the intensity of earthquakes. Although most people define a quake by its Richter scale rating, assumptions of the scale's superiority are patently false.

The Richter scale measures an earthquake's shockwaves with the use of a logarithmic scale on which each unit represents a tenfold increase in energy (a 7.0 is ten times more powerful than a 6.0). This measurement, which gives a general idea of an event's magnitude, is limited because it cannot completely describe the impact. This is particularly true of quakes that measure above 6.8, where the scale saturates and "sees" each earthquake as the same.

The Moment Magnitude Scale

For more than a decade, scientists have used the moment magnitude scale to measure severe earthquakes. Taking into account such factors as geological properties and ground slippage (displacement), the moment magnitude scale—devised by seismologists Hiroo Kanomori and Tom Hanks—expresses a tremor's total energy. Like the Richter scale, it operates logarithmically. Unlike the Richter scale, it doesn't saturate in its upper reaches and can more closely pinpoint real-world effects and overall destruction. Perhaps newscasters will get on board some day.

Yankee Doodle Duds

Is it possibly true that television ratings plummet when the New York Yankees fail to reach the World Series?

* * * *

IN 2004, THE series between the Boston Red Sox and the St. Louis Cardinals garnished a 15.8 rating, among the highest in the previous 11 seasons. The 2005 through 2007 Fall Classics did not reach the peaks of the late 1990s, but they hardly plummeted without the presence of the Yankees. When New York does make the Series, the ratings are merely mediocre. The 1998 World Series between the Yankees and San Diego Padres attracted a 14.1 share of the viewing public, the worst TV rating in post-season baseball history up to that time. Furthermore, the much-anticipated 2000 Subway Series between the Yankees and cross-town-rival New York Mets garnished only a 12.4 rating, the worst television percentage of any Yankee World Series appearance. The Sultans of Swat may have the cash, but they don't have the clout.

How Many Words for Snow?

The notion that the Inuit have dozens of words for snow is widespread, completely false, and still taught in schools. How did this unfounded myth gain such momentum?

* * * *

From Ivory Towers to Pop Culture

Most people have been told at least once that "Eskimos" (explanation to follow) have many words for snow. This pearl of wisdom is usually shot off in an academic setting as an example of how different cultures adapt their language to the specifics of their environment. Few know that this "fact" is not only false but makes no sense, given a basic understanding of Eskimo languages.

The myth got started in 1911 when renowned anthropologist Franz Boas pointed out that Eskimos have four distinct root words for snow, translating as "snow on the ground," "falling snow," "drifting snow," and "snowdrift." It is unclear where Boas collected this linguistic data. Eskimos speak a polysynthetic language, meaning they take a root word, such as *snow*, and then add on to it a potentially endless number of descriptors. For example, Eskimos could take their root word for snowdrift and tack on to it their words for *cold*, *high*, *insurmountable*, and *frightening*, thus creating one very long and descriptive word. Because the language works in this way, there are, technically, an infinite number of "Eskimo words for snow."

Because Boas was widely read in academic circles, textbooks soon started to make seemingly random claims about the number of ways Eskimos refer to snow. According to Roger Brown's

Words and Things, Eskimos have just three words for snow. Carol Eastman, in *Aspects of Language and Culture,* claims they have "many words" for snow. Once these academic postulations drifted into the mainstream, the number of Eskimo words for snow inexplicably skyrocketed. A 1984 *New York Times* article put the number at 100, while a 1988 article in the same paper marveled at the "four dozen" different words for snow.

What Is an "Eskimo" Word, Anyway?

The Eskimo-words-for-snow myth becomes even more nonsensical when you consider that there is no such thing as one "Eskimo" language. "Eskimo" has become a popular blanket term for the indigenous peoples of eastern Siberia, Alaska, Canada, and Greenland. Eskimos are generally divided into the Inuit and Yupik. In some places, the term "Inuit" has come to be used as a replacement term for "Eskimo," but in other regions this is not accepted, as not all Eskimos are Inuit. The groups collectively referred to as Eskimo or Inuit speak many different languages, though there are commonalities among them.

The popularity of the language myth is fueled by an "Oh wow, aren't they strange!" factor that often comes with fast facts about different cultures. Even if some Inuit languages do have more words for snow, that fact in itself isn't terribly provocative. Linguist Geoffrey Pullum points out in his essay *The Great Eskimo Vocabulary Hoax:* "Botanists have names for leaf shapes; interior decorators have names for shades of mauve; printers have many names for different fonts . . . would anyone think of writing about printers the same kind of slop we find written about Eskimos in bad linguistics textbooks?"

Perhaps at some point in the future, a rumor will spread among the Eskimos that "Americans" have a dizzying number of words for snow, including but not limited to *flurry, blizzard, nieve, neve, slush, schnee, snowball,* and *snowflake.*

No Room at the Inn?

In the traditional Christmas story, an innkeeper turns Mary and Joseph away from his at-capacity establishment but offers them the use of his stable. It's a nice tale, but there are problems with the details.

* * * *

No Vacancy

IN DESCRIBING THE event, the gospel writer Luke says, "And she brought forth her firstborn son, and wrapped him in swaddling clothes, and laid him in a manger; because there was no room for them in the inn." To modern ears, the word *inn* conjures images of a cozy bed-and-breakfast. But in Luke's original Greek, the word *kataluma* can also be translated as "lodging place." He uses this word later, when referring to the room in which Jesus ate the Passover meal with his disciples just before his death. At the time, most houses had such rooms, usually on an upper level and often with a separate entrance, which could be used by guests. It may be that Mary and Joseph were headed to such a guest room at the house in Bethlehem, but they arrived late and the room was already occupied.

Away in a What?

The word *kataluma* can also refer to a covered shelter with open cooking fires where travelers came with their family members and animals—a crowded, noisy place that Mary and Joseph might have wanted to avoid at the time. In any event, there was no room, so they went to a stable, correct? Sorry—the gospels don't mention a stable, either. After Jesus was born, Mary laid him in a manger, or feeding trough. If the *kataluma* was in a private house, the manger may have been in the lower level where the family's animals were kept. Or the manger may have been in a nearby cave where animals could stay out of the elements—not exactly the three-sided wooden shelter of nativity scenes, but at least a more private place in which to give birth.

Who Built the Pyramids?

The Great Pyramids of Egypt have maintained their mystery through the eons, and there's still a lot we don't know about them. But we do know this: Slaves, particularly the ancient Hebrew slaves, did not build these grand structures.

⁕ ⁕ ⁕ ⁕

IT'S EASY TO see why people think slaves built the pyramids. Most ancient societies kept slaves, and the Egyptians were no exception. And Hebrew slaves did build other Egyptian monuments during their 400 years of captivity, according to the Old Testament. Even ancient scholars such as the Greek historian Herodotus (fifth century B.C.) and the Jewish historian Josephus (first century A.D.) believed that the Egyptians used slave labor in the construction of the pyramids.

Based on evidence of the lifestyles of these ancient builders, however, researchers have discredited the notion that they were slaves (Nubians, Assyrians, or Hebrews, among others) who were forced to labor. They had more likely willingly labored, both for grain (or other foodstuffs) and to ensure their place in the afterlife. What's more, we now know that the Great Pyramids were built more than a thousand years before the time of the Hebrews (who actually became enslaved during Egypt's New Kingdom).

Archaeologists have determined that many of the people who built the pyramids were conscripted farmers and peasants who lived in the countryside during the Old Kingdom. Archaeologist Mark Lehner of the Semitic Museum at Harvard University has spent more than a decade studying the workers' villages that existed close to the Giza plateau, where the pyramids were built. He has confirmed that the people who built the pyramids were not slaves—rather, they were skilled laborers and "ordinary men and women."

Out of This World!

The original Star Trek *series brought us two popular quotes: Captain James Kirk's "Beam me up, Scotty," and Dr. Leonard "Bones" McCoy's "Damn-it, Jim, I'm a doctor, not a . . ." But any Trekkie worth his or her dilithium crystals knows that neither quote is exact.*

* * * *

A Universe of One-Liners

CAPTAIN KIRK SAID a lot of things in the original series (when he wasn't busy smooching green-skinned space vixens), but he never said, "Beam me up, Scotty." He did, however, utter a number of variations on that statement over the course of the series and in subsequent movies. These included: "Beam me up," "Beam us up, Scotty," "Beam them out of there, Scotty," and "Scotty, beam me up." It's a minor point, to be sure, but one of great importance to the legions of die-hard *Star Trek* fans.

The quote most often attributed to Dr. McCoy has had its variations, too. Most people put a "Damn-it" in front of the line, but Bones never uttered that expletive in the TV series (a product of the 1960s, *Star Trek* was almost devoid of curse words). However, the doctor did mutter, "Damn-it, Jim!" and "Damn-it, Spock!" on a number of occasions in various *Star Trek* movies.

The "I'm a doctor, not a . . ." routine was used a couple of times during the original series, most evidently in the episode titled "The Devil in the Dark," in which Captain Kirk orders Dr. McCoy to attend to an injured Horta, a creature that is essentially a sentient rock. McCoy's response is typical of his character: "I'm a doctor, not a bricklayer!" Bones made similar sarcastic comments whenever he was required to perform a task that was outside his expertise. In the case of the Horta, he did as he was instructed and ably patched up the wounded creature with cement.

Knowing a good quote when it hears one, the *Star Trek* franchise used Dr. McCoy's popular catchphrase throughout later series and motion pictures. For example, the holographic doctor in *Star Trek: Voyager* (played by Robert Picardo) used the phrase on a couple of occasions. And so did others. In one episode in which Picardo's doctor asks another holographic physician for help after their ship has been taken over by Romulans, the second doctor replies, "I'm a doctor, not a commando!"

Likewise, in the series *Star Trek: Deep Space Nine* (the episode titled "Trials and Tribble-ations"), when Dr. Bashir (Alexander Siddiq) is asked about events in the 23rd century, he quips, "I'm a doctor, not a historian."

There are additional variations of the "I'm a doctor, not a . . ." statement. In the movie *Star Trek: First Contact,* a holographic doctor says, "I'm a doctor, not a doorstop." And in the video game *Star Trek: Bridge Commander,* players who try to engage engineer Brex in too much chitchat are eventually scolded with, "Damn-it, Jim! I'm an engineer, not a conversationalist!"

Long-Lasting Legacy

Star Trek has given viewers much over the 40-plus years it has been around, including innovative scientific concepts that are actually starting to become reality. Its influence even reached NASA, which named its prototype space shuttle *Enterprise,* after the starship featured in the show.

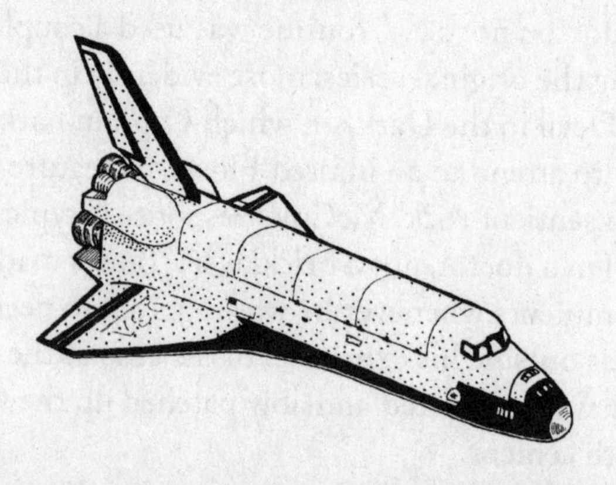

"Rum, Sodomy, and the Lash"

Is it possible that Prime Minister Winston Churchill, whose favorite port in a storm was the one with the most alcohol, would dish out a disparaging dictum about the British Royal Navy?

❋ ❋ ❋ ❋

WHEN GREAT BRITAIN was a dominant naval power, it was said that Britannia ruled the waves with a navy rich in resources and tradition. Therefore, it came as a shock to the British population when it was widely reported that Winston Churchill was of the opinion that the only true traditions that the Royal Navy observed were "rum, sodomy, and the lash."

In fact, Churchill's dissenters perpetrated the origin of this myth-quote. In the 1940s, while he was serving as prime minister and his country was fighting for its very survival, Churchill's political foes concocted an amusing smear campaign that focused on his apparent disdain for and distrust of the navy. According to Churchill's competitors, young Winston had been denied entry to the Royal Naval College because he suffered from a speech impediment, and the scars from that snub never healed.

That wound still riled him when he allegedly rose in the House of Commons and delivered a scathing speech that ridiculed the Royal Navy and its traditions, which he summarized as the equivalent of alcohol, sex, and torture. But the entire incident proved to be fabricated. Records show that Churchill never attempted to join the navy, and documents concur that he never used the House of Commons as a platform to voice his opinions on the Admiralty. Yet, the line remains one of the most popular quotes attributed to Churchill. Its fame was cemented when he supposedly confided to his assistant, Anthony Montague-Browne, that although he had never spoken those words, he certainly wished he had.

For Crying Out Loud!

Every new parent dreams about getting a good night's sleep, but there's a lot of disagreement over the best way to help a baby sleep soundly.

* * * *

OF COURSE, NO one expects a newborn or young infant to sleep through the night. But according to some experts, how parents put their infants to bed and how they tend to them when they awaken may make a difference in their sleep habits in the long run.

There are two main schools of thought on the subject. Respected medical experts continue to weigh in with differing theories and studies. Well-meaning—and experienced—friends and family add to the confusion by offering their surefire bedtime strategies.

To Soothe or Not to Soothe?

On one side of the debate are proponents of routines that purportedly train a baby to fall asleep by himself. This approach advises parents to let the baby cry for short intervals of time. On the other side are those who don't believe a crying infant should ever be left alone and that it's a parent's responsibility to make the baby feel secure enough to fall asleep.

The controversy over sleep routines escalated in 1985 with the publication of *Solve Your Child's Sleep Problems,* by pediatrician Richard Ferber. He advocates allowing babies to soothe themselves to sleep, starting around four to six months of age. The basic idea of "Ferberizing," as it has come to be known, is to put the baby in the crib while still awake, after a calming bedtime routine. Parents are told to not pick up the baby or feed him, even if he cries. Instead, Ferber advocates letting the baby cry for a few minutes before returning to the room to comfort him. Parents are instructed to gradually increase the amount of time the baby is left alone. Ferber calls this approach "progressive waiting," and he believes it eventually teaches the baby to fall asleep without parental intervention. Successfully Ferberized babies will be able to fall back asleep without crying out for their parents when they awaken during the night.

Critics of the Ferber method (which was somewhat revised in the 2006 edition of his book) argue that letting a baby "cry it out" only teaches an infant that nobody cares about him. Two Harvard researchers have gone so far as to say that leaving crying babies alone in their cribs can be traumatic, leading to emotional problems later in life.

Parental Prerogative

There is much anecdotal evidence from parents to support both points of view. Many moms and dads praise Ferber's system, saying that their babies began sleeping through the night after following the routine for a few days. Other parents tell horror stories of letting a baby cry himself to sleep, only to later discover that the child was suffering from, say, a painful ear infection.

The American Academy of Pediatrics says that it is perfectly normal for babies to wake up during the night. A newborn is hungry every one and a half to three hours and needs nourishment in order to fall back asleep. Around the age of six months, most babies can sleep a span of six to eight hours without nursing or having a bottle. However, some doctors insist that it is unreasonable to expect an infant to sleep soundly through the night before he turns a year old.

A Filling Idea

A common misconception is that introducing solid food early will help solve a baby's sleep problems. But some pediatricians say babies should be given only breast milk or formula until they are six months of age. Younger babies cannot properly digest solid food and may end up getting a stomachache instead of a good night's sleep.

Cut the Fat

Exercise enthusiasts and gym-goers everywhere fear It: Take too long a break from a workout regimen, and those hard-earned muscles will turn into fat. Luckily for the lazy, this change is actually impossible.

* * * *

EVEN IF YOU lie on the couch all day eating bonbons, moving only slightly to reach the remote control, your muscles will not turn into fat. However, you will most certainly gain weight, and your muscle strength will diminish.

There are hundreds of different types of cells in the human body—muscle and fat are just two of them. Like spaghetti and meatballs, muscle and fat cells are often grouped together, but they don't have much in common, and one will never turn into the other. Muscle cells are long, fibrous, and mostly striated, while fat cells are round and globular. Each type of cell has a different function: Fat cells store energy, and muscle cells burn energy.

When you stop exercising, your muscles begin to shrink, or atrophy. You don't lose muscle cells; the ones you have just get smaller and flabbier. If you continue to consume the same number of calories as you did when you were active, your fat cells will grow and store the excess calories that are not being burned. The engorged fat cells will take up residence in the territory that muscles used to occupy, creating the illusion that muscle has magically converted to fat.

Hitler's Dance Fever

When France surrendered to the Nazis during the second World War, did Adolf Hitler really perform an odd little jig? The Führer was guilty of many things, but inappropriate dancing wasn't among his crimes.

* * * *

IN JUNE 1940, as Adolf Hitler prepared to accept the surrender of the French government at Compiegne, France, he gave the sidewalk a single stamp with his boot, seeming to punctuate the astonishing turn of history. But people in Allied countries saw something much different in newsreels: Hitler danced a childish two-step, seeming to gloat over the German victory. Played over and over again in movie theaters, the clip ridiculed the Nazi leader while raising the fighting spirit of the public for the many battles to come. Although still recorded as a historical fact in many history books, Hitler's jig was in fact the invention of crafty British propagandists who had simply looped the footage of his single step so that he appeared to be doing a victory dance.

Hitler was also the target of a few less-remembered hoaxes. In 1933, a picture supposedly showing him as a toddler—with a scowling mouth and menacing eyes—was published widely in Great Britain and the United States. The German consulate in Chicago wrote a letter to the Chicago Tribune protesting against the photo's veracity. Five years later, the image was identified by a woman named Harriet Downs, who said it was actually a baby photograph of her own son, John, which had been obviously retouched and darkened to make the child appear more sinister. The photo was then officially retracted by Acme Newspictures.

Feed a Cold, Starve a Fever

Don't worry if you can't remember whether you're supposed to feed a cold and starve a fever, or the other way around. Neither approach will cure you—but one could make you feel better.

⁕ ⁕ ⁕ ⁕

NO ONE KNOWS for sure where this oft-repeated advice originally came from. But some myth busters have traced the adage back to the Middle Ages, when people believed illnesses were caused either by low temperatures or high temperatures. Those caused by low temperatures, including the common cold, needed fuel in the form of food, so eating was the treatment of choice. To the medieval mind, fever—or any other illnesses that caused a high temperature—was fueled by food, so the recommended treatment was to eat nothing or very little to help the body cool down.

Some evidence of this line of thought can be found in the writings of a dictionary maker named Withals, who in 1574 wrote, "Fasting is a great remedie of fever." But if it actually worked for people back then, it was probably a placebo effect.

Today, most medical experts (except for practitioners who promote fasting for healing) totally disagree with the notion of overeating or fasting to treat viral infections that cause colds and flu. When you have a cold or the flu, you actually need more fluids than usual. Drink plenty of water, juice, soup, and tea, and eat enough food to satisfy your appetite. Hot fluids will soothe a cough, ease a sore throat, and open clogged nasal passages. Food will supply nutrients that help bolster your immune system.

So stock up on chicken soup and tea and honey when the inevitable cold or fever strikes. And if a pint of mint chocolate chip ice cream helps you endure the aches and sniffles, why not indulge?

Which Witches Burned?

Contrary to popular belief, no witches were burned at the stake during the Salem witch trials, and men (and even dogs!) were not immune from punishment.

✻ ✻ ✻ ✻

Blame it on Tituba

According to 17th-century residents of Salem, Massachusetts, Satan was always seeking to tempt God-fearing locals into witchcraft. In 1692, the town pastor owned a slave from Barbados named Tituba, who entertained the local children with fortune-telling. No one knows the actual reason, but the girls among the group soon began to claim that they were being spiritually tormented. They also began to exhibit such strange behaviors as hysteria, seizures, and apparent hallucinations. Some people identified the "illnesses" as a condition known as ergotism, which is caused by a rye fungus. But the more likely explanation is simpler. These were just children being children—eager for attention, imitating one another, and aware that the attention ends once the charade does.

Nonetheless, Tituba was quickly identified as the source of the beguiling "spells," and she ultimately confessed under pressure. But rather than being burned at the stake, her punishment was that she was indentured for life to pay the costs of her jailing. Her arrest was the snowball behind an avalanche of accusations. Those subsequently charged with being witches were either "proven guilty" or "soon to be proven guilty."

Witches Take Many Forms

About 150 people (and two dogs) were arrested during the trials, and 19 people (including six men) were hanged. One gent who refused to enter a plea was subjected to "pressing," a form of torture in which rocks are slowly placed atop a person's body until he or she finally suffocates—a process that can take as long as three days. Suddenly, incineration doesn't look so bad.

The Beatles' Ode to LSD

It's got a trippy melody and mind-blowing lyrics. Despite what the drug-devouring flower-power generation believed, however, "Lucy in the Sky with Diamonds" is not a song about LSD.

❋ ❋ ❋ ❋

"LUCY IN THE Sky with Diamonds" is the third track on the Beatles' magnificent album *Sgt. Pepper's Lonely Hearts Club Band*, which was released in April 1967. That was the beginning of the Summer of Love, when hippies, freaks, and counterculture types were experimenting with all kinds of mind-expanding hallucinogens, including the trip-inducing drug lysergic acid diethylamide. To some chemically altered minds, "Lucy in the Sky with Diamonds" was a cleverly coded reference to LSD, evidenced by the first letter of each of the key words in the song's title. Furthermore, believers were convinced that John Lennon's evocative lyrics and airy vocals were the perfect musical expression of an acid trip. The myth of "Lucy in the Sky with Diamonds" as the Beatles' ode to LSD was born.

Each of the Beatles has readily admitted to using acid during this period in their lives, but all of them have denied that "Lucy in the Sky with Diamonds" was inspired by the drug. The true inspiration for the song, as consistently stated by Lennon, was a drawing made by his then-four-year-old son, Julian, depicting a little girl surrounded by twinkling stars. The drawing, Julian explained to his father, was of his schoolmate Lucy O'Donnell—who was floating through the sky among diamonds. Lennon said the image reminded him of Lewis Carroll's *Through the Looking-Glass*, which in turn inspired the lyrics in the iconic Beatles song.

"Lucy in the Sky with Diamonds" isn't, as many believe, a song about LSD. Given the context of the times, however, it's not inconceivable that Lennon dropped a hit while writing it.

Myth Conceptions

Combine a partial fact with a believable untruth and you have the kind of myth that won't go away.

* * * *

Myth: Twins skip a generation—if your grandmother had twins, so will you.

Fact: While twins can run in families, the whole "skip a generation" thing isn't true. Identical twins result from one fertilized egg randomly splitting, creating two siblings with identical DNA. But there is no known gene that influences this process, so it's really just a rare coincidence when a big family has more than one set of twins.

Myth: The chupacabra lives!

Fact: Sorry, folks, but there is simply no such animal known in Spanish as the *chupacabra*, or "goat sucker." This ferocious little animal, found in the American southwest, is a figment of overactive imaginations. While we're at it, there's no such thing as a jackalope, either.

Myth: Eating carrots will keep you from going blind.

Fact: While carrots contain a lot of Vitamin A, which is good for eyes, skin, teeth, and bones, eating carrots isn't going to keep you from needing glasses. This rumor came about during World War II and has stuck around ever since.

Myth: Eating turkey makes you sleepy.

Fact: Not quite. It's been proven that turkey, chicken, and minced beef contain nearly equivalent amounts of tryptophan, the chemical that gets blamed for the sleep-inducing factor of turkey. Other common sources of protein (like cheese, for example) contain more tryptophan per gram than turkey does.

Myth: Coca-Cola is an effective contraceptive.

Fact: Please, don't try this at home. Scientists studied the spermicidal properties of Coke in 1985, but their slim findings were refuted years later by another scientific team in Taiwan. Our question is, who decided to try this in the first place?

Hockey's Other Hat Trick

One might logically assume that NHL great Gordie Howe invented the "Gordie Howe Hat Trick." Then again, when you assume…

* * * *

MR. HOCKEY DID not invent the three-pronged feat that bears his name. In fact, the term used to describe the art of recording a goal, an assist, and a fight in a single game didn't enter the sport's lexicon until 1991. That's a full ten years after the game's longest-serving veteran hung up his blades.

Honoring a Hockey Great

Make no mistake: Gordie Howe was more than capable of achieving all three elements necessary to complete the celebrated triple. He was a wizard at putting the biscuit in the basket, a magician at deftly slipping a pass through myriad sticks and skates and putting the disc on the tape of a teammate's stick, and he wasn't opposed to delivering a knuckle sandwich to a deserving adversary. However, the tattered pages of the NHL record books show that he recorded only one Howe Hat Trick in his 32-year career in the NHL and the World Hockey Association. On December 22, 1955, in a game against the Boston Bruins, Howe (playing for the Detroit Red Wings) scored the tying goal, set up the winning 3–2 tally, and bested Beantown left winger Lionel Heinrich in a spirited tussle.

The Record Holder

The Gordie Howe Hat Trick isn't an official statistic—in fact, the San Jose Sharks are the only franchise that lists the achievement in its media guide—but it is a widely acknowledged measurement of a skater's ability to play the game with both physical skill and artistic grace. The New York Rangers' Brendan Shanahan is the NHL's all-time leader in "Howe Hats." According to *The Hockey News*, Shanny scored a goal, recorded an assist, and had a fight in the same game nine times.

What about the Football Game?

Most people were taught that Thanksgiving originated with the Pilgrims when they invited local Native Americans to celebrate the first successful harvest. Here's what really happened.

❋ ❋ ❋ ❋

THERE ARE ONLY two original accounts of the event we think of as the first Thanksgiving, both very brief. In the fall of 1621, the Pilgrims, having barely survived their first arduous year, managed to bring in a modest harvest. They celebrated with a traditional English harvest feast that included food, dancing, and games. The local Wampanoag Indians were there, and both groups demonstrated their skill at musketry and archery.

So that was the first Thanksgiving, right? Not exactly. To the Pilgrims, a thanksgiving day was a special religious holiday that consisted of prayer, fasting, and praise—not at all like the party atmosphere that accompanied a harvest feast.

Our modern Thanksgiving, which combines the concepts of harvest feast and a day of thanksgiving, is actually a 19th-century development. In the decades after the Pilgrims, national days of thanksgiving were decreed on various occasions, and some states celebrated a Thanksgiving holiday annually. But there was no recurring national holiday until 1863, when a woman named Sarah Josepha Hale launched a campaign for an annual celebration that would "greatly aid and strengthen public harmony of feeling."

Such sentiments were sorely needed in a nation torn apart by the Civil War. So, in the aftermath of the bloody Battle of Gettysburg, President Lincoln decreed a national day of thanksgiving that would fall on the last Thursday in November, probably to coincide with the anniversary of the Pilgrims' landing at Plymouth. The date was later shifted to the third Thursday in November, simply to give retailers a longer Christmas shopping season.

Not Just For Nerds: Role-Playing Games

Many myths abound about role-playing games (RPGs), the increasingly popular tabletop adventures in which players create characters whose fates are decided by a roll of the dice.

* * * *

Gamers are all nerds who live in their moms' basements. A wide variety of people from all walks of life answer to the siren song of a 20-sided die (d20). Men aren't the only ones playing, either—the number of women gamers is growing rapidly. Gone are the days of the stereotypical geeks dominating the gaming tables as more and more people—from teens to college kids to well-employed adults—discover the fun in gaming.

Live-action role-playing (LARP) is sick—people bite each other and hit each other with weapons. There are always a few groups who prefer to play things out with props, but most standard LARP rules don't allow it. In fact, the rules for combat are strict, for the players' safety. A common way to play combat is that the participants have index cards—one for each "weapon" they use—rather than a prop. When they "attack," they tap the target with the card. Physical contact between players is extremely limited, if not strictly forbidden.

Dungeons & Dragons (D&D) is satanic. This rumor most likely started because of the large image of a demon on the cover of the original *Dungeon Masters Guide*. If it has a devil on the cover, that must mean it's satanic, right? In fact, the first couple of modules (suggested campaigns) were designed so the players would eventually fight and destroy the devil. So why is there a demon on the cover? The answer is simple: design and marketing.

A big nasty demon presents the possibility of a more interesting challenge than a low level "kobold" or a simple skeleton.

It's obvious that game developers were eager to get away from the satanic stereotype—the second-edition *Monster Manual* is noticeably thinner in the "D" section (all references to demons and devils were removed). However, the use of demonic nasties as villains was popular with players, so the characters made a comeback in the third edition, released in 2000.

RPGs teach kids how to cast spells. If you think reading the *D&D Players Handbook* will teach you how to cast a real magic missile, you're in for a surprise. Waving your arms around and saying some variation of "Abracadabra, fireball!" won't accomplish anything but knocking over your soda.

The *Harry Potter* books give more details on how to cast a spell than D&D does. The rules lay out whether you need special components—such as an identifying spell that requires a pearl worth 100 gold pieces—or if you need to be able to speak and/or make certain motions (since you could be silenced or held immobile and therefore unable to make the required verbal or movement-based parts of the spell). What those specific words or gestures are is not usually mentioned, though a few gestures were included in earlier editions as in-jokes. For instance, to cast a spell that caused small flames to spout from the caster's fingers, the caster had to make a hand gesture that was the same as one would make to light a lighter. To cast a spell in the game, the player would say, "I cast [insert spell name]." No rituals involved.

Spreading Fear

Surgery carries risks, but spreading cancer is not one of them.

* * * *

ONE OF THE most unfortunate medical misconceptions—that surgery causes cancer to spread throughout the body—has persisted for years. Specialists in cancer surgery say there is no truth to this misconception, but it sometimes prevents cancer patients from seeking life-saving treatment.

A 2005 survey by the American Cancer Society found that 41 percent of respondents believed cancer spreads throughout the body as a result of surgery. And 37 percent of those responding to a 2003 survey by the Philadelphia Veterans' Affairs Medical Center survey thought cancer spreads when simply exposed to air.

Experts surmise that misconceptions about cancer surgery were launched before the advent of effective diagnostic tools such as magnetic resonance imaging and ultrasound. Back then, the only conclusive way to know if a patient had cancer was through exploratory surgery. In many cases, by the time the patient was opened up in the operating room, the cancer already had metastasized, and there was little doctors could do to stop the advancement of the disease. Ailing patients mistakenly blamed the surgeon for their declining health, believing it was the operation that had caused the cancer to spread.

Today, doctors are able to identify cancer at much earlier stages, when patients have better treatment options. Surgery, often in combination with chemotherapy and radiation, is an important part of the journey back to health. Surgeons are able to safely perform biopsies and remove tumors without introducing cancer cells into other parts of the body. Cancer spreads three ways: through the blood, through the lymphatic system, or by invading tissue near the tumor.

Ready, Set, Cycle!

From the sandbox to the sorority house, gal groups are powerful. Guys might even say magical. But can women exert a kind of chemical alchemy over one another?

* * * *

A 1971 STUDY OF 135 MEMBERS of an all-female college dormitory showed that women who lived together tended to have menstrual periods within days of one another. In the study, roommates whose periods averaged more than six days apart in October were less than five days apart six months later.

The theory of menstrual synchrony—called the "McClintock Effect" after the author of the study—was long accepted as established fact, especially among women. McClintock, a graduate student who then became a University of Chicago psychologist, speculated that pheromones (chemical messengers received through the sense of smell) were responsible for the phenomenon.

Recent studies have cast doubt on the theory. In 1992, H. Clyde Wilson published a report accusing a slew of studies—McClintock's included—of faulty research and shoddy methodology. And researchers have pointed out that synchrony is impossible when women have cycles of different lengths.

In 2007, psychologist Jeffery Schank published a study involving 186 Chinese women who lived together in a college dorm. Though he uncovered some interesting menstruation patterns among the women, he found no evidence that their cycles were in sync. The researchers reviewed McClintock's original study and went so far as to say that the results in 1971 could be chalked up to chance.

The strongest evidence against menstrual synchrony is that, in all studies, women's cycle lengths still vary radically, even if their start dates get closer over a designated period of time.

Did Nero Fiddle While Rome Burned?

Over the ages, the phrase "Nero fiddled while Rome burned" has become a euphemism for irresponsible behavior in the midst of a crisis. But as a matter of historical fact, legend has it wrong.

* * * *

IN A.D. 64, much of Rome burned to the ground in what is known as the Great Fire. According to legend, the reigning emperor, Nero, purposely set the blaze to see "how Troy looked when it was in flames." From atop a palace tower, he played his fiddle and sang as the fire raged and consumed the capital.

Nero, a patron of the arts who played the lyre, wrote poetry, and fancied himself a great artist, often performed in public, challenging the beliefs of Rome's political class who believed such displays were beneath the dignity of an emperor. But music was, in fact, the most dignified of Nero's interests. Under the influence of a corrupt adviser who encouraged his excesses, his life became a series of spectacles, orgies, and murders. A few months after his first public performance, the Great Fire ravaged Rome for five days. Roman historian Suetonius, who hadn't even been born at the time of the fire, describes Nero singing from the Tower of Maecenas as he watched the inferno. Dio Cassius, a historian who lived a hundred years later, places him on a palace roof, singing "The Capture of Troy."

However, the historian Tacitus, who actually witnessed the fire, ascertained that the emperor was at his villa in Antium, 30 miles away. Many contemporary historians agree that Nero was not in Rome when the fire broke out. According to Tacitus, Nero rushed back to Rome to organize a relief effort and, with uncharacteristic discipline and leadership, set about rebuilding and beautifying the city he loved.

Do Real Cats Hate Water?

One of humankind's most beloved creatures is mysterious in many ways, and a popular feline myth is that cats hate water.

* * * *

SOME CATS MAY fear water because of how we use it around them: Many a noisy tomcat has had a bucket of water thrown its way, and mischievous kids might tease Mittens with a spray from the garden hose. Forcing a bath on a cat is a sure way to get it to loathe the water, but that's no different for any other animal. There's actually a lot of evidence that cats love water. Many cats don't hesitate to jump into a filled sink or running shower and actually seem amused as water from a faucet drips on their heads. One reason for the positive reaction is that cats are attracted to the motion and sound of water.

Among the larger cats, climate makes a difference. Tigers, lions, jaguars, and ocelots from the hot savannas are likely to fancy a plunge into cool streams and ponds to get a break from the heat. Logically, cats that live in cold environments—including snow leopards, lynx, bobcats, and cougars—show little interest in getting wet.

Cats are, of course, creatures of habit, so a pet that has been exposed to water since it was young will tolerate a bath much better than one whose human companion shielded it from water out of fear for its safety.

Tree Thief

It's the unkindest cut of all! The holidays are stressful. Between buying gifts, planning dinners, and visiting relatives, it's understandable that some things get pushed off to the last minute. Maybe the tree hasn't been bought yet, and the ones remaining are spindly little things.

* * * *

Idiot Ruins Christmas

ONE SCROOGE SNUCK into Seattle's Washington Park Arboretum in December 2009 and cut down a rare Chinese keteleeria tree worth over $10,000. David Zuckerman, the horticulturist for the University of Washington Botanic Gardens, was shocked that this particular tree was cut. "It's a Charlie Brown tree," he said. Many firs, which are more traditional Christmas trees, are prevalent throughout the arboretum. The keteleeria is a sparsely branched tree, not as bushy as firs. It's also much more rare. The keteleeria tree, native to China, is endangered. It was planted at the arboretum so that its seeds could be used to help propagate the species.

The theft wasn't even a last-minute panic decision on Christmas Eve. The perpetrator cut down the tree sometime on December 9. That's 15 solid tree-shopping days that don't necessitate cutting down the nearest endangered keteleeria. Those who worked at the arboretum to cultivate the tree from a sapling to a fine seven-foot specimen were particularly hurt by the act. The local botanic gardens might have to watch their flowers around Mother's Day.

It was not even the first time a tree had been stolen from this arboretum. Several years previously, a fir was cut down and displayed in a restaurant. The perpetrator was caught, but trees that are cut down can't be repaired. Staffers are considering covering the trees with a nontoxic paint during December that can be washed off after the holidays are over.

Christmas Myths!

Be careful what you believe—the Christmas holidays are rife with urban legends, very few of which are true.

⁕ ⁕ ⁕ ⁕

Christmas has spawned a variety of urban legends over the years, myths and incredible stories that, upon closer examination, almost always prove to be false. Here is a sampling of the most popular:

1. **Myth:** The number of suicides jumps dramatically over the Christmas holidays.

Fact: Proponents of this urban legend believe that the joy of the Christmas season exacerbates the hopelessness felt by many, causing them to take their own lives. However, numerous studies have found this not to be true. One of the most compelling was a Mayo Clinic survey of suicides over a 35-year period that failed to find even a small spike in self-inflicted deaths before, during, or after the Christmas holidays.

2. **Myth:** *The Twelve Days of Christmas* was written as a coded reference to Catholicism during a period in British history when the religion was illegal.

Fact: According to this myth, *The Twelve Days of Christmas* is a "catechism song" chock full of hidden meaning. "Two turtle doves," for example, refers to the Old and New Testaments of the Bible, and "three French hens" means the theological virtues of faith, hope, and charity. However, there is absolutely no historical evidence that this claim, which dates back only to the 1990s, is true. At best, it's mere speculation.

3. **Myth:** Salvation Army bell ringers get to keep a portion of the money placed in their kettles.

Fact: Absolutely untrue, Salvation Army officials state. Most bell-ringers are volunteers, though some are paid seasonal employees, often recruited from homeless shelters and the Salvation Army's own retirement homes. Those hired receive a salary, while all of the money dropped into their kettles goes toward a variety of charitable endeavors.

4. **Myth:** A dad who was supposed to miss Christmas because he was on a business trip decided to cancel at the last minute and surprise his kids by dressing up as St. Nick and sliding down the chimney. Unfortunately, the man became stuck and died of asphyxiation. His family knew nothing about his plans until they lit a fire in the fireplace.

Fact: This is one of the oldest and most oft-repeated holiday urban legends to make the rounds. It's certainly a great story, but a single real-life case has yet to be verified. You can find a variation on this urban legend in the movie *Gremlins* (1984); it's how Phoebe Cates's character lost her dad.

5. **Myth:** The common abbreviation for Christmas—Xmas—is an attempt to remove Christ from the holiday.

Fact: Everyone needs to calm down. The use of "Xmas" as an abbreviation for Christmas is eons old and based on the fact that the Greek word for Christ begins with the letter *chi,* which is represented in the modern Roman alphabet by a symbol that closely resembles an "X."

6. **Myth:** The candy cane was invented as a tribute to Jesus. The shape represents the letter "J" and the red and white stripes symbolize purity and the blood of Christ.

Fact: A variation on this urban legend also suggests that candy canes were created as a form of secret identification among Christians during a period of persecution; both stories are false. The truth is that this popular Christmas candy has been around at least since the late 17th century (when there was no persecution of Christians in Europe), but the color striping is strictly a 20th century invention developed for decoration and flavor.

Don't Be Fooled

Abraham Lincoln is reputed to have said, "You can fool all of the people some of the time and some of the people all the time, but you cannot fool all the people all the time." But any fool knows that the true sources of famous quotes are often difficult to track.

⁂ ⁂ ⁂ ⁂

ALTHOUGH THIS LEGENDARY quote may make a wise point about lies and politics, it is unwise to assume it came from the sage mouth of our sixteenth president. Historians have been unable to confirm that Lincoln ever spoke these words. The first written reference that attributes the quote to Lincoln is in 1901's *"Abe" Lincoln's Yarns and Stories,* by Alexander K. McClure, in which the author claims the president made the remark in casual conversation. The quote pops up again in a 1905 book called *Complete Works of Abraham Lincoln.*

According to the most common story, Lincoln spoke these words on September 2, 1858, in a speech made in Clinton, Illinois. This was during the famous Lincoln–Douglas debates, when the two contenders for an Illinois seat in the U.S. Senate took part in eloquent political wrangling, mostly over the issue of slavery. A complete record of the speech does not exist, so no one can prove or disprove whether the future president ever said anything about the relative ease of foolery. The quote definitely does not appear in any of Lincoln's writings.

Lincoln scholars generally deny that he is the true source of the quote, but others believe he may have said something like it in one of his speeches, and the words then became a sound bite. Whether Lincoln said it or not, the confusion over this quote does prove that, unless you have it in writing, nowadays it's difficult to fool anyone about anything.

Monkee Business

He now sits in a cage at Corcoran State Prison, but is it possible that Charles Manson, the wild-eyed cult leader who ruled over a posse of deranged, drug-addled followers, almost became a member of the Monkees?

* * * *

IN THE MID-1960S, perceptive TV producers Bob Rafelson and Bert Schneider decided to cash in on the Beatles' success in *A Hard Day's Night* by formulating their own recipe for a boy-band. Borrowing from the Liverpool lads' madcap antics and animalistic moniker, Rafelson and Schneider concocted a half-hour comedy show that featured music, skits, and general goofiness. They called it *The Monkees*. In September 1965, an open audition attracted 437 musicians, actors, acrobats, hippies, has-beens, and wannabes, including future Buffalo Springfield guitarist Steven Stills, Lovin' Spoonful founder John Sebastian, and Three Dog Night vocalist Danny Hutton.

Another member of the Tinseltown music fraternity rumored to have staked a claim to fame in front of the producers was a scruffy mite named Charles Manson, who would gain infamy for masterminding a bizarre, ritualistic murder spree in August 1969. That event eventually put the former felon back into prison for a lifetime stay, but a popular myth placed Manson among the throng of Hollywood hopefuls trying out for a part. Although the tale has some credibility—Manson did have musical ability and befriended Beach Boys drummer Dennis Wilson, who helped the maniac record demo tapes—he could not have attended the audition. At the time, he was in a prison at McNeil Island, Washington, and wasn't released until 1967. Even if he had somehow wrangled a weekend pass, Manson was 30 years old at the time, making him ineligible to be one of the "4 Insane Boys, Age 17–21," as the posting indicated. In the end, though, it appears he did meet one of those requirements.

The Dastardly Draw

A deeply ingrained image of the Wild West is that of two men approaching each other with hands on their gun butts, determined to prove who can draw and fire faster. Exciting stuff, but it seldom happened that way.

❋ ❋ ❋ ❋

IN THIS CLASSIC showdown, usually outside the town saloon, the two scowling scoundrels stand motionless. After a blur of movement and two nearly simultaneous gunshots, one man drops, hit by the first bullet. Variations of this scenario include one gunman shooting his adversary in the hand or, with even greater sportsmanship, letting his opponent draw first.

In 1865, Wild Bill Hickok and Dave Tutt met in front of an expectant crowd for a prearranged pistol duel in Springfield, Missouri. At a distance of 75 yards, Tutt fired and missed. Hickok steadied his revolver with his left hand and triggered a slug into Tutt's heart. It's a stereotypical depiction, but this sort of face-off was actually uncommon.

In Western gunfights, the primary consideration was accuracy, not speed. Gunfighters usually didn't even carry their weapons in holsters. Pistols were shoved into hip pockets or waistbands, and a rifle or shotgun was usually preferred over a handgun. A study of almost 600 shootouts indicates that in one gunfight after another, men emptied their weapons at their adversaries without hitting anyone (except, perhaps, for a luckless bystander) or inflicting only minor wounds.

During the first decade of the 20th century, Westerns became a staple of the burgeoning film industry. Hollywood pounced on the handful of duels such as Hickok versus Tutt, and soon the fast-draw contest became an integral part of the genre. Like modern detective films without car chases, Westerns were incomplete without fast-draw duels.

Alleged Celebrity UFO Sightings

It's not just moonshine-swilling farmers in rural areas who claim to have seen UFOs hovering in the night sky. Plenty of celebrities have also reportedly witnessed unidentified flying objects and have been happy to talk about their experiences afterward.

* * * *

Jimmy Carter

NOT EVEN PRESIDENTS are immune from UFO sightings. During Jimmy Carter's presidential campaign of 1976, he told reporters that in 1969, before he was governor of Georgia, he saw what could have been a UFO. "It was the darndest thing I've ever seen," he said of the incident. He claimed that the object that he and a group of others had watched for ten minutes was as bright as the moon. Carter was often referred to as "the UFO president" after being elected because he filed a report on the matter.

David Duchovny

In 1982, long before he starred as a believer in the supernatural on the hit sci-fi series *The X-Files,* David Duchovny thought he saw a UFO. Although, by his own admission, he's reluctant to say with any certainty that it wasn't something he simply imagined as a result of stress and overwork. "There was something in the air and it was gone," he later told reporters. "I thought: 'You've got to get some rest, David.'"

Jackie Gleason

Jackie Gleason was a comedian and actor best known for his work on the sitcom *The Honeymooners* and his role as Minnesota Fats in *The Hustler* (1961). He was also supposedly a paranormal enthusiast who claimed to have witnessed several unidentified objects flying in the sky. Gleason's second wife, Beverly McKittrick, claimed that in 1974 President Nixon took Gleason to the Homestead Air Force Base in Florida, where he saw the wreckage of a crashed extraterrestrial spaceship and the bodies of dead aliens. The incident had such a profound affect on him that it curtailed his famous appetite for alcohol, at least for a while. Gleason was so inspired by the visit that he later built a house in upstate New York that was designed to look like a spaceship and was called "The Mother Ship."

John Lennon

In 1974, former Beatle John Lennon claimed that he witnessed a flying saucer from the balcony of his apartment in New York City. Lennon was with his girlfriend May Pang, who later described the craft as circular with white lights around its rim and said that it hovered in the sky above their window. Lennon talked about the event frequently and even referenced it in two of his songs, "Out of the Blue" and "Nobody Told Me," which contains the lyric, "There's UFOs over New York and I ain't too surprised . . ."

Ronald Reagan

Former actor and U.S. president Ronald Reagan witnessed UFOs on two occasions. Once during his term as California governor (1967–1975), Reagan and his wife Nancy arrived late to a party hosted by actor William Holden. Guests including Steve Allen and Lucille Ball reported that the couple excitedly described how they had just witnessed a UFO while driving along the Pacific Coast Highway. They had stopped to watch the event, which made them late to the party.

Reagan also confessed to a *Wall Street Journal* reporter that in 1974, when the gubernatorial jet was preparing to land in Bakersfield, California, he noticed a strange bright light in the sky. The pilot followed the light for a short time before it suddenly shot up vertically at a high rate of speed and disappeared from sight. Reagan stopped short of labeling the light a UFO, of course. As actress Lucille Ball said in reference to Reagan's first alleged UFO sighting, "After he was elected president, I kept thinking about that event and wondered if he still would have won if he told everyone that he saw a flying saucer."

William Shatner

For decades, the man who played Captain Kirk in the original *Star Trek* series claimed that an alien saved his life. When the actor and a group of friends were riding their motorbikes through the desert in the late 1960s, Shatner was inadvertently left behind when his bike wouldn't restart after driving into a giant pothole. Shatner said that he spotted an alien in a silver suit standing on a ridge and that it led him to a gas station and safety. Shatner later stated in his autobiography, *Up Till Now*, that he made up the part about the alien during a television interview.

Outrageous Media Hoaxes

Fair and balanced hasn't always been the mantra of the media. In fact, some newspapers used to pride themselves on the outlandish stories they could come up with. Here are a few of the most outrageous hoaxes in journalism.

* * * *

Man on the Moon

IN 1835, IN one of America's earliest media hoaxes, *The New York Sun* reported that a scientist had seen strange creatures on the moon through a telescope. The story described batlike people who inhabited Earth's neighbor. Readers couldn't get enough of the story, so other publishers scrambled to create their own version. When faced with criticism, *The Sun* defended itself, stating that the story couldn't be proven untrue, but eventually the stories were revealed as hoaxes.

Hoaxer Ben Franklin

For nearly a decade, Ben Franklin perpetrated a hoax continually claiming that Titan Leeds, the publisher of the main competitor to Franklin's *Poor Richard's Almanac*, was dead. This greatly decreased Leeds's circulation, since no one wanted to read the ramblings of a dead man. Leeds protested, but year after year, Franklin published annual memorials to his "deceased" competitor. When Leeds really did pass away, Franklin praised the man's associates for finally admitting he was dead.

Anarchy in London

In 1926, a dozen years before *The War of the Worlds,* the BBC staged a radio play about an anarchic uprising. The "newscast" told of riots in the streets that led to the destruction of Big Ben and government buildings. The population took the play so seriously that the military was ready to put down the imaginary rioters. The following day the network apologized and the government assured the public that the BBC would not be allowed such free range in the future. The British were ridiculed worldwide, especially in the United States, where the public had not yet been introduced to a young actor named Orson Welles.

Wild Animals on the Loose in New York City

In 1874, the *New York Herald* published stories detailing how animals at the city zoo had escaped and were rampaging through the streets. The mayor ordered all citizens to remain in their homes while the National Guard grappled with the situation. The problem was that the stories weren't true. In fact, the final line of the article read, "Of course, the entire story given above is pure fabrication." Apparently, no one read that far as the city was thrown into a panic. When the smoke cleared, the editor wasn't fired...he was given a bonus for raising the newspaper's circulation.

Mr. Hearst's War

Media mogul William Randolph Hearst had no problem with manipulating the truth to sell newspapers. One of his most famous hoaxes was a series of misrepresentations of what was really occurring in Cuba during the lead-up to the Spanish–American War. He sent artist Frederic Remington to the island to capture the atrocities, but the artist found none. "You furnish the pictures, I'll furnish the war," Hearst replied. But Hearst's misuse of pictures was not limited to that event. Consumed with a passion to defeat the communists, he once ordered his editors to run pictures showing an imaginary Russian famine. However, on the same day, they unwittingly published truthful stories about the rich harvest Russia was enjoying.

Poe's Prank

Though Edgar Allan Poe is best known for his macabre works of fiction, he had his hand in a few works of journalistic fiction as well. One of his best known was a piece that ran in *The New York Sun* in 1844, the same year he wrote his classic poem "The Raven." The article claimed that daring adventurer Monck Mason had crossed the Atlantic Ocean in a hot air balloon. Mason had only intended to cross the English Channel but had been blown off course and arrived 75 hours later in South Carolina. When readers investigated the claim, Poe and *The Sun* conveniently admitted they had not received confirmation of the story.

Millard Fillmore's Bathtub Bunk

Everyone seems to know that Millard Fillmore was the first president to have a bathtub installed in the White House. The only problem is, it isn't true. The story, along with a detailed history of the bathtub, was all a hoax perpetrated by writer H. L. Mencken when he worked for *The New York Evening Mail*. "The success of this idle hoax, done in time of war, when more serious writing was impossible, vastly astonished me," Mencken later wrote. The excitement around his piece and the public's inability to accept the truth affected Mencken, and he began to wonder how much of the rest of history was indeed, in his words, "bunk."

✳ Chapter 9

Odd Quaffs and Unforgettable Edibles

Drink Up!

There are worse things you can do than gulp down urine. In fact, the practice of urine consumption, or urophagia, dates back to the ancient Egyptians, Chinese, Indians, and Aztecs, who imbibed in this very personal nectar for health purposes.

✳ ✳ ✳ ✳

It Quenches Your Thirst in a Pinch

Of course, "health purposes" is a relative term. In dire circumstances in which no fresh water is available, drinking urine can supposedly help prevent dehydration. Given that urine is composed mostly of water, the first golden drink may be fairly harmless. If you were to go back to the urine well repeatedly, however, you'd run into the law of diminishing returns—the percentage of usable water in your pee would decrease and harmful flushed waste products would increase.

These harmful substances could include drugs or other chemicals from the environment that exit the body in a hurry via the kidneys, which would take a beating as they continuously tried to recycle compounded toxins. In addition, the high salt content in urine would eventually lead to dehydration, not stave it off.

Urine as a Hallucinogen

However, if it's psychotropic trips you're after, perhaps scoring the Koryak tribe of Siberia as your new drinking buddies would be the thing. Koryak tribesmen swig each other's urine to prolong highs after consuming mind-altering mushrooms during rituals.

Some cultures believe that supplemental consumption of urine keeps illnesses at bay. The holistic approach of Indian Ayurvedic medicine known as Amaroli uses urine to treat asthma, arthritis, allergies, acne, cancer, heart disease, indigestion, migraines, wrinkles, and other afflictions. There is no proof that these treatments work, but since urine is sterile (as long as it isn't contaminated with the Koryak's psychedelic mushrooms), it does have antibacterial, antifungal, and antiviral qualities.

Urine Therapy

Some people praise pee as a power-packed elixir brimming with vitamin, hormone, and protein goodness. Apparently, rocker Jim Morrison and actor Steve McQueen participated in urine therapy.

Like most anything, however, moderation is the key. According to the Chinese Association of Urine Therapy, negative side effects include diarrhea, fatigue, fever, and muscle soreness; these problems worsen as you drink more. You may also want to beware if you ever see someone clad in a T-shirt that's emblazoned with the words, "Koryak Chug-a-lug Champ."

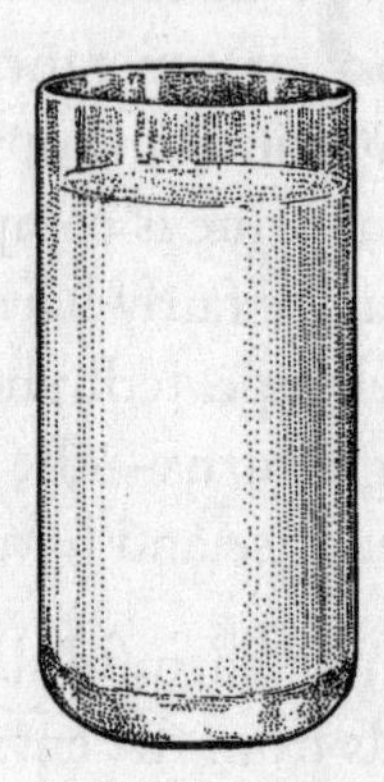

Hints of Earwax and Vomit? Pour Me a Glass!

Chicagoans are known for their flinty wintertime resolve. But what many people don't realize is that the city has a unique secret weapon: they can heat up from the inside out with a bitter—some say vile—swill known as Malört.

* * * *

VARIOUSLY COMPARED TO nail polish remover and earwax, vomit and formaldehyde, Malört's sharp, stinging taste has given it a reputation for seediness only matched by that of its biggest fans, the bikers and derelicts who inhabit Chicago's many backroom taverns and dive bars. Malört schnapps is named for the Swedish word for its principal ingredient, wormwood, the bitter herb that allegedly gave hallucinogenic properties to absinthe—which, in turn, rotted the brains of a generation of bohemians in 19th-century Montmartre. And, though Malört's only public distributor, Carl Jeppson Liquors, uses a variety of wormwood that doesn't contain any dangerous hallucinatory chemicals, its taste may make you feel like you're having a bad acid trip.

Carl Jeppson, the Swedish immigrant who began selling his concoction after the repeal of Prohibition in the 1930s, may have been a master of understatement in addition to having a taste for the bizarre. He liked to brag that Malört was brewed "for that unique group of drinkers who disdain light flavor or neutral spirit." On the other hand, Jeppson knew that his brew wasn't just another regurgitated variation on a theme, and that there were enough adventurous drinkers in Chicago to make his company a hit.

The grimace-inducing spirit doesn't simply taste, in the words of one bar patron, like drinking bug spray. Oh no—it tastes like drinking bug spray for hours at a time. (Mmm, sign us up!)

Which is to say, the flavor isn't simply potent, but long-lasting, as well. But many of the brave souls who consume enough of the vile brew become converts—among the earliest of these were Chicago's North Side Swedish and Eastern European populations, who became especially fond of Malört and made it, over the years, a staple at many bars in diverse pockets of the city.

But don't expect to find Malört everywhere. Its esoteric following, along with its acquired taste, has also made the drink something of a bottom-shelf novelty. Some bars may boast 30-page liquor menus and keep extra bottles of Blue Label under the counter, but that doesn't mean they'll have Malört. And don't even bother requesting the toxic potion anywhere outside the city; it is only distributed in Chicago. Today, Jeppson Liquors is still the only known producer of Malört. And, though the distillery traded in Chicago's bitter-cold winters for the sun and warmth of Florida long ago, the brand, operated out of a Chicago apartment, has remained loyal to the city through and through. The label still says "Chicago, U.S.A." in big, bold type, and is emblazoned with a shield bearing the sky-blue stripes and red stars of the Chicago flag.

Toad-ally High

Bored with all of the traditional ways of getting high: marijuana, cocaine, and Ecstasy? Looking for a new disgusting and unhygienic way to tune out for a while? Look no further than the banks of the Colorado River, dude.

* * * *

HOPPING ALONG THE river's shores in southern Arizona, California, and northern New Mexico, the *Bufo alvarius* (also called the "Cane Toad" and "Colorado River Toad") would normally be in danger of being the main course for a wolf or Gila monster. That is, it would if it weren't for a highly toxic venom that this carnivorous toad produces whenever it gets agitated: the same venom that can get you high as a kite if properly ingested.

The toad's venom is a concentrated chemical called bufotenine that also happens to contain the powerful hallucinogen 5-MeO-DMT (or 5-methoxy-dimethyltryptamine). Ingested directly from the toad's skin in toxic doses (such as licking its skin), bufotenine is powerful enough to kill dogs and other small animals. However, when ingested in other ways—such as smoking the toad's venom—the toxic bufotenine burns off, leaving only the 5-MeO-DMT chemicals. Those can produce an intense, albeit, short-lived rush that has been described as 100 times more powerful than LSD or magic mushrooms, even if it takes a lot more work to get it.

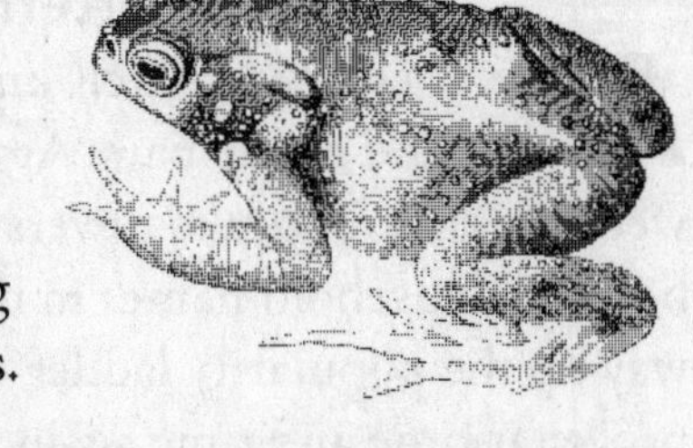

As one of the few animals that excrete 5-MeO-DMT, *Bufo alvarius* are leathery, greenish-gray or brown critters that can grow up to seven inches long. They have four large glands that are located above the ear membranes and where their hind legs meet their bodies. Toad-smokers first milk the venom from the amphibian by rubbing its glands, which causes it to excrete the bufotenine. Then they catch the milky white liquid in a glass dish or other container. After the bufotenine has evaporated into a crystalline substance, it is collected using a razor blade or other sharp instrument and put in a glass-smoking pipe, and then lit and inhaled. Sounds, uh, fun!

This Ain't Your Grandma's Wedding Cake

One more wedding tradition has inevitably fallen victim to pomp, circumstance, and apparently, bridal wizardry.

* * * *

Alterna-cakes

Among the most well-known cake alternatives are cupcakes and candy bars. Arguably rivaling the traditional wedding cake, cupcake towers and candy bar buffets have really become household names in the wedding world. Working their way up the popularity ladder, these customized creations have proven it time and time again: What they lack in tradition, they gain in delightful designs. Because both choices can be dipped, dolloped, shellacked, sprinkled, and stacked any way a bride sees fit, they are easily taking over the market of goodies miniature in size but powerful in taste.

Kreme Is the New Cake

Not to be outdone by the more popular wedding cake alternatives, places like Baskin-Robbins and Krispy Kreme Doughnuts have also jumped on board the wedding dessert bandwagon. Each has started customizing dishes for couples looking for something original and nontraditional. Krispy Kreme will refine, ruffle, and raise up rows of doughnuts in the name of "holey" matrimony, while Baskin-Robbins likes to give their newlyweds-to-be the option of choosing their favorite flavors to be merged and quaffed into a giant ice cream creation . . . giving ice sculptures and "cold feet" reassurance that it's super cool to be a part of wedded bliss. (What? Too much?)

Sweet Snack Cakes

While chilled cream cakes and ice sculptures give couples a reason to freeze their assets, stacking their favorites in tiers is yet another alternative that will leave their guests with full bellies and warm hearts. Everything from fruit-packed pastries to mousse-filled toasting flutes—and even pancakes and pork pies (yes, pancakes and pork pies!)—have been displayed in reception halls across America. Not only can this cut down on service charges, but it also allows newlyweds the opportunity to have more than one choice. One of the most popular "tier"-ing choices is a combination of prepackaged confectioneries, such as Twinkies, Ding Dongs, and Sno Balls, assembled and stacked on cake plates and platters. These are obviously done by a bride looking to be the "Hostess with the Most-est!"

Cultural Confections

Aiming to please isn't the only reason people are leaning toward wedding cake alternatives. More often than not, dessert options are decided by someone's preferences, culture, heritage, or beliefs. Thus, many couples will look to other parts of the world for inspiration. For instance, while many Americans revel in the fruitcake as a holiday gag gift, places such as the British Isles, the Caribbean, Ireland, and Scotland revere it as a wedding-day "must have."

Couples looking for something different might look to Europe where France venerates its caramel covered, cream-filled pastry towers *(croquembouche)* or Lithuania, where *sakotis*—a cookie-like dessert shaped into a Christmas tree—are nothing short of blue ribbons. Even newlyweds who don't have a sweet tooth can look for inspiration in Korea, where they cover ground steamed rice in red bean powder, or Japan and India, where they use "dummy cakes." While guests might not like eating cardboard, the money saved will leave at least two people smiling!

Bottom line, wedding cakes and all their alternatives really don't have specific rules to follow. In fact, since everyone knows calories don't count when someone's getting married (or that's what we tell ourselves at least), it may even be encouraged to go for broke and try them all!

Out to Lunch

Human beings are social creatures, and what brings us together most often is food. Our early ancestors shared their kill around a fire, but when we don't feel like cooking, we just run out for a bite. Here are a few nuggets of knowledge on restaurants and restaurant culture.

* * * *

Oldest Restaurant: Casa Botin. Situated on Calle de Cuchilleros in the heart of Madrid, Spain, Casa Botin was founded in 1725 and still cooks up such famous food as roast pig, baby eels, and caramel custard in its 18th-century stove.

Most Expensive Sandwich: the Von Essen Platinum Club Sandwich. This sandwich, thought to be a favorite of Edward, Duke of Windsor, and his wife, Wallis Simpson, is available at the Cliveden House in Berkshire, England. For £100, or roughly $160, diners can delight in Iberico ham, poulet de Bresse, white truffles, quail eggs, and sun-dried tomatoes on fermented sourdough bread.

Most Expensive Restaurant: Aragawa. Admission to this tiny steak house in Tokyo is by invitation only. The succulent, locally raised Kobe beef is presented simply (with pepper and mustard only) at the price of nearly $400 per person.

Most Popular Occasion for Dining Out: a birthday. Mother's Day follows close behind, and Valentine's Day takes third place.

Percentage of an Individual's Meals Eaten in Restaurants: 24 percent. Though restaurants continue to beckon more people to their tables, the average family still consumes the majority of its meals at home.

Number of Restaurant Employees in the United States: 12.8 million. Only the government employs more people.

Taste of Wisconsin

The Seven Wonders of Wisconsin cuisine are beer, cheese, and brats, brats, brats, brats, and brats. Whether at festivals, picnics, tailgating, or in the backyard, the Badger State has a love affair with the bratwurst.

* * * *

The Sausage of Champions

IF YOU ARE what you eat, Wisconsinites are bratwurst. And they eat lots of brats. For example, every year, during a single Memorial Day weekend, residents of the capital city, Madison, eat around 20 miles of brats at the World's Largest Brat Fest. That's roughly 200,000 sausages.

Wisconsin loves brats, even if no one is too certain what they are. "Translated, a bratwurst means a sausage to be fried," says Debra Usinger, director of retail operations and corporate services for the 129-year-old Usinger's Famous Sausage, based in Milwaukee. And she should know, since she's also a great-granddaughter of the founder.

The grilled brat served as a sandwich, Usinger says, is a distinctly American creation. In Germany, the sausages are browned in a pan and served on a plate, with nary a bun in sight. The spirit if not the intent of old-style Germanic cooking style is preserved in the Sheboygan area, where the act of grilling brats on a barbecue is still called a "brat fry." And the history of the brat goes back farther than that.

Did bratwurst fill the bellies of the builders of the pyramids? Not quite, but Ancient Egypt is where the first sausage was born. Even 4,000-plus years ago, it was immigrant food. The constant movement of nomads and roaming armies brought sausages up to Europe through the Mediterranean. Long after Alexander the Great, after all, Napoleon said that an army travels on its stomach. What travel food is more convenient?

"You couldn't bring steak with you, but you could take sausages that were smoked and dried," says Usinger. "It was portable."

As sausages spread, their flavors changed. "You might have a region that did not have a particular spice that was available in another region," she says. New tastes were incorporated as sausages traveled the globe. The bratwurst itself became a distinct member of the sausage family in Germany. In Nuremberg, a restaurant named Zum Goldenen Stern calls itself "the world's oldest bratwurst kitchen." Because the restaurant dates back to 1419, few dispute its claim.

In Germany, and especially Austria, a primary recipe developed for what came to be called the bratwurst: a mild, seasoned sausage consisting of pork, salt, pepper, nutmeg, mace, and other spices. Some were prepared with veal. Using rare cuts of meat was not the bratwurst's original reason for being. It was just the opposite.

"Sometimes less palatable parts were used," says Usinger. "Nobody wanted to waste anything because nobody could *afford* to waste anything." The parts were ground up. But how did they keep it all in one piece? Enter digestion—a pig or sheep's digestion, rather, and not your own. Sausages are contained in casings made from the cleansed parts of the animal intestines used for digestion, not for waste. This was an ideal solution because it was an otherwise unusable part of the animal.

In these modern times, you can find brats made from beef, chicken, turkey, spinach, tofu, and other ingredients. "People are very creative with newer trends," says Usinger. "That's what I like about sausage. You can be creative in so many different ways."

As for brat preparation, before grilling them Usinger recommends bringing water to a boil. Put in the brats, and turn off the heat. Keep the pot covered. Continued boiling may burst the casing, shriveling the brats; the meat will be exposed, and the brats will overcook. On the other hand, her brother Fritz, the company's president, just plops them on the grill. But when it comes to boiling in beer, the Usinger siblings agree that it's a bad brat idea. Beer can impart a bitter taste, changing the spicing profile of the brat.

If you don't pre-cook your brats, you should ideally grill them about 25 to 30 minutes, turning often to keep the casing from splitting. Keeping the juices inside retains flavor. Don't cook brats like steak, searing the outside. When they're firm to the touch and reach about 180 degrees internal temperature, they're done.

Some people even use brats as ingredients in other dishes, slicing them for soups and salads, or removing the casings to use them as loose sausage. Some even make brat patties—sausage burgers!

If you happen to be in the area on Memorial Day, you can take part in the annual World's Largest Brat Fest in Madison. Billed as "the biggest picnic on the planet," the weekend features volunteers attempting to set new world records—by serving brats to raise funds for more than 70 groups and charities. Brat Fest started in 1982 as a customer appreciation day at Madison's Hilldale Shopping Center. Sentry Foods owner Tom Metcalfe set out one 19-inch grill, one table, and three chairs. Things have changed somewhat.

Festivities have moved to the Dane County Fairgrounds, at the Alliant Energy Center, and a 19-inch grill isn't big enough to serve the necessary 88 brats every minute of the four-day festival. The event is now sponsored by Johnsonville, which donates most of the brats and also provides a gargantuan grill: 53,000 pounds, 6 feet in diameter, 20 feet tall, and 65 feet long. It has the capacity to serve about 2,500 brat-lovers per hour.

* In 2004, the people of Campbellsport grilled and served a 48-foot-long bratwurst made from 25 pounds of pork. Topped with nearly 25 pounds of ketchup and mustard, more than a pound of onion, and four gallons of relish, it was cut into 160 pieces and sold to benefit the fire department.

Pride of the Lone Star Pitmasters

Barbecue has emerged as a culinary art form in Texas—as serious as religion and as controversial as politics. Anyone and everyone has an opinion about what it takes to make good barbecue and what it means to be called a legend.

* * * *

ASK ANY GROUP of typical Texans about how barbecue got started in the Lone Star State, and they'll answer back with as many different stories as someone can slice cuts of meat from a 1,600-pound steer.

Some historians cite that the Texas tradition of barbecuing meats began during the 1850s, when German and Czech immigrants first settled the "German Belt," a swath of land stretching from Houston all the way to the rolling Hill Country. As they did in their homeland, the immigrant butchers sold their fresh meat in a storefront market and cooked the less savory cuts in smokers out back.

The lesser-quality meat provided cheap eats to itinerant African American and Hispanic cotton pickers who took advantage of the bargain. No sooner had they purchased a hunk of smoked pork loin or sausage link than they set upon devouring it, directly out of the butcher paper it came wrapped in. For them, this was "barbecue," much to the astonishment of the butchers.

Alternate Theories

Others claim that African Americans imported barbecue-style cooking from the traditional American South. After Texas joined the Union as a slave state in 1845, Southern cotton planters eyed it as a place to buy cut-rate land. Of course, the wealthy plantation owners moving into the area brought their large slave families with them. After slavery was abolished in 1865, the freedman's knack for turning inferior cuts of meat into mouth-watering fare influenced backyard cooks far beyond the plantation.

Still others date the beginnings of Texas barbecue to the 1800s, when the Anglo cowboys and Mexican vaqueros of the Rio Grande Valley cooked game such as rabbit, squirrel, and venison in open pits out on the range. In the tradition of Mexican *barbacoa,* they wrapped the meat in agave leaves and buried it under coals, where it slowly cooked for hours. So tender was the final product that it was said to have reduced hardened range riders to tears.

A Pillar of Texas Cuisine

Whatever the origins, one thing is certain: Barbecue has emerged as a culinary art form in Texas—as serious as religion and as controversial as politics. Every Bubba and Bobby Joe has an opinion to offer.

Should it be beef, pork, chicken, or turkey? Brisket, ribs, or sausage? Sauce or no sauce? Serve it with white bread? What type of side orders should be served with it, if any? What type of wood should it be cooked over? How low and how slow should it be cooked? A traditional pit or a smoker box? Prepackaged rubs or homemade? Some barbecue aficionados will even argue over whether a fork and knife should be used . . . or if people should just go natural by eating with their hands.

Rules of the Road

Although there is no universal agreement on these and other aspects of Texas 'cue, there are a few accepted standards that have emerged. First and foremost is the way the meat is cooked, and that means always over real wood or coals. Texas pitmasters who hope to pass off electric or gas-fired barbecue had better watch their backs. In some areas hereabouts, they still ride people out of town on a rail . . .

Second, any true Texan knows that beef is the meat of choice. Brisket—the normally tough cut of meat that's taken from the lower chest of the cow—is the common commodity of most pitmasters of this region. Locally prepared sausage (from Elgin) is also a favorite, made with beef, including the tripe, with natural casings. Texas pit bosses cook up a fair amount of beef ribs, too—something a pit-master in Memphis or North Carolina wouldn't even consider (in those regions, pork rules).

Location, Location, Location

So, where can people find the best Texas barbecue joints? Sage advice suggests a drive straight to Lockhart, the nerve center of the state's barbecue scene. Located just 30 miles south of Austin off Highway 183, the town's quartet of BBQ restaurants serves more than 5,000 visitors each week and more than adequately answers the question, "Where's the beef?" In the fall of 2003, the Texas state senate even passed a resolution proclaiming Lockhart "The Barbecue Capital of Texas."

The town square should be a first stop, where Smitty's Market sets the standard for the stereotypical meat-market format. According to *Texas Monthly* magazine, Smitty's serves some of the best barbecue in Texas and, by the magazine's count, is at least in the top five barbecue restaurants in the state. Brisket, pork chops, and sausage are among the daily fare, with pork ribs on the weekend. There's a large dining area at one side of the store, and out back, visitors will find an area where pitmasters toil over massive stone pits, covered by metal lids. The pits are well seasoned, thick with years of smoke, as are the restaurant's walls.

If their appetites remain up to the task, culinary explorers should head over to Kreuz Market next, another legend in the making spun off by a Smitty's family member. For barbecue lovers, this is the Disneyland of smoked fare, a cavernous building that houses immense wood-fired smokers packed with every imaginable cut of meat. Learned pitmasters—armed with an impressive arsenal of butcher knives—slice the meat by the pound, putting on a show that is eclipsed only by the taste of the mouth-watering food. Here, forget the sauce and the utensils. It's a communal food fest, with diners elbowing up to each other on long tables outfitted with copious rolls of paper towels.

Those into a more intimate setting should seek out Black's Barbecue. Owned and operated by the Black family since 1932, it's billed as Texas's oldest major barbecue restaurant continuously owned by the same family. Here, they smoke meats over hardwood for hours to create a flavor that *Gourmet* magazine touted as "the best BBQ in the heart of Texas, and therefore the best on earth."

It's definitely a friendly, family atmosphere here, with knotty-pine paneling, gingham table cloths, and food that will make diners swear off the chain joints.

Chisholm Trail Lockhart Bar-B-Q & Hot Sausage completes the foursome of Lockhart classics, an up-and-comer that started operations in 1978. Here, diners will be happy to find the usual standards, including brisket, beef and pork ribs, pork chops, chicken, ham, and turkey. The fajitas and the sausage from the restaurant's own recipe are also worth a try. Unlike some of the other local barbecue eateries that are light on sides, Chisholm Trail lays it on thicker than any of the other contenders. To wit, they feature a large cafeteria-style hot food bar with everything one might want—including pinto beans, green beans, fried okra and squash. There's also a salad bar with traditional coleslaw, potato salad, and numerous special salads.

It's All Good

The truth of the matter is that those who make the trip to Lockhart and taste for themselves the stuff that barbecue legends are made of really won't care if Texas barbecue originated at the early butcher shops, from inventive slave cooks, or from vaqueros. The only thing on their minds will be the wonderful taste, the intoxicating smell, and scheduling the next opportunity to swing through town again to sample more real Texas 'cue: the pride of the Lone Star pitmasters.

* Early Texas explorer Álvar Núñez Cabeza de Vaca's last name means head of a cow in Spanish. Why? One of his ancestors left a cow's head at a mountain pass to guide Christian forces trying to attack the Moors in the 13th century.
* For more than 30 years, Lufkin has hosted the Southern Hushpuppy Cookoff. Past recipes have included such nontraditional ingredients as shrimp, crabmeat, onions, vegetables, chili, and even whiskey.
* Anyone who follows Tejano music at all has heard of San Antonio's Flaco Jiménez. His sub-genre, conjunto, has long been the music of the Tejano working class, and his five Grammys attest to his success.

New York City's Oldest Restaurants

Ten places to feast on New York history.

❋ ❋ ❋ ❋

BECAUSE 70 PERCENT OF restaurants in New York close or change hands after just five years, the ten restaurants noted below have performed a miraculous feat: Each opened for business prior to 1900. Savor the past along with a steak, a pastrami sandwich, a burger, or a pint at:

Fraunces Tavern® (1762): New Yorkers owe many debts to Samuel Fraunces, founder of The Queen's Head Tavern. Not only did the Fraunces kitchen establish a take-out service (to go to Gen. George Washington's camp during the Revolution), the place also provided a home to colonialists who wanted to knock back a pint while griping about the Stamp Act and other hot political issues of the day. In a particularly big moment, the Fraunces hosted Washington's postwar gala and farewell speech to officers of the Continental Army. Just as significant is the structure itself, the oldest in Manhattan. Today, diners enjoy a colonial setting and browse the Revolutionary War artifacts in the museum upstairs.

Bridge Café (1847): The space currently occupied by the Bridge Café was built in 1794 and has offered some type of food or drinking enterprise ever since. However, it wasn't officially designated as a drinking establishment until 1847. In 1879, the place got into some trouble when it was classed as a "disorderly house" (a brothel, that is). The Bridge took its current name in 1979.

Pete's Tavern (1864): Pete's Tavern has been in the same location since it opened, making it the longest-lived continuously operating bar and restaurant in the city. The booths haven't changed in 100 years, so lucky (and imaginative) diners can share a table with the spirit of O. Henry, who wrote *The Gift of the Magi* at the booth by the front door. The original bar spans 30 feet and survived Prohibition disguised as a flower shop.

Old Homestead Steakhouse (1868): The Old Homestead has seen many changes in its neighborhood since it first opened. The formerly grimy meatpacking district is now home to hipsters and even an elevated park that was constructed on an abandoned freight rail. Amidst all this upscale development, the giant cow tethered to the restaurant's marquee ("We're the King of Beef") makes Old Homestead hard to miss, as does the neon sign notifying the public of its status as New York's oldest steakhouse. This was one of the first American restaurants to offer Kobe beef, and because that apparently went over big, the Homestead is now comfortable offering a $41 hamburger.

Landmark Tavern (1868): Farther up the west side, the Landmark Tavern originally perched on the shores of the Hudson River (there was no such thing as 12th Avenue in 1868). Its Irish founders tended to the needs of the waterfront workers on the first floor, serving nickel beer on a bar carved from a single mahogany tree. Those original owners raised their kids and subsequent generations on the two upstairs floors until Prohibition, when the tavern closed for 30 minutes while the whiskey barrels were relocated to a speakeasy on the third floor. The speakeasy is long gone, but the Landmark's bar, stamped tin ceiling, floor tiles, and maybe even a Confederate soldier remain.

P. J. Clarke's (1884): Although the two-story building surrounded by high-rise towers on the northeast corner of Third Avenue and 55th Street has proudly stood in the neighborhood since 1864, the first restaurant opened on the site 20 years later. Serving celebrities (Frank Sinatra "owned" Table 20), businesspeople, and everyday folks alike, the checkered tablecloths give the bar and restaurant a homey feel. It hosted genius, too: Johnny Mercer supposedly worked out the melody for "One for My Baby (And One More for the Road)" while sitting at the bar. Clarke's is infamous (or beloved, depending on who you ask) for the gigantic urinals that dominate the men's room.

Keens Steakhouse (1885): This English chophouse is famous for its muttonchops and, in its early years, for the now-defunct Herald Square Theater District actors who hung out there. Women were not permitted inside until 1905, when actress Lillie Langtry sued for admittance. In keeping with British tradition, Keens allowed Pipe Club members to store their fragile clay pipes on its premises. The Club has had more than 90,000 members and now houses the world's largest collection of churchwarden pipes, as well as pipes that once belonged to Albert Einstein, John Pierpont (J. P.) Morgan, Babe Ruth, and Teddy Roosevelt.

Peter Luger Steakhouse (1887): The German steakhouse that was to become Peter Luger opened in Williamsburg before there was a Williamsburg Bridge. The wood-paneled décor of ledges lined with steins hasn't changed much since then, but the steaks have been voted the best in the city by the Zagat Survey for more than 25 years in a row. Cows are selected by the present owner's granddaughter, who personally attends the meat market, looks each beast in the eye, and gives a thumbs up or down.

Katz's Delicatessen (1888): When the Katz family opened its delicatessen after arriving from Russia, they had to compete with dozens of Jewish delis that crowded the teeming streets of the Lower East Side. It was a tough fight, but in the end, the Katz's special way of smoking and pickling won out. During World War II, the owners sent salamis to their sons serving in the army, launching what has been Katz's slogan ever since: "Send a Salami to Your Boy in the Army." According to local lore, the U.S. Army bombarded the enemy with nicely aged salamis, which were then so enjoyed by the besieged that they surrendered.

Old Town Bar and Restaurant (1892): Chock full of original fixtures, the Old Town Bar and Restaurant has the oldest functioning dumbwaiter system in New York City. Its marble and mahogany bar is 55 feet long and is backed by a beveled-edge mirror that fills more than 250 square feet. The Old Town's tin ceilings tower 18 feet above patrons' heads. The enormous urinals, however, weren't installed until 1910. The bar lived through Prohibition as a speakeasy protected by Tammany Hall, New York's powerful Democratic political machine.

Food Stylist

Q: People have hair stylists, but what's a food stylist?

A: My job description is one sentence long: I get food ready for its close-up. As you may have heard, people eat first with their eyes. In Japan, in fact, people consider food presentation an art form and will hesitate to eat something if it looks unattractive.

Q: Do you cook the food as well as style it?

A: Yep, everything—I have the training of a professional chef. My team shops for the ingredients, cooks the dish, and then makes sure it looks as it should on film. And filming could happen at any time, so we often have to perform a lot of maintenance to keep the dish looking fresh. When photographers have to shoot a banquet scene, I know I've done my job well if I've prepared a table full of food that is six hours old and too cold to eat—and it's still irresistible.

Q: Talk about your tool kit. What equipment does it take to style food?

A: In addition to standard kitchen utensils, I use art supplies such as palette knives and brushes, even beauty supplies such as eyebrow tweezers and bobby pins. A lot of food styling involves sculpture. I use glue, oil, cotton swabs, paper towels, and a lot of other inedible materials to create an "exhibit" that looks like a pile of food.

Q: Can people eat your styled food?

A: Some of it, some of the time—mainly when I first prepare it. Since vegetables need to look fresh, I will blanch them rather than cook them. I doubt you'd want to eat my ice cream, which is a mixture of shortening, corn syrup, and powdered sugar that doesn't melt under the lights. My clear acrylic ice cubes don't melt, but they also won't chill your drink. If I were you, I would stay away from the pancakes, because I pour motor oil on them to simulate syrup.

Sushi: The Hallmark of Japan's Fast-food Nation

While the combination of fish and rice has been a mainstay in Asian cuisine for millennia, sushi as we know it today began with an entrepreneurial mind, lots of street carts, and a little bit of sumo wrestling.

⁂ ⁂ ⁂ ⁂

Complexities in Fish Fermentation

MODERN SUSHI CONSUMERS are accustomed to sushi of the fast-food variety—throw rice on seaweed, add a strip of fish, and presto, you have a snack that is both nutritious and delicious. Yet the earliest type of sushi, known as *Narezushi*, took over six months to prepare and was so smelly that it was eventually replaced by the stink-free sushi that we know today.

The sushi prototype actually originated in China and was then perfected in Japan. Before the days of refrigeration, innovations in fish preservation abounded. One popular method was to press fish between layers of salt for months at a time. At some point, it was discovered that fish would ferment faster if it was rolled in rice *and* pickled with salt. After the fermentation was complete, the rice was discarded, and the fish was eaten alone.

This method of fish preservation was popular in China and Southeast Asia, but it was only in Japan that the process eventually evolved into the snack-size morsels of sushi. The original Narezushi was created by a complex process that involved salting and pickling fish for more than a month, piling the pickled fish in between layers of cool rice, sealing everything into a barrel, then pickling some more. This drawn-out exercise created a sourness that was awful to smell but delectable to the taste buds.

It's difficult to popularize a food that's six months in the making, but this all changed with the invention of rice vinegar, which decreased fermentation time. Sushilike dishes were popular during Japan's Muromachi Period (1338–1573) and again during the Azuchi-Momoyama period (1574–1600). The first type of sushi to gain widespread popularity was the Oshizushi of Osaka, which was rectangular in shape and consisted of alternating layers of rice, fish, and sometimes pickled vegetables. Yet even this popular sushi, quite literally, stank.

Sushi Served Up Fast

The process of fermentation had to be abandoned completely before sushi could come out smelling like roses—or at least not like old fish. It wasn't until the rise of a large urban city, where fish could be caught and consumed in plenty within a 24-hour time frame, that modern sushi became possible. In late 18th-century Edo (Tokyo), it became common to place a strip of raw fish on a mound of rice. This is called *nigiri-zushi*, or hand-made, sushi.

The popularization of nigiri is usually attributed to early 19th-century entrepreneur and chef Yohei Hanaya. Legend has it that he was throwing a dinner party in 1824 when he realized he didn't have enough fish to go around. As a solution, he placed small slabs of fish on large mounds of rice. The entrepreneur in Hanaya realized he'd just discovered a gold mine.

Hanaya transformed his off-the-cuff innovation into a fast-food phenomenon. He sold his nigiri sushi in carts throughout the streets of Edo, and soon others opened sushi carts of their own. Hanaya opened his first cart in front of Ryukoku Temple, where frequent sumo wrestling tournaments created an ongoing glut of pedestrians. The idea was that sushi is a finger food that can quickly be prepared and eaten, right on the street. The fish in these early sushi were often cooked, marinated, or heavily salted, so dipping in soy sauce was not necessary. Sushi went from food carts to restaurants throughout Edo to restaurants throughout Japan and eventually gained its current status as a worldwide food favorite.

The word *sushi* actually refers to the pickled or vinegared rice, not the fish itself. *Sashimi* refers to the raw fish.

11 Facts About Pizza

Since 1987, October has been officially designated National Pizza Month in the United States.

⁕ ⁕ ⁕ ⁕

1. Approximately three billion pizzas are sold in the United States every year, plus an additional one billion frozen pizzas.
2. Pizza is a $30 billion industry in the United States.
3. Pizzerias represent 17 percent of all U.S. restaurants.
4. 93 percent of Americans eat pizza at least once a month.
5. Women are twice as likely as men to order vegetarian toppings on their pizza.
6. About 36 percent of all pizzas contain pepperoni, making it the most popular topping in the United States.
7. The first known pizzeria, Antica Pizzeria, opened in Naples, Italy, in 1738.
8. More pizza is consumed during the week of the Super Bowl than any other time of the year.
9. On average, each person in the United States eats around 23 pounds of pizza every year.
10. The first pizzeria in the United States was opened by Gennaro Lombardi in 1895 in New York City.
11. The record for the world's largest pizza depends on how you slice it. According to Guinness World Records, the

record for the world's largest circular pizza was set at Norwood Hypermarket in South Africa in 1990. The gigantic pie measured 122 feet 8 inches across, weighed 26,883 pounds, and contained 9,920 pounds of flour, 3,968 pounds of cheese, and 1,984 pounds of sauce. In 2005, the record for the world's largest rectangular pizza was set in Iowa Falls, Iowa. Pizza restaurant owner Bill Bahr and a team of 200 helpers created the 129-foot by 98.6-foot pizza from 4,000 pounds of cheese, 700 pounds of sauce and 9,500 sections of crust. The enormous pie was enough to feed the town's 5,200 residents ten slices each.

Smoking Bishop

Don't be dumbfounded over this Dickensian drink.

* * * *

MOST EVERYONE IS familiar with Charles Dickens's *A Christmas Carol.* Ebenezer Scrooge bah-humbugs his way through the holiday until a quartet of ghosts appear in his bedroom. Near the end of the story, the now-reformed Scrooge happily cheers Bob Cratchit, "A merrier Christmas, Bob, my good fellow, than I have given you for many a year! I'll raise your salary, and endeavor to assist your struggling family, and we will discuss your affairs this very afternoon over a bowl of Smoking Bishop, Bob!" According to Dickens's great-grandson Cedric, the Victorians named several drinks after members of the clergy. Some common names include Pope for burgundy, Cardinal for champagne or rye, Archbishop for claret, and Bishop for port.

The Recipe (Bring Your Insulin)

The recipe isn't complicated. Bake five oranges and a grapefruit until brown. Push whole cloves into the fruit. Add a quarter pound of sugar and two bottles of red wine. Dickens says to leave this in a warm place for a day, but that's Victorian-era nonsense. You can get a fine result simmering the mixture on a stove on low heat for an hour. Squeeze the juice out of the oranges and grapefruit, then add the port. Again, Dickens says not to boil it, but that's more lack of understanding than anything else. Presumably the fear is that all the alcohol will boil off, but the truth is it takes hours for that to happen. Bring to a boil and then reduce to low heat. Serve warm. By serving warm, the drink is said to be "smoking," hence Smoking Bishop. Although Dickens makes no mention of it, you can add cinnamon sticks and/or brown sugar (or other seasonal spices such as nutmeg) to taste. If you're accustomed to making mulled wine, you can pick and choose spices and juices to add. Your taste buds will guide you. And don't forget to raise a glass to Tiny Tim's health!

Tamale Time!

In Texas, it wouldn't be Christmas without delicious tamales!

✻ ✻ ✻ ✻

A VARIETY OF DIFFERENT foods have come to be associated with Christmas, including fruitcake, ham, eggnog, and that proverbial favorite, figgy pudding. But tamales? Absolutely! Especially if you're a Texan. For generations, dining on Christmas tamales has been a popular tradition in the Lone Star State and its southwestern neighbors. Some people hand-prepare the zesty Latin American dish, while others special-order them from neighborhood shops. The meal is typically eaten on Christmas Eve. Tamales are made using a starchy, corn-based dough called masa, which is filled with seasoned meat, vegetables, chilies, or cheese, then wrapped in cornhusks and steamed. (The cornhusks are discarded before eating.) Tamales are fairly time-consuming to prepare, which is why some folklorists believe tamales have come to be associated specifically with Christmas—it's the one special occasion most people are willing to put in the extra effort. As a result, Christmas tamales have become a fond family custom for many.

It allows relatives to gather, visit, and catch up while everyone participates in the creation of the flavorful feast.

An Ancient Dish

According to historians, tamales can be traced as far back as 8000 B.C. They were a staple among the ancient Aztecs and were as popular then as peanut butter sandwiches are today. Spanish explorers discovered tamales upon their arrival in Mexico, and they quickly took to the yummy dish, bringing it with them to all of their colonies. It was an ideal food because tamales transported well. So this Christmas, add a spicy bit of ancient history to your holiday meal by making your own tamales. (You can find a variety of easy-to-follow recipes on the Internet.) If it sounds like too much work, any Mexican restaurant will be happy to oblige you.

Setting Sake Straight

Most Americans consider sake a Japanese rice wine, but it is actually more akin to beer. Furthermore, a look back in time suggests that sake may have originated in China, not Japan.

* * * *

What Is Sake?

THE JAPANESE WORD for sake, *nihonshu*, literally means "Japanese alcoholic beverage" and does not necessarily refer to the specific rice-based beverage that foreigners exclusively call *sake*. What differentiates sake from other alcoholic beverages is its unique fermentation process. Although all wines are the result of a single-step fermentation of plant juices, sake requires a multiple-step fermentation process, as does beer. The requisite ingredients are rice, water, yeast, and an additional substance that will convert the starch in the rice to sugar. People have always found ways to make alcohol with whatever ingredients are available, so it is likely that beverages similar to sake emerged soon after rice cultivation began. The most popular theory holds that the brewing of rice into alcohol began around 4000 B.C. along the Yangtze River in China, and the process was later exported to Japan.

The Many Ways to Ferment Rice

The sake of yore was different from the sake that's popular today. At one time it was fermented with human saliva, which reliably converts starch to sugar. Early sake devotees chewed a combination of rice, chestnuts, millet, and acorns, then spit the mixture into a container to ferment. This "chew and spit" approach to alcohol production has been seen the world over in tribal societies. Subsequent discoveries and technological developments allowed for more innovative approaches to fermentation. Sometime in the early centuries A.D., a type of mold called koji-kin was discovered to be efficient in fermenting rice. In the 1300s, mass sake production began in Japan, and it soon became the most popular national beverage.

Freaky Foods from Around the World

People around the globe eat many things that seem bizarre and distasteful to American palates. And we're not just talking about snails, raw herring, and spices so hot they nearly melt your face. These foods are downright strange!

❋ ❋ ❋ ❋

Balut

A FAVORITE IN SOME parts of Southeast Asia, balut consists of a fertilized duck egg that's been steamed at the brink of hatching. Diners, who usually pick up one from a street vendor on the way home from a night of imbibing, chow down on the entire contents of the egg: the head, beak, feet, and innards. Only the eggshell is discarded.

Bird's Nest Soup

The key ingredient of this famous Chinese dish is the saliva-rich nest of the cave swiftlet, a swallow that lives on cave walls in Southeast Asia. This delicacy is so popular that it has endangered the bird's population. Some enterprising suppliers have started farming nests by providing birds with houses in which to build. Still, wild nests remain the most highly prized.

Century Eggs

When Sam I Am was expressing his disgust for green eggs in Dr. Seuss's classic book, he was probably talking about the Chinese specialty known as century eggs, also called thousand year eggs or preserved eggs. To make the dish, chicken, duck, or goose eggs are preserved in a mixture of salt, lime, clay, ash, and rice husks or tea leaves, then allowed to ferment for several weeks or months. The process turns the egg whites into gelatinous, transparent dark-brown masses, while the yolks become pungent yet intensely flavored orbs of variegated shades of green. What's not to like?

Cobra's Blood Cocktails

Those looking for an exotic drink in Indonesia can quench their thirst with a fresh cocktail of cobra's blood, served either straight or mixed with liquor. Depending on the establishment, consumers can choose between different types of cobras to supply the blood, with the King Cobra being the most prized and expensive.

Fugu/Pufferfish

Diners must have the utmost confidence in the chef preparing this Japanese dish because eating improperly prepared pufferfish can result in sudden death. As such, Japan requires extensive training and apprenticeship, as well as special licensing, for the chefs preparing this highly sought-after delicacy.

Kopi Luwak Coffee

Java junkies may need to save up to indulge in this exotic Indonesian coffee. What makes it so special? For kopi luwak coffee, only the sweetest, tastiest coffee beans are picked by experts—not human experts, but rather wild civets who live in Indonesia's coffee-producing forests. The animals nibble on the fruity exterior before swallowing the hard inner beans. During the digestive process, their gastric juices remove some of the proteins that ordinarily make coffee bitter. Humans don't intervene in the coffee-making process until it's time to separate the undigested beans from the civet dung. Experts claim the result is an exceptionally smooth and balanced cup of Joe.

Seal Flipper Pie

A specialty of Newfoundland, this maritime favorite is made from the chewy cartilage-rich flippers of seals, usually cooked in fatback with root vegetables and sealed in a flaky pastry crust or topped with dumplings.

The Good Old Days of Soda Fountains

In an age when drive-through coffee shops are serving up iced mochachinos, it's easy to forget that chrome-topped soda fountains once held a place of distinction in American culture.

* * * *

The Golden Age

IN 1819, THE first soda fountain patent was granted to Samuel Fahnestock. This nifty invention combined syrup and water with carbon dioxide to make fizzy drinks—and they instantly caught on.

The first soda fountains were installed in drugstores, which were sterile storefronts originally intended only to dispense medicines. To attract more business, pharmacists started to sell a variety of goods, including drinks and light lunch fare. That way, customers could come in to shop, take time out for some refreshment, and possibly do extra shopping before they left.

Typical soda fountains (the name for both the invention and the shops where the fountains could be found) featured long countertops, swivel stools, goose-neck spigots, and a mirrored back bar, all of which helped attract the attention of young and old alike. Soda fountains were also installed in candy shops and ice-cream parlors. Before long, freestanding soda fountains were being built across the country.

Two of the world's most popular beverages got their start at soda fountains. In 1886, Coca-Cola was first sold to the public at the soda fountain in a pharmacy in Georgia. Pepsi's creator, Caleb Bradham, was a pharmacist who started to sell his beverage in his own drugstore in 1898.

Soda fountain drinks had to be made to order, and this was typically done by male clerks in crisp white coats. Affectionately referred to as "soda jerks" (for the jerking motion required to draw soda from the spigots), these popular, entertaining mixologists were the rock stars of the early 1900s. Think of a modern-day bartender juggling bottles of liquor to make a drink: Soda jerks performed roughly the same feats, except that they used ice cream and soda.

Birth of the Brooklyn Egg Cream

In Brooklyn, New York, candy shop owner Louis Auster created the egg cream, a fountain drink concoction that actually contained neither eggs nor cream.

You make an egg cream any way you like it, but a basic recipe combines a good pour of chocolate syrup with twice as much whole milk, along with seltzer water to fill the glass. (In New York, an egg cream isn't considered authentic unless it's made with Fox's "U-Bet" chocolate syrup.)

The foam that rises to the top of the glass appears similar to egg whites, which may be how the drink got its name. Some claim that the original chocolate syrup contained eggs and cream; others say "egg cream" comes from the Yiddish phrase *ekt keem*, meaning "pure sweetness"; still others believe that when little kids ordered "a cream" at the counter, it sounded like they were saying "egg cream." Whatever the etymology, the drink is legendary among soda fountain aficionados. Auster claimed that he often sold more than 3,000 egg creams a day. With limited seating, this meant that most customers had to stand to drink them, prompting the traditional belief that if you really want to enjoy an egg cream, you have to do so standing up.

Several beverage companies approached Auster to purchase the rights to the drink and bottle it for mass distribution, but trying to bottle an egg cream was harder than they thought: The milk spoiled quickly, and preservatives ruined the taste. Thus, the egg cream remained a soda fountain exclusive.

Sip and Socialize

Prohibition and the temperance movement gave soda fountains a boost of popularity during the 1920s, serving as a stand-in for pubs. Booze became legal again in 1933, but by that time, fountains had become such a part of Americana that few closed shop. During the 1950s, soda fountains became the hangout of choice for teenagers everywhere.

It wasn't until the 1960s that the soda fountain's popularity began to wane. People were more interested in war protests and puka beads than Brown Cows and lemon-lime-flavored Green Rivers. As more beverages were available in cans and bottles and life became increasingly fast-paced, people no longer had time for the leisurely pace of the soda shop.

Some fountains survived and still serve frothy egg creams to customers on swivel stools, and many of these establishments at-tempt to appeal to a wide audience by re-creating that old-fashioned atmosphere.

Recipes

Brown Cow

4 scoops ice cream (chocolate for a Brown Cow, vanilla for a White Cow)
4 tablespoons flavored syrup (usually chocolate)
$1^1/_2$ cups milk
Whirl in a blender until smooth. Share, or not.

Green River

3 ounces lemon-lime syrup
10 ounces seltzer water
Stir. Add ice, if desired.

Hoboken

$^1/_2$ cup pineapple syrup
A splash of milk
Seltzer water
Chocolate ice cream
Blend and enjoy!

Catawba Flip

1 scoop vanilla ice cream
1 large egg
2 ounces grape juice
Shaved ice
Seltzer water
Blend first four ingredients until smooth.
Pour into a tall glass and fill with seltzer water.

Junk Food All-Stars

Master showman P. T. Barnum may have inspired more creativity in other people than any other entrepreneur in the past century. One item still keeps bakery ovens hot, not because of the thing itself, but because of its packaging: animal cracker boxes.

✻ ✻ ✻ ✻

Barnum's Biscuits

THESE TINY, TASTY biscuits were around long before Barnum arrived on the scene. However, his traveling circuses stirred a pioneering success in packaging small boxes colorfully illustrated as animal circus wagons —large enough to hold a good portion of cookies and small enough to use as an ornament or toy.

A Crackerjack Idea

Pre-Columbian peoples likely discovered popcorn when someone overcooked corn kernels. Confectioners mixing popcorn with molasses or melted marshmallows produced popcorn balls. However, Cracker Jack was the snack that stormed the North American market in 1900. Movie-house buttered popcorn came much later.

Cracker Jack, featuring sweet-meets-salty caramel-coated popcorn, was a wild success—even beyond the tiny free prizes found in every box. First, the creators, Fred and Louis Rueckheim, figured out a way to prevent all the components from sticking together in a big gooey ball. They went for bite-size pieces, and then added peanuts. Their real coup over the competition came when they packaged it all in a sealed box that preserved their product.

Of course, their crackerjack of a name helped too. The branding was so successful that Cracker Jack was listed in the Sears Catalogue without even a line of product description.

Candy Bars That No Longer Exist

The first candy bar was manufactured in Norway in 1906. Within a few decades, candy lovers on both sides of the Atlantic had more than 5,000 different bars to choose from. Some, such as Baby Ruth (introduced in 1921) and Mr. Goodbar (1925), are still around. Others have not been so fortunate.

* * * *

- The Seven Up Bar, manufactured by the Pearson's Candy Company of Minnesota from 1951 to 1979, had seven individual sections, each with a different creamy filling covered in milk or dark chocolate.
- Chicken Dinner certainly wins the award for the candy bar with the strangest name. A chocolate-covered caramel peanut roll, it was made by the Sperry Company of Milwaukee from around 1920 until the company was sold in 1962. Sperry also produced a bar named Cold Turkey, but the ingredients are, alas, unknown.
- Powerhouse, a quarter-pound bar of peanuts, caramel, and fudge, even had its own comic strip superhero, Roger Wilco, who offered kids prizes in return for a wrapper and 15¢. He always reminded his fans that "Candy is delicious food." The bar, originally manufactured by Walter Johnson Candy, sold from the 1920s until 1988.
- The Marathon Bar introduced by the Mars Company in 1974 also had its own cartoon spokesmen—Marathon Mike and the Pirates. The eight-inch-long braided roll of caramel-covered chocolate was discontinued in 1981.
- Nestle's Triple-Decker—three layers of dark, milk, and white chocolate in one bar—sold from the 1940s to the 1970s. The company reportedly stopped making it because the cost was too high.

* Last but not least, the Good News bar was saved in the nick of time by loyal fans in Hawaii. Introduced in the late 1930s, this combination of chocolate, peanuts, and caramel really caught on in the islands. When the company tried to discontinue it, the natives rose up in protest. Thanks to them, it is still sold today, but only in Hawaii.

Inspiration Station

The chocolate chip cookie: accidental inspiration

* * * *

OKAY, MAYBE IT'S a stretch to call a cookie a "work of art," but there are millions of people who would argue that a great chocolate chip cookie gives any masterpiece a run for its money. The inspiration behind the beloved sweet was born out of necessity. Ruth Graves Wakefield was a dietician and co-owner of a Massachusetts tourist lodge in the 1930s along with her husband. Their lodge was dubbed the Tollhouse Inn, and it quickly gained a reputation for serving fantastic homemade desserts.

One day, Wakefield was making a batch of chocolate cookies when she realized she was out of her usual baker's chocolate. She substituted chunks of semi-sweet chocolate instead; rather than melting into the cookie, the chocolate stayed in chunks after baking. Lo, the chocolate chip cookie was born. Handily, the chocolate she used had been a gift from her friend Andrew Nestlé. Nestlé's chocolate manufacturing company bought Wakefield's recipe and supplied her with a lifetime supply of chocolate chips. The Tollhouse recipe can still be found on the back of every bag of Nestlé chocolate morsels.

Cookie Monster: Proust's Madeleine

Leave it to Marcel Proust to make a mountain out of a molehill—or a seven-book, six-volume, 4,300-page story out of a cookie. *In Search of Lost Time* is a loosely autobiographical account of Proust's life that took him 13 years to write. The story begins when the protagonist bites into a madeleine, a pound-cake-like cookie in the shape of a seashell that was popular in France. The taste of the madeleine dipped in a cup of tea instantly sends the author back to the halcyon days of his childhood. From there, it's a bonanza of time, space, memory, and decidedly Proustian motifs—all from an afternoon snack.

The Gift That Keeps On Giving

We decided to find out how much beef jerky you can get out of a cow. Our results are certain to delight jerky lovers.

* * * *

The Wonders of Jerky

THERE'S A COMMON misconception that men would starve to death if they were left to their own devices. This is patently untrue—as long as there's a convenience store nearby that sells beef jerky. Indeed, a beer, a baseball game, and a bag of beef jerky might make for the perfect afternoon for the average male. A serious beef jerky habit can get pretty expensive, though—even the cheap-o gas-station variety costs a few dollars for just a couple of ounces of dried, peppery goodness. It makes you wonder whether it wouldn't be easier just to cut out the middleman and go straight to the cattle auction.

As it turns out, you can get quite a bit of beef jerky from a single cow, but determining just how much requires a little agricultural mathematics. A large "beef animal" can weigh over 1,200 pounds—this counts everything, including the guts, bones, and other inedible material. Before being turned into jerky, the cow needs to be butchered and trimmed into lean cuts (the best jerky is made from boneless steaks and roasts).

Pull Out Your Calculator

According to the agricultural school at South Dakota State University, the yield of an average cow that's butchered for lean beef is about 38 percent of its original weight—which means that a twelve hundred pound feeder will give you about 456 pounds of steaks, roasts, and ground beef with which to work. Next, the high-quality meat is cured and seasoned before being sliced into strips and dried. Meanwhile, the ground beef can be made into lower-quality jerky, which is labeled as "ground and formed" on the packaging. During the drying process, the meat loses up to three-quarters of its original weight; however, this still leaves us with a very generous portion of beef jerky from a single cow: roughly 115 pounds. Considering that an average bag of beef jerky weighs 1.8 ounces, we're talking more than 1,000 bags of beef jerky from a single cow.

Of course, this is only a hypothetical situation—in the real world, the fattier cuts of meat are almost never used for jerky, because they're more difficult to dry and are (in their jerky form) much more perishable. Besides, as delicious as jerky is, steakhouse patrons know that there are other, more satisfying uses for the loin, tenderloin, and rib-eye cuts.

At any rate, when you consider the massive amount of work that's involved in butchering, cleaning, curing, slicing, and dehydrating a cow, it's probably better to leave it to the professionals. Hey, we've got baseball to watch.

That's Offal! 7 Dishes Made From Animal Guts

For anyone raised on the standard American hamburger, the idea of eating the brains, guts, or feet of a cow may seem gross and totally out of the question. But in many parts of the world, those are the parts of the animal that are considered delicacies. The next time you make a trip to the butcher, ask for offal (the entrails and intestine) and rustle up something a little different for dinner.

* * * *

1. Kokoretsi

In Greece, it's all about lamb. Skewered, baked, roasted—if it's lamb, the Greeks are cooking it. But these folks would never waste good meat: Kokoretsi is a traditional Balkan dish often served for Easter that includes lamb intestine, heart, lung, and kidneys, or a combination of any of the above. Chunks of these organs are speared onto a skewer and then wrapped up in the small intestine, which forms a kind of sausage casing. The spear is set over a fire and sprinkled with oregano and lemon juice. In a few hours, *opah!* Grilled guts for everyone!

2. Haggis

On paper, it just doesn't look appetizing: mix sheep's heart, liver, and lung meat with oatmeal and fat, and stuff the mixture into its own stomach; boil for three hours and enjoy. Still, that's exactly what haggis is, and the Scottish have been enjoying this dish for centuries. Traditionally served in a sauce with turnips and potatoes, this dish is used also in that popular national pastime, "haggis throwing."

3. Tripe

Is your stomach making those growly hungry noises? Try some stomach! Served in many countries across the globe, the stomach, known as tripe, is the main ingredient in many regional dishes. Beef tripe is the most commonly used stomach, but sheep, goat, and pig stomachs are often on the menu as well. Tripe is used often in soups and in French sausages, fried up in Filipino dishes, and used as a relish in Zimbabwe. In Ireland and Northern England, tripe is simply served up with onions and a stiff drink.

4. Khash

If you're a cow, you'd better watch your step. Folks in Armenia are crazy about khash, a dish primarily made from cow's feet. First, the hooves are removed. Next, the feet are cleaned and cooked in plain boiling water overnight. By morning, the mixture is a thick broth and the meat has separated from the bone. Brain and stomach bits can be added for extra flavor. Armenians are careful about when they serve this favorite dish—it reportedly has strong healing properties—and it is usually served only on important holidays. Peppers, pickled veggies, and cheese go well with kash, but the favorite accompaniment is homemade vodka.

5. Yakitori

Who doesn't like a juicy grilled chicken skewer? Look closely if you purchase one from a Japanese street vendor, however. Yakitori are chicken skewers that contain more parts of the chicken than you may care to taste, including the heart, liver, gizzard, skin, tail, small intestine, tongue, and wing.

6. Rocky Mountain Oysters

There's not a lot on a farm animal you *can't* eat, as evidenced by Rocky Mountain Oysters, or bull testicles. Once the testicles of the bull (or lamb or buffalo) are removed, they're peeled, dipped in flour, and deep-fried to a golden crunch. This dish is commonly found in the American West, where bulls are prevalent, and also in bull-populated Spain.

7. White Pudding

To make white pudding (and if you live in Scotland, Ireland, Nova Scotia, or Iceland, you might), you'll first need a big bowl of suet, or pork fat. It's the main ingredient in this dish, which also includes meat and oatmeal, and, in some earlier recipes, sheep brains. The pudding is similar to blood pudding, but without the blood. Sometimes formed into a sausage shape, white pudding can be cooked whole, fried, or battered and served in place of fish with chips.

A Feast for the Fearless: The World's Most Revolting Foods

Turning our attention briefly to some of the most curious delicacies of the world, from maggot-infested cheese to mouse-flavored liquor, we present the following culinary adventures.

❋ ❋ ❋ ❋

Baby Mice Wine

THOSE BRAVE MEN and women who enjoy eating the worm from the bottom of a tequila bottle and want to advance to spirit-soaked vertebrates might be interested in baby mice wine, which is made by preserving newborn mice in a bottle of rice wine. This traditional health tonic from Korea and China is said to aid the rejuvenation of one's vital organs. Anecdotal evidence, however, suggests that the sight of dead baby mice floating helplessly in liquor is more likely to break your heart than rejuvenate it.

Casu Marzu

The Sardinian delicacy *casu marzu* is a hard sheep's milk cheese infested with *Piophila casei*, the "cheese fly." The larvae eat the cheese and release an enzyme that triggers a fermentation process, causing their abode to putrefy. The cheese is not considered true casu marzu until it becomes a caustic, viscous gluey mass that burns your mouth and wriggles on your tongue when you eat it. *Nota bene:* The cheese fly is also called "the cheese skipper," because its larvae have the amazing ability to leap up to six inches in the air when disturbed. Since the larvae rightfully consider it disturbing to be eaten, it is suggested that consumers of casu marzu make use of protective eye gear during the repast.

Cobra Heart

This Vietnamese delicacy delivers precisely what it promises: a beating cobra heart, sometimes accompanied by a cobra kidney and chased by a slug of cobra blood. Preparations involve a large blade and a live cobra. If you find yourself in the uncomfortable situation where the snake has already been served but you feel your courage failing, ask for a glass of rice wine and drop the heart into it. Bottoms up!

Escamoles

Escamoles are the eggs, or larvae, of the giant venomous black *Liometopum* ant. This savory Mexican chow, which supposedly has the consistency of cottage cheese and a surprisingly buttery and nutty flavor, can be found both in rural markets and in multi-star restaurants in Mexico City. A popular way to eat escamoles is in a taco with a dollop of guacamole, but it is said that they are also quite delicious fried with black butter or with onions and garlic.

Hákarl

Hákarl, an Icelandic dish dating back to the Vikings, is putrefied shark meat. Traditionally, it has been prepared by burying a side of shark in gravel for three months or more; nowadays, it might be boiled in several changes of water or soaked in a large vat filled with brine and then cured in the open air for two months. This is done to purge the shark meat of urine and trimethylamine oxide. Sharks have an extra concentration of both to maintain essential body fluid levels, but the combination makes the meat toxic. Since rancid shark meat is not considered all that tasty, native wisdom prescribes washing it down with a hearty dose of liquor.

Lutefisk

If the idea of rotten shark meat does not appeal, consider lutefisk, or "lye fish"—possibly the furthest from rotten that food can get. This traditional Scandinavian dish is made by steeping pieces of cod in lye solution. The result is translucent and gelatinous, stinks to the high heavens, and corrodes metal kitchenware. Enjoy it covered with pork drippings, white sauce, or melted butter, with potatoes and Norwegian flatbread on the side. (As a side note: The annual lutefisk-eating contest in Madison, Minnesota, is scheduled right before an event called the Outhouse Race. This might not be entirely a coincidence.)

Pacha

This dish can be found everywhere sheep can be found, especially in the Middle East. To put it simply, pacha, which is the Iraqi name for it, is a sheep's head stewed, boiled, or otherwise slow-cooked for five to six hours together with the sheep's intestines, stomach, and feet. Other meats might also be added to the broth. Something to keep in mind: If you are served this dish in Turkmenistan, where it is called *kelle-bashayak*, this means two things—one, you're the guest of honor at the gathering, and two, you will be expected to help consume the head or else risk offending the hosts.

Sweet Lemons? It's a Miracle!

"When life gives you lemons, make lemonade." A great maxim made even easier if you have a supply of miracle fruit.

* * * *

Change Your Taste

MIRACLE FRUIT (*SYNSEPALUM dulcificum*) is a small berry that can literally change the way common foods taste. After eating a miracle berry, stout beer tastes like a chocolate milkshake and cheese tastes like cake frosting. It can make hot sauce taste like a glazed donut, vinegar taste like apple juice, and oysters taste like chewing gum.

Miracle fruit is indigenous to West Africa and grows on trees that can reach 20 feet high. Although the trees produce crops only twice a year, the berries can be freeze dried or refrigerated indefinitely. The fruit can also be ordered online in tablet or granulated form. After eating one small berry, its glycoproteins bind to the tongue's taste buds, producing miraculin, which makes bitter and sour foods taste sweet. The effects last from 30 minutes to 2 hours.

Miracle fruit first gained popularity in the United States during the 1970s when it was marketed as a dieting aid, so that people could enjoy low-calorie menus without feeling deprived. So why aren't we all eating the fruit? Allegedly, the Food and Drug Administration folded to pressure exerted by the sugar industry and stopped allowing its import.

Miracle Fruit for the Masses

If you would like to have your own miracle fruit, you can order berries or seeds from vendors online. There are five steps to enjoying miracle fruit:

1. Buy a selection of foods like citrus fruits, rhubarb, bleu cheese, stout beer, and cheap tequila.
2. Wash the miracle fruit and put a berry into your mouth. Swirl it around for about a minute.
3. Bite into the berry, liberally coating your tongue with the juice.
4. Taste a lemon wedge—it should taste like sweetened lemonade.
5. Proceed with the other foods, throwing caution to the wind.

12 Items That Would Need a Name Change to Sell in America

The meanings of many foreign words get lost in translation when converted into English. But the English names of the following products would probably benefit from a name change if they want to be successful in the United States.

1. **Cream Collon:** Glico's Cream Collon is a tasty cookie from Japan. The small cylindrical wafers wrapped around a creamy center actually do resemble a cross section of a lower intestine filled with cream. An ad says to "Hold them between your lips, suck gently, and out pops the filling." Yum! Glico's Cream Collon can be ordered on the Internet.

2. **Ass Glue:** Ass glue is made from fried donkey skin and is considered a powerful tonic by Chinese herbalists, who use it to fortify the body after illness, injury, or surgery. If you have a dry cough, a dry mouth, or are irritable, you can find ass glue at most Chinese herb shops.

3. **Mini-Dickmann's:** Mini-Dickmann's, a German candy made by Storck, is described as a "chocolate foam kiss." Available in milk, plain, or white chocolate, Mini-Dickmann's are only an inch and a half long. Too embarrassed to be seen with a box of Mini-Dickmann's? Try Super-Dickmann's, the four-inch variety. Both sizes are available from Storck USA.

4. **Kockens Anis:** If you think anis sounds funny, you'll laugh even harder when you see Kockens Anis in Swedish grocery stores. Anis is aniseed, a fragrant spice used in baking, and Kockens is the brand name. While aniseed is found in most U.S. grocery stores, don't ask for the Kockens brand because it's not available and could get you coldcocked.

5. **Aass Fatøl:** On those rare hot days, Norwegians like to quench their thirst with a cold bottle of Aass Fatøl beer. The word fatøl appears on many Scandinavian beer labels and means "cask." This beer comes from the Aass Brewery, the oldest brewery in Norway. If you'd like to get some Aass, it's imported in the United States.
6. **Big Nuts:** Big Nuts is a chocolate-covered hazelnut candy from the Meurisse candy company in Belgium. For those who like candy that makes a statement, Big Nuts is available online.
7. **Dickmilch:** Dairy cases in Germany are the place to find dickmilch, a traditional beverage made by Schwalbchen. In German, dickmilch means "thick milk" and is made by keeping milk at room temperature until it thickens and sours. Called sour milk in the United States, it's a common ingredient in German and Amish baked goods.
8. **Pee Cola:** If you're asked to take a cola taste test in Ghana, one of the selections may be a local brand named Pee Cola. The drink was named after the country's biggest movie star Jagger Pee, star of Baboni: The Phobia Girl. Don't bother looking for a six-pack of Pee to chug because it's not available in the United States.
9. **Golden Gaytime:** Street's Golden Gaytime is toffee-flavored ice cream dipped in fine chocolate and crunchy cookie pieces and served on a stick. Also available in a cone, Golden Gaytime is one of many Street's ice cream treats sold in Australia but not in the United States. One memorable advertising slogan remarks: "It's so hard to have a Gaytime on your own."
10. **Piddle in the Hole:** Take a Piddle in the Hole at a pub in England and you'll be drinking a beer from the Wyre Piddle Brewery. Made in the village of Wyre Piddle, the brewery also makes Piddle in the Wind, Piddle in the

Dark, and Piddle in the Snow. Before you run out for a Piddle, it's only available in the UK.

11. **Shito:** Shito is a spicy hot chili pepper condiment that, like ketchup in the United States and salsa in Mexico, is served with most everything in Ghana. There are two versions: a spicy oil made with dried chili pepper and dried shrimp; and a fresh version made from fresh chili pepper, onion, and tomato. Shito appears as an ingredient in Ghanaian recipes but hasn't found a market in the United States.

12. **Fart Juice:** While it may sound like an affliction caused by drinking it, Fart Juice is a potent potable in Poland. Made from the leftover liquid from cooking dried beans, this green beverage could pass for a vegetable juice and is probably a gas to drink, but it's not available in the United States.

Exotic Fruit Explosion

As U.S. consumers grow increasingly fond of "superfruits," the time is ripe for some exotic new flavors.

* * * *

PRIOR TO ABOUT 2005, few U.S. consumers had heard of "superfruit." Coined by marketers to describe fruit that is nutrient- and antioxidant-rich (as well as somewhat of a novelty), the term quickly gained widespread (perhaps even *super*) usage. Some of the first fruits marketed as ultra-healthy superfruits included blueberry, mango, and pomegranate, followed by açaí, mangosteen, and goji berry. Using these flavors, food and beverage companies have launched a variety of health products and trendy treats.

So what other flavors are likely to achieve superfruit status? Some of the following exotic fruits have already been showing up in energy drinks, beauty products, and fancy fortified water.

Camu-camu: This fun-to-say fruit is native to Peru and is very acidic in taste. Organically prepared camu-camu is said to contain more vitamin C than any other fruit on the planet. It is popular in Japan as a food and as an ingredient in beauty products. Camu-camu also is in energy drinks, juice, and dietary supplements.

Cherimoya: This heart-shape fruit from South America tastes like a tropical blend of pineapple, banana, and strawberry. Its healthy properties include vitamin C, calcium, and fiber, and it is particularly used in beverages, smoothies, or ice cream.

Lulo: Also known as "naranjilla," or "little orange," this is a small, tangy fruit hailing from Columbia, Ecuador, and Peru. Some say the green juice of lulo tastes like a combination of lime and rhubarb. The fruit contains vitamins B and C, calcium, and iron, and is popular in beverages and purees.

Rambutan: This distinctive-looking fruit comes from Southeast Asia. *Rambutan* translates to "hairy" in some Asian languages, and indeed, the fruit has a hairy exterior. Often used in desserts, the rambutan's sweet taste is similar to that of a litchi, and the fruit is a reliable source of vitamin C and calcium.

Victuals and Libations

If it's true that you are what you eat, then you might want to stay away from slimehead, in the hole, or bang belly...except that the first is a fish we also call orange roughy, the second is sausage baked in batter, and the last one is a dessert. Looking at these and other examples, it's easy to see that the English language serves up a veritable buffet of interesting words for what—and how—we eat and drink.

* * * *

YERD-HUNGER IS AN overwhelming desire for food, just as *dipsomania* is a craving for alcohol. A *libation* is a beverage—generally an alcoholic one. Total abstention from alcohol is called *nephalism*. On the other hand, if you've had too much to drink, you might be feeling *crapulous*, or tipsy. When you've overimbibed, you might be *adrip, anchored at Sot's Bay, bosko absoluto, capernoited, impixlocated, incog, lappy, pruned, swozzled, whooshed*, or *zissified*.

To *gourmandize* is to overeat. To *englut* or to *ingurgitate* is to gulp excessively. When you hear your stomach growling, that's *borborygmus*. Following a Thanksgiving feast, you might feel *farctate*, which is the state of having overeaten. *Lurcation, gulosity*, and *edacity* all refer to gluttony. The opposite of a glutton is a *jejunator*, which is a person who is fasting. To break your fast, you *genticulate*, or eat breakfast. To *degust* is to savor good food, and *abligurition* is the practice of spending lavishly for food. *Magirology* is the art or science of cooking. *Commensality* (*mensa* is the Latin word for "table") is the act of eating with other people. A *knabble* is a bite or nibble of food. People who *fletcherize* chew their food at least 30 times before swallowing. To *masticate* is to chew.

The mouth-filling word *brychomnivorous* describes someone who eats in a furious manner. *Equivores, erucivores, nucivores*, and *ranivores* have diets of horsemeat, caterpillars, nuts, and frogs, respectively. An onion eater is a *cipovore*, and an *alliaphage* dines on garlic. Newborn babies are *lactiphagous*, meaning that their diets consist solely of milk. Animals that feed on grasses are *ambivores* or *graminivores*. Finally, the proper—and only correct—pronunciation of *victuals*, which means "food" or "food provisions," is "vittles" (VIT-uhls). The *c* and *u* are not pronounced.

Bon appétit, and cheers!

✳ Chapter 10

Infamous Achievements, Saves, and Solutions

Audacious Prison Escapes

When you have nothing to lose, you have everything to gain. Such appears to be the thinking behind these wild escapes.

✳ ✳ ✳ ✳

Alcatraz, San Francisco Bay, California

IN THE MOVIE *Escape from Alcatraz* (1979) a determined Clint Eastwood (playing Frank Morris) escapes from the famously "escape-proof" penitentiary. In reality, Morris and two accomplices rode a makeshift raft off the "the Rock" on June 11, 1962. They were never seen again. The facility closed in 1963.

Libby Prison, Virginia

After digging a 50-foot tunnel, 109 Union soldiers broke free on February 9 and 10, 1864. Over half made it to safety behind northern lines.

Brushy Mountain State Prison, Tennessee

The convicted killer of Martin Luther King Jr., James Earl Ray used a makeshift ladder to scale the prison's 14-foot-high walls in 1977. He eluded authorities for 54 hours in what's been described as "one of the greatest manhunts in modern memory."

Colditz P.O.W. Camp, Germany

During World War II, British inmates built a glider to escape from this Nazi P.O.W. camp. Before they could use it, they were liberated. Tests later proved that the contraption could indeed fly.

Pascal Payet, France

Payet pulled off not one but two daring prison escapes, both times with the use of accomplices flying hijacked helicopters. After a 2007 escape from France's Grasse Prison, the fugitive underwent cosmetic surgery. Despite his proactive attempt at eluding authorities, Payet was soon recaptured.

Salag Luft III, Germany

Immortalized in *The Great Escape* (1963), the March 24, 1944, escape from the Nazi P.O.W. camp featured three tunnels (Tom, Dick, and Harry) that reached beyond the prison's fences. In all, 76 men crawled to freedom. Sadly, only three evaded recapture.

Bragging Rights

Highest Scores Ever Racked Up on Popular Video Arcade Games

The days of begging your parents for quarters to use in the arcade are over. Now, video games are big business, raking in an estimated $35 billion annually and spawning professional leagues around the globe. Here are some high scores racked up by the world's best players.

* * * *

Pac Man: Billy Mitchell,

Ft. Lauderdale, Florida,

Score: 3,333,360

In 1999, video game legend Billy Mitchell (yes, there are video game legends) took more than six hours to become the first person to get a "perfect" score on *Pac Man*—the most famous arcade game of all time—completing 256 boards while eating every dot, power pellet, ghost, and piece of fruit. Mitchell achieved his score at the Funspot Family Fun Center in Weirs Beach, New Hampshire, the unofficial proving ground for world-record–seeking gamers. According to Twin Galaxies, the official record-keeping institution of arcade gaming, it was the first perfect score in more than ten billion attempts worldwide since the game's invention.

Donkey Kong:

Billy Mitchell, Ft. Lauderdale, Florida,

Score: 1,050,200

Not content with the world's only perfect *Pac Man* score, Mitchell has been in a heated battle with rival Steve Wiebe for the *Donkey Kong* crown. (Wiebe has the second highest total ever at 1,049,100.) The rivalry has been so intense and has sparked such interest that an award-winning documentary, *The King of Kong,* was made about the pair in 2007.

Ms. Pac Man:

Abdner Ashman, Queens, New York,

Score: 933,580

Ashman is no Billy Mitchell, but he holds the world records for *Ms. Pac Man, Jr. Pac Man,* and *Robotron: 2084.*

Space Invaders:

Donald Hayes, Windham, New Hampshire,

Score: 55,160

Donald Hayes not only has the world record for *Space Invaders,* he also ranks in the top ten for dozens of other arcade games, including the classics *Galaga, Centipede, Dig Dug,* and *Q-bert.*

A Serious Track Record

Nobody in America moves more people daily than New York's Metropolitan Transportation Authority.

* * * *

BUSES ALONE TRUCK along 2.3 million, while the subway carries more than three times as many (7.8 million). As a result, many New Yorkers don't own cars or even learn how to drive. The city's extensive public transit gives it the distinction of having the lowest per capita emissions from automobiles of any metropolitan area in the nation. In fact, the city's petrol consumption compares to the national average of the 1920s.

When the system originally opened on October 27, 1904, it was made up of only 28 subway stations, all of them in Manhattan. Now, according to the MTA, the station count of 468 is just 35 short of the combined total of every other subway system in the United States. Approximately 660 miles of track are used for commuter service, out of 840 miles in total—if laid end to end, those tracks would span the distance from New York to Chicago.

The longest ride you can take without changing trains is the A line from 207th Street in Manhattan to Far Rockaway in Queens for a total of 31 miles, all for the same fare (unlike some other systems that charge by the zone). It should come as no surprise that, as of 2009, the busiest station of all was Times Square (with a staggering annual total of nearly 59 million passengers), followed by Grand Central, just crosstown east on 42nd Street, with more than 42 million. New York has the fourth most populated subway system in the world (super-crowded Tokyo tops the list), with an annual ridership of 1.563 *billion.*

And while tourists often fear becoming a crime statistic, especially on that very same subway (which runs 24/7/365, unlike—say—London's tube, which closes down after midnight), the odds have been consistently reduced ever since the time of Rudy Giuliani as mayor. In 2004, Mayor Michael Bloomberg was quoted as saying, "The subway system is safer than it has been at any time since we started tabulating subway crime statistics nearly 40 years ago." (Nonetheless, it's still advisable to avoid flashing cash.)

Dynamite! The Life and Times of Alfred Nobel

It was a legacy that was hard to live down. But Alfred B. Nobel—inventor of the explosive dynamite—took what many considered to be a negative and used it to give back to humanity. Nobel's story is one of lifelong learning and perseverance, as well as isolation and loneliness.

* * * *

The Initial Spark

ALFRED NOBEL WAS born in Stockholm, Sweden, in 1833, the son of Immanuel and Andriette Nobel. His father was a natural-born inventor who received three mechanical patents at the young age of 24. Yet, due to a series of financial disasters, Nobel was born into a family mired in bankruptcy. Frail and sickly, he excelled in academics during a brief stint in grade school. The family moved to St. Petersburg, Russia, when Nobel was only nine years old. He showed an amazing facility with languages, eventually becoming fluent in five: Swedish, Russian, French, English, and German.

Nobel's formal education concluded with a series of tutors, including a chemistry professor named Ninin who introduced a teenage Nobel to the volatile liquid nitroglycerine. Discovered by Italian chemist Ascanio Sobrero in 1847, the oily and highly unstable fluid was extremely sensitive to shock, making it a dangerous explosive. Sobrero was convinced that the substance could never be tamed. But Nobel's father had become an expert in manufacturing armaments, such as land and sea mines, so Nobel's interest in nitro served the family business as well.

The end of the Crimean War sent the Nobel and Sons munitions company into bankruptcy. Returning to Stockholm, Nobel and his three brothers formed a new company, appropriately named Brothers Nobel, and continued working with nitroglycerine. The Swedish military showed great interest in the brothers' efforts, so a demonstration was arranged. Nobel filled an iron barrel with half gunpowder and half nitro and urged everyone to step back. The result was a bit *too* successful—a thundering explosion rocked the area, prompting the military to deem the mixture much too risky. They immediately ceased all further contact with the Brothers Nobel.

Stockholm Becomes a Boomtown

Nobel was obsessed with calming the wild nitroglycerine, working nonstop for days at a time. By 1863, he succeeded in igniting gunpowder and nitro with a fuse. The triumph resulted in new manufacturing plants in Sweden, Germany, and Norway. But his efforts were not without disaster.

In the fall of 1864, Nobel's younger brother Emil was mixing chemicals at the plant in Stockholm. More than 40 gallons of nitroglycerine were stored there, so when a tremendous explosion rocked the area, it literally blew its six victims, including Emil, to pieces. For safety's sake, Nobel moved his experiments to a barge in Bockholm Bay, Sweden.

A Flash of Ingenuity

In an 1889 letter, Nobel noted his greatest invention by writing, "I prepared and detonated the first dynamite charge . . . in November 1863," although most sources note the patent date of 1867. Nobel finally calmed nitro by mixing it with kieselguhr—a chalky mineral made mostly of silica. The result was a paste that could be safely handled and rolled into paper tubes. He called his new creation *dynamite,* from the Greek word for "power." Dynamite became a safe means of removing large amounts of earth for mining and tunneling, as well as construction. Establishing worldwide manufacturing and distribution of dynamite, Nobel became wealthy and successful.

But there was a problem with this new invention: When it was stored, the nitro would slowly leak from the dynamite, once again resulting in an unstable and dangerous liquid. Nobel studied the problem and came up with the solution in 1876. He mixed nitro with wood pulp and other minerals, creating a compound Nobel called blasting gelatin. When ignited with a separate detonator, it became an even better method for excavation.

The Final Blast

Nobel knew that his invention would not only be used for constructive purposes. A substance as devastating as dynamite would also be used for violence and crime. Nobel had no wish to be remembered solely as the inventor of dynamite—after all, he amassed more than 350 patents in his career, many in the field of synthetic materials. He also wrote a tragic play, *Nemesis*, late in his life.

An incorrectly placed obituary in an 1888 French newspaper prematurely told of Nobel's passing, referring to him as "the merchant of death." Such a sobriquet did not suit Nobel. In fragile and failing health, he wrote a will in November 1895; the will stipulated that his fortune be used to establish a fund for making annual awards to individuals who "shall have conferred the greatest benefit on mankind." Five prizes would be given, one each in the fields of chemistry, physics, medicine, literature, and peace— "the best work for fraternity between the nations, for the abolition or reduction of standing armies, and for the holding and promotion of peace." (A sixth prize, for economics, was established in 1969.) Nobel, satisfied that he had left the world with a sense of hope and promise, died on December 10, 1896, following a stroke.

Prize Winners

The first Nobel Prizes were given in 1901. In the time since, U.S. presidents Teddy Roosevelt, Woodrow Wilson, and Jimmy Carter have received the peace prize, along with other notables including Mother Teresa, Albert Schweitzer, Mikhail Gorbachev, Al Gore, and Martin Luther King, Jr. Winners in literature include Rudyard Kipling, George Bernard Shaw, Sinclair Lewis, Eugene O'Neill, Pearl S. Buck, William Faulkner, Winston Churchill, Ernest Hemingway, Boris Pasternak, John Steinbeck, Jean-Paul Sartre, and Toni Morrison. Recipients in the sciences and medicine include Marie Curie (twice; as well as daughter Irene, some 24 years later), Linus Pauling, Frederick Sanger (twice), Wilhelm Roentgen, Guglielmo Marconi, Max Planck, Albert Einstein, Niels Bohr (as well as son Aage, 53 years later), Enrico Fermi, and Ivan Pavlov. Recipients in economics include Milton Friedman and James Tobin. Those who many believe should have won a Nobel Prize but never did include Thomas Edison, Nikola Tesla, Mohandas Gandhi (nominated five times), Robert Oppenheimer, and Dmitry Mendeleyev.

The World's Oldest Things

Stop complaining about your knees. You've got nothing on these true ancients.

* * * *

* Biologists believe that *turritopsis nutricula,* a type of hydrozoan, has the longest lifespan of any animal. Like some other hydrozoans, it begins life as a polyp and later changes into a jellyfish. But it can also turn back into a polyp, rebooting its life cycle. It seems to be able to do this indefinitely, which means its lifespan is essentially infinite. Nobody knows how old any specific *turritopsis nutricula* is, however.
* In 2007, a 405-year-old quahog clam claimed the title for oldest specific animal. The scientists who found it off the

coast of Iceland fixed its age by counting its growth rings.

- The oldest known living tree is a spruce on Fulu Mountain in Sweden, discovered in 2004. According to carbon dating, the tree is 9,550 years old, which means it took root soon after the last ice age.
- The oldest known building site is an arrangement of ten post holes on a hill near Tokyo. Archeologists believe the holes were the foundation for two huts, half a million years old, built by the extinct human species *Homo erectus*.
- The oldest structure built by modern humans is Gobekli Tepe, a temple of elaborately carved pillars unearthed in Turkey. Archeologists believe the temple is 11,500 years old, which predates Stonehenge by about 7,000 years.
- Some Wyoming residents claim the oldest structure is their own Fossil Cabin. It wasn't built until 1933, but the construction material—a collection of dinosaur fossils—is between 100 million and 200 million years old.
- The oldest known object on Earth is a microscopic zircon crystal, found in western Australia. Through chemical analysis, scientists have determined it is 4.4 billion years old—only 100 to 200 million years younger than Earth itself.

Damming The Red River

A combination of ingenuity and sheer brawn saved the day for Union soldiers stuck in a jam on Louisiana's Red River in 1864.

⁕ ⁕ ⁕ ⁕

GREAT BATTLES ARE decided in many different ways. Acts of bravery and courage under fire often determine the outcome of armed conflict. Other struggles end in defeat through the folly or indecision of a commander. Only 14 people were given the official Thanks of Congress for their services to the Union during the Civil War, and all but one—Lieutenant

Colonel Joseph Baile—were commanders in the armed forces. As the chief engineer of the 19th Corps, Bailey saved a Union gunboat fleet trapped on Louisiana's Red River in May 1864 using a single formidable weapon—his brain.

Revered River

The Red River, a tributary of the Mississippi, was highly prized by Union forces as a channel by which to capture Shreveport and establish their control over northern Louisiana. The plan was that Shreveport and the surrounding area would then serve as a springboard to launch excursions into Texas and Arkansas. In April 1864, General Nathaniel Banks led a Union force of 32,000 soldiers and 13 gunboats, including 6 formidable iron-clads, northward up the Red River.

Stuck in the Mud

A series of skirmishes with a determined Confederate opposition slowed the Union advance. At Alexandria, Confederate defenders had been busy digging channels to divert the river. They had already succeeded in lowering the water level to a depth of less than four feet, which created quite a problem for the Northern fleet. The smallest Union gunboats required seven feet of water to travel, and the larger craft needed a depth of more than ten feet.

The Union advance was stalled. The gunboats were at risk of becoming stranded on the river as the waters continued to recede, which would make them sitting ducks for either capture or destruction by Confederate artillery. On May 1, Chief Engineer Bailey was authorized to take any actions necessary to free the boats and permit the advance to continue. To the amazement and consternation of the Union leadership, he resolved to build a dam.

Dam It All

Bailey had worked in the Wisconsin woods in his youth, and he sought out lumbering experience among the 3,500 soldiers he had at his disposal. He then directed the felling of hundreds of trees in the Red River's adjacent forests and had the cut timber fashioned into cribbing. Other soldiers were detailed to gather rock and earth to fill the cribbing that would be used in the makeshift dam. All the while, Bailey's lumberjacks and laborers were subject to continuous Confederate sniper fire.

To complete the dam, Bailey intended to have four barges deliberately overloaded and sunk midstream above the rapids. He would then connect the cribs to the barges, creating two 300-foot wing dams jutting into the river from each bank. This backup would force the water level up to a point at which the gunboats could maneuver. The remaining open water in the middle of the river would form an ingenious spillway through which the gunboats could pass over the rapids.

Cleverness Lauded

The dam was built in ten days. The river's water level rose to almost 13 feet because of Bailey's dam, which allowed the gunboats safe passage into navigable water. The last boat passed over just as the dam was breaking due to the water pressure in the river. In his report to the War Department, General Banks lauded Chief Engineer Bailey as the person who both saved a Union fleet worth more than $2 million and permitted the Red River Campaign to continue.

Magical Moments

Don Larsen summoned greatness when it mattered the most.

* * * *

The Setting: Yankee Stadium; October 8, 1956

The Magic: Don Larsen's perfect World Series game is one for the ages.

IN 1954, DON Larsen, then a member of the Baltimore Orioles, had a record of three wins and 21 defeats. He was one of seven Orioles swapped for ten Yankees that off-season, and he obviously took to his new surroundings, sporting a 9–2 record in a partial '55 season and an 11–5 record as a swingman in '56. That year he started 20 games and relieved in 18 others.

But on one October afternoon, the solid but until then unremarkable Larsen gave one of the greatest pitching performances of all time, retiring all 27 Brooklyn Dodgers he faced. There were three turning points in his perfecto (the first perfect game in the majors in 34 years). One was a heads-up piece of defensive work on a second-inning Jackie Robinson grounder, which Andy Carey deflected to shortstop Gil McDougald, who threw to first in time to nail Robinson by an eyelash. Second was Mickey Mantle's fourth-inning home run (the first hit allowed by Sal Maglie that day), and third was a nifty running grab of a Gil Hodges line drive by Mantle the next inning. Beyond that, Larsen was utterly dominant: He used only 97 pitches, and only one batter ever saw a three-ball count. Larsen lasted ten more years in the bigs, mostly as a reliever and without much success, but his one perfect day of World Series glory is an achievement that remains unmatched.

Soundly Whipped

Was U.S. Air Force captain Chuck Yeager first to break the sound barrier? The answer may surprise, mesmerize, or antagonize.

* * * *

MANY BELIEVE THAT celebrated test pilot Chuck Yeager was first to punch through the elusive sound barrier. They are correct—at least in part. When Captain Yeager's Bell X-1 rocket plane hit Mach 1.07 (just beyond the speed of sound) on October 14, 1947, the grinning West Virginian became the first *human* to conquer the sound barrier in level flight. Although the feat was a crowning achievement for humankind, Yeager's X-1 was far from the first *object* to surpass the speed of sound.

That title goes to something far less glamorous than a celebrated aviator—a plain old bullwhip. As far back as 5000 B.C., the Chinese were producing these long, handheld implements to drive cattle. When the whip uncoiled to its fully extended length, a noticeable "crack" would encourage the animals to fall in line. The question has always been: Why does the whip issue such a sound, even when it is striking nothing more than thin air?

In 1927, high-speed photography of a whip-stroke solved the mystery. The pictures showed that a sonic boom was being created as the extreme tip of the whip "snapped" out to its full length. Because the speed of sound is dependent upon elevation (greater speed is required to break through at sea level than at higher altitudes), the whip had to be traveling at an even faster speed than Captain Yeager.

Greatest Shortcuts

The construction of great canals such as Suez and Panama cost vast amounts of money and required incredible feats of engineering to link seas and oceans with only a few miles of artificial waterway. But it sure beat sailing thousands of miles around entire continents.

✻ ✻ ✻ ✻

IN SEPTEMBER 1513, Vasco Núñez de Balboa left the Spanish colony of Darién on the Caribbean coast of the narrow Panamanian isthmus to climb the highest mountain in the area—literally, to see what he could see. Upon reaching the summit, he gazed westward and became the first European to see the eastern shores of the vast Pacific Ocean.

From there, he led a party of conquistadores toward his discovery. They labored up and down rugged ridges and hacked their way through relentless, impenetrable jungle, sweating bullets under their metal breastplates the entire way. It took four days for Balboa and his men to complete the 40-mile trek to a beach on the Pacific.

Of course, Balboa's journey would have been much easier if someone had built the Panama Canal beforehand.

For 2,500 years, civilizations have carved canals through bodies of land to make water transportation easier, faster, and cheaper. History's first great navigational canal was built in Egypt by the Persian emperor Darius I between 510 and 520 B.C., linking the Nile and the Red Sea. A generation later, the Chinese began their reign as the world's greatest canal builders, a distinction they would hold for 1,000 years—until the Europeans began building canals using technology developed centuries earlier during the construction of the world's longest artificial waterway: the 1,100-mile Grand Canal of China.

It's been a long time since a "great" canal was built anywhere in the world. Here's a brief look at the most recent three.

The Erie Canal—Clinton's Big Ditch

At the turn of the 19th century, the United States was bursting at its seams, and Americans were eyeing new areas of settlement west of the Appalachians. But overland routes were slow and the cost of moving goods along them was exorbitant.

The idea of building a canal linking the Great Lakes with the eastern seaboard as a way of opening the west had been floated since the mid-1700s. It finally became more than wishful thinking in 1817, when construction of the Erie Canal began.

Citing its folly and $7-million price tag, detractors labeled the canal "Clinton's Big Ditch" in reference to its biggest proponent, New York governor Dewitt Clinton. When completed in 1825, however, the Erie Canal was hailed as the "Eighth Wonder of the World," cutting 363 miles through thick forest and swamp to link Lake Erie at Buffalo with the Hudson River at Albany. Sadly though, more than 1,000 workers died during its construction, primarily from swamp-borne diseases.

The Erie Canal fulfilled its promise, becoming a favored pathway for the great migration westward, slashing transportation costs a whopping 95 percent, and bringing unprecedented prosperity to the towns along its route.

The Suez Canal—Grand Triumph

The centuries-old dream of a canal linking the Mediterranean and the Red Sea became reality in 1859 when French diplomat Ferdinand de Lesseps stuck the first shovel in the ground to commence building of the Suez Canal.

Over the next ten years, 2.4 million laborers would toil—and 125,000 would die—to move 97 million cubic yards of earth and build a 100-mile Sinai shortcut that made the 10,000-mile sea journey from Europe around Africa to India redundant.

De Lesseps convinced an old friend, Egypt's King Said, to grant him a concession to build and operate the canal for 99 years. French investors eagerly bankrolled three-quarters of the 200 million francs ($50 million) needed for the project. Said had to kick in the rest to keep the project afloat because others, particularly the British, rejected it as financial lunacy—seemingly justified when the canal's final cost rang in at double the original estimate.

The Suez dramatically expanded world trade by significantly reducing sailing time and cost between east and west. De Lesseps had been proven right and was proclaimed the world's greatest canal digger. The British, leery of France's new backdoor to their Indian empire, spent the next 20 years trying to wrest control of the Suez from their imperial rival.

The Panama Canal—Spectacular Failure

When it came time to build the next great canal half a world away in Panama, everyone turned to de Lesseps to dig it.

But here de Lesseps was in over his head. Suez was a walk in the park compared to Panama. In the Suez, flat land at sea level had allowed de Lesseps to build a lockless channel. A canal in

Panama, however, would have to slice through the multiple elevations of the Continental Divide.

Beginning in 1880, de Lesseps, ignoring all advice, began a nine-year effort to dig a sea-level canal through the mountains. This strategy, combined with financial mismanagement and the death of some 22,000 workers from disease and landslides, killed his scheme. Panama had crushed the hero of Suez.

The Panama Canal—Success!

The idea of a Panama canal, however, persevered. In 1903, the United States, under the expansionist, big-stick leadership of Theodore Roosevelt, bought out the French and assumed control of the project. Using raised-lock engineering and disease-control methods that included spraying oil on mosquito breeding grounds to eliminate malaria and yellow fever, the Americans completed the canal in 1914.

The Panama Canal, the last of the world's great canals, made sailing from New York to San Francisco a breeze. A trip that once covered 14,000 miles while circumnavigating South America was now a mere 6,000-mile pleasure cruise.

Achieving The Gratitude Of Your Country

Now the military's highest, most coveted honor, the Medal of Honor was generously passed out during the Civil War.

❋ ❋ ❋ ❋

DURING THE FOUR years Americans fought in World War II, 464 men—out of about 16 million who served—won the Medal of Honor. Only 245 received it during the Vietnam War. Just one member of the Coast Guard has ever received one, and only 17 members of the Air Force. It's a rare, highly respected honor, given only to service members who do something dramatically courageous during combat.

At least, that's the way it's been for the past century. During the Civil War, the story was a little different.

Establishing the Medal

No medal for valor existed when the Civil War started. Army General-in-Chief Winfield Scott didn't like medals—he felt they smacked of show-off European armies. Others felt individual medals discouraged unity among the troops.

But Navy leaders believed that recognizing bravery was motivational, so they proposed a medal of valor "to be bestowed upon such petty officers, seamen, landsmen, and Marines as shall most distinguish themselves by their gallantry and other seamanlike qualities during the present war." Lincoln agreed and authorized the medal on December 12, 1861. When General Scott resigned from the Army, the new Army leaders decided they wanted a medal, too. The Army version of the Navy medal was approved on July 12, 1862.

The new medal was handsome—a five-pointed star hanging from a ribbon resembling the American flag. Soldiers wore it around their necks.

The First Honorees

The stage was set for recognizing outstanding soldiers. So to whom did the first medals go? Well, they were brave men, but the event didn't involve dramatic hand-to-hand combat. Rather, they were soldiers led by Union spy James J. Andrews who snuck behind Confederate lines in Georgia to destroy railroad tracks, bridges, and telegraph lines. They failed and most were captured, but the six who survived the raid were awarded with Medals of Honor in 1863.

At least those honorees did something fearless and dramatic. Because it was the only medal around, leaders started handing the Medal of Honor out somewhat indiscriminately. Consider the case of the 27th Maine Infantry Regiment. All 864 members won the medal. That's more than all the winners from World War I, World War II, and the Vietnam War combined! One might think those Maine boys did something heroic—won a major battle against tremendous odds, captured the Confederate capital, or something along those lines. Well, one would be wrong. This regiment received the medal because it did a stellar job sitting around the barracks in forts near Washington, D.C. Offered as an incentive for troops to stay to protect the capital after their enlistment was up, a clerical error resulted in the entire regiment sharing the honor.

Of course, many Civil War Medal of Honor winners did actually risk their lives. A number of winners were cited for capturing Confederate battle flags. These flags inspired and guided troops in battle, so a soldier risked serious personal damage sneaking be hind enemy lines to snag one of these prizes. Of course, some battle flags were just dropped in the heat of combat or lay with dead or wounded color bearers, so the first lucky Union soldier to come across one and hurry it back to his lines could be awarded a shiny new Medal of Honor.

The rules for winning a Medal of Honor were tightened in the early 20th century. Now soldiers must risk their lives in combat "above and beyond the call of duty." When the restrictions were enacted in 1917, a panel of generals reviewed all previous winners and revoked 911 medals, including those given to the brave protectors from Maine. The final total of Civil War Medals of Honor awarded is 1,522.

Spread the Glove

A future glove manufacturer started the trend, while the need for protection and some ingenuity resulted in the baseball glove we know today.

* * * *

Albert Spalding was not the first pro baseball player to sport a glove, but he was probably the first star of the game to wear one. He saw his first baseball glove in Boston on the hands (one on each) of first baseman Charles C. Waite in 1875. Spalding started wearing one when he moved to first base in 1877, and soon everyone was asking about them. Conveniently, Spalding and his brother owned a sporting goods store in Chicago, a company that would go on to produce innumerable gloves over the next 130 years and counting.

In 1883, a hard ball broke two of shortstop Arthur Irwin's fingers. Since he wanted to continue to play, Irwin purchased a large buckskin glove, added padding, and sewed the third and fourth fingers together. Within a few years most players were using the "Irwin glove." The baseball glove was improved upon again in 1919 thanks to St. Louis spitball pitcher Bill Doak. He suggested to the Rawlings Sporting Goods Company that a webbed thumb and forefinger be added to create a pocket. The result was the revolutionary "Doak model," the standard baseball glove used to this day.

One player whose glove couldn't be found on the field was Bid McPhee. The son of a saddlemaker, McPhee eschewed the leather on his hand long after almost everyone else had relented. Instead he soaked his hands in brine every spring to toughen them. It worked. His 529 putouts at second base for the 1886 Cincinnati Reds of the American Association has never been surpassed, glove or no glove. Yet with his fielding slipping and his finger sore to start the 1896 season, McPhee finally donned a glove in his 15th season. The future Hall of Famer led second basemen in fielding, and his .978 fielding percentage stood as a record for 23 years.

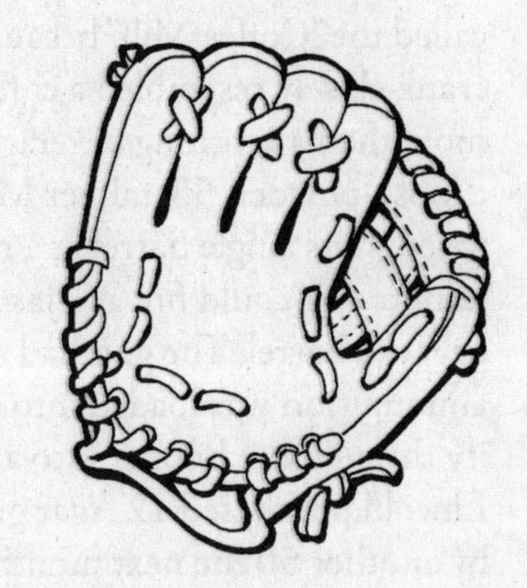

What Kind of Weapon Is That?

How did Johnny Reb and Billy Yank fight a modern war? With new-fangled weapons! Inventors from North and South struggled to create new devices that could fill the bill.

⁂ ⁂ ⁂ ⁂

THE CIVIL WAR is commonly known as the "first modern war" because of the introduction of so many "firsts." For instance, great measures were taken to improve the conditions for soldiers on both sides, including innovations such as the use of railroads and ships to move soldiers, supplies, and the wounded from one place to another; food preservation methods such as canned goods; and improved medical procedures. Inventors from the Union and the Confederacy also tirelessly worked to create weapons that would give their troops the advantage on the battlefield. They successfully introduced ironclad ships, submarines, and land and sea mines. The need to provide maximum firepower, however, resulted in a collection of weapons that were innovative—perhaps too innovative.

Rapid Fire in the North

The Agar "Coffee Mill" gun, invented by Northerner Wilson Agar, was used ineffectively by both sides during the war. It was called the "Coffee Mill" because its top-mounted hopper and crank closely resembled a coffee grinder. It percolated much more than a morning beverage, however. The gun fired either a loose-powder .58-caliber Minie-type ball or a paper round through its single barrel at a rate of about 120 rounds per minute—it could fire at a faster rate, but that tended to overheat the barrel. The gun had a range of about 1,000 yards. The ammunition was loaded into steel containers and fed by gravity through the hopper into a crank-driven revolving chamber. Lincoln purchased 12 Agar guns in November 1861, followed by another 50 the next month.

Southern-Style Firepower

Ingenuity was not limited to Northern inventors. Some of the most creative designs were developed in the South. Unfortunately, the designs proved to be less than efficient on the battlefield.

On April 22, 1862, several men stood near Newton's Bridge outside of Athens, Georgia, to test an invention of John Gilleland, who was a carpenter prior to the war. The newly forged cannon—a double-barreled six-pounder gun—was rolled into position. The barrels were aligned with a three-degree divergence to fire shells at a slight angle from each other. Lying on the ground next to the gun were pairs of shells attached to each other by a chain. If the gun worked, the two shells would propel the chain to cut down anything in its path. Gilleland stepped up and loaded a shell in each barrel, leaving a portion of the chain draped between the muzzles. But when he fired both barrels, the shells didn't fire simultaneously. Instead, the first traveled erratically while the other followed behind it. Subsequent tests ended with the same result. When the gun was sent to the Confederate government for evaluation, it was returned to be reconfigured so that instead of shells connected by a chain, each barrel was loaded with a canister. The gun eventually saw action on August 2, 1864, when Northern forces approached Athens. It successfully drove away Union troops and now stands outside Athens City Hall.

Rapid Fire That Worked

One successful Confederate invention was the Williams rapid-fire gun, created by Captain R. S. Williams. A breech-loaded artillery piece, it had a four-foot-long barrel. This weapon differed from other artillery in that it was hand-cranked and capable of firing up to 20 rounds per minute with a maximum range of 2,000 yards. The Williams rapid-fire gun proved to be very effective in its first action on May 31, 1862, at the Battle of Seven Pines. Although it was discovered that rapid fire caused the breech to expand and fail to lock, the gun was used in both the Eastern and Western Theaters throughout the war. "We had heard artillery before," one Union officer observed, "but we had never heard anything that made such a horrible noise as the shot from these breechloaders."

The Ingenious Astrodome

The first covered baseball and football stadium—nicknamed the "Eighth Wonder of the World"—is a tribute to American ingenuity.

❋ ❋ ❋ ❋

THE STORY OF the Astrodome began in 1962 when Major League Baseball expanded to include the Houston Colt .45s, owned by Houston Judge Roy Hofheinz. The city's subtropical weather made scheduling difficult—extreme heat and humidity were a challenge for players, and tropical downpours were always a risk.

What About a Roof?

However, after a trip to Rome's Colosseum, Judge Hofheinz had an idea. He'd learned that the original Colosseum had a retractable fabric canopy called a *velarium*. With more modern technology, Hofheinz speculated, a modern sports arena could be enclosed within a dome and air-conditioned.

In November 1964, his dream became a reality as the Harris County Domed Stadium was completed. It stood 18 stories tall and covered nearly ten acres of land, about six miles from downtown Houston. The ceiling was made of clear Lucite plastic; sunlight lit the interior well enough that the playing field could be natural Bermuda grass, bred for indoor use.

Around the world, Houston's domed stadium was acclaimed as an engineering miracle. Almost immediately, other cities launched plans to enclose their existing stadiums or build entirely new ones based on the Houston design.

The Space Age

Hofheinz soon renamed the stadium the Astrodome to highlight Houston's connection with the space industry. At the same time, the Houston Colt .45s became the Astros. During an exhibition game, Mickey Mantle hit the first home run in the Astrodome. Everything seemed perfect, until the first official games in the new dome. Players complained that they couldn't see fly balls due to glare from the Lucite panes in the ceiling.

Two sections of the panes were painted white, but the grass died when it couldn't get enough sunlight. For most of the 1965 season, teams played on dying grass and dirt that had been painted green.

Installing the Astroturf

Researchers at Monsanto invented artificial grass, which they called *ChemGrass,* that same year, but it wasn't in full production at the start of the 1966 baseball season. So, although most of the infield was covered with the ChemGrass—soon renamed *AstroTurf*—the outfield remained painted dirt until more AstroTurf arrived in July.

The stadium continued to adapt to challenges, and its popularity grew with fans and teams alike. In 1968, the Houston Oilers football team made the Astrodome their stadium, following the arrival of the annual Houston Livestock Show and Rodeo in 1966. Basketball games were also featured at the Astrodome, including the 1971 NCAA Final Four games and the 1989 NBA All-Star Game. In 1973, the Astrodome was also the site of the famous Battle of the Sexes tennis match in which Billie Jean King beat Bobby Riggs.

Declining Fortunes

However, by 1996, the age of the Astrodome became evident. The Oilers demanded a new stadium, but when Houston turned them down, owner Bud Adams moved the team to Tennessee. Next, the Astros insisted on a new ballpark or they'd leave the city as well. In 2000, a new park, now called Minute Maid Park, was built in downtown Houston. And this ballpark has a retractable roof.

In 2002, Reliant Stadium opened next to the Astrodome, as the home of Houston's new NFL team, the Texans. Like Minute Maid Park, it features a retractable roof.

The Astrodome became known as the "lonely landmark" due to the fact that so few events have been scheduled there since the new stadiums were built. The stadium did, however, serve one more noble act of kindness in 2005, when its doors were opened to welcome displaced survivors of Hurricane Katrina. For two weeks, more than 13,000 people found shelter inside the Astrodome, which once again became the focal point of world headlines and applause.

Tunneling to Water—and Safety

When the Assyrians were threatening to capture Jerusalem, King Hezekiah acted quickly to fend off the attack. He ended by tunneling his way to invulnerability.

* * * *

After the Assyrians defeated the northern kingdom of Israel in 721 B.C., they returned in 701 to conquer the southern kingdom of Judah. As the invaders approached Jerusalem, brutally destroying several other cities of Judah along the way, King Hezekiah of Judah built up Jerusalem's defenses to better fend them off. And it worked.

When the Assyrians finally reached Jerusalem, they surrounded and besieged the city but were unable to break through its walls. The Assyrians settled in, assuming that even if they couldn't get in, the people of Jerusalem wouldn't be able to get out to reach their water supply. The people of Jerusalem would surrender when they got thirsty enough. But the Assyrians were in for a surprise.

While the Assyrians were brutalizing other Judean cities, King Hezekiah had exhibited great foresight. Until then, all of Jerusalem's fresh water had come from the underground Gihon Spring outside the city walls and flowed through unprotected canals into storage pools inside the city. Because these canals could be blocked by the enemy, cutting off the city's water supply, Hezekiah had workers close off the canals and dig a six-foot-high tunnel under the city walls through 1,749 feet of solid bedrock. The work was done mainly with hand tools by two sets of men who worked from opposite directions and met in the middle. Hezekiah then used the completed tunnel to divert the water from the spring through the tunnel, which sloped gently downward, allowing the water to flow naturally into a large reservoir, called the Pool of Siloam, inside the city walls. And so the people of Jerusalem weren't trapped without water. The Assyrian siege failed. Archaeologists have found remains of this tunnel, including a plaque written by its engineers that describes the meeting of the two groups of workers as they completed their gargantuan task.

Infamous Zoo Escapes

On Monday, November 9, 1874, New Yorkers awoke to shocking news. According to the New York Herald, enraged animals had broken out of the Central Park Zoo the preceding Sunday. At that very moment, a leopard, cheetah, panther, and other beasts of prey roamed the park in search of hapless victims. Men rushed out with rifles, prepared to defend their families. But it was all for naught—the whole story was a hoax perpetuated by a reporter irate at what he thought was lax security at the zoo. As it turns out, though, that reporter may have been on to something.

❋ ❋ ❋ ❋

FAST-FORWARD TO SUNDAY, July 5, 2009. More than 5,000 visitors were evacuated from Great Britain's Chester Zoo in the city of Liverpool. "Chimps Gone Wild!" the headlines screamed the next morning. Apparently, 30 chimpanzees had escaped from their island enclosure. This great breakout was certainly no hoax, but neither was it cause for alarm. The chimps got no further than the area where their food was kept. They gorged themselves until they had to lie down and rub their aching bellies. A little later, zoo wardens rounded them up, and the escapees returned to their island peacefully.

When it comes to zoo escapes, primates are often among the prime offenders. One of the greatest of all nonhuman escape artists was the legendary Fu Manchu of Omaha's Henry Doorly Zoo. Back in 1968, this orangutan confounded his keepers by repeatedly escaping from his cage no matter how well it was secured. Only when a worker noticed Fu slipping a shiny wire from his mouth did the hairy Houdini's secret come out. The orangutan had fashioned a "key" from this wire and was using it to pick the lock. What's even more impressive is that he had the sense to hide it between his teeth and jaw—a place no one was likely to look. Once officials realized what the cagey animal had been up to, they stripped his cage of wires. Though Fu Manchu never escaped again, he was rewarded for his efforts with an honorary membership in the American Locksmiths Association.

Oliver, a capuchin monkey in Mississippi's Tupelo Buffalo Park and Zoo, went Fu Manchu one better. In 2007, he escaped from his cage twice in three weeks. Both times he traveled several miles before being apprehended. Zookeepers suspected him of picking the lock, but they never figured out how he did it. Their solution was to secure his cage with three locks, a triple threat that has so far kept him inside. Word of Oliver's escapades drew so many visitors to the zoo that officials decided to capitalize on the capuchin culprit. A best-selling item at the zoo's gift shop is a T-shirt emblazoned with "Oliver's Great Escape" along with a map of his routes.

Evelyn, a gorilla at the Los Angeles Zoo, didn't need to pick a lock. She escaped on October 11, 2000, via climbing vines, à la Tarzan. After clambering over the wall of her enclosure, she strolled around the zoo for about an hour. Patrons were cleared from the area, and Evelyn's brief attempt to experience the zoo from a visitor's point of view ended when she was tranquilized and returned to her enclosure without further incident.

Juan, a 294-pound Andean spectacled bear at Germany's Berlin Zoo, had a much more amazing adventure on August 30, 2004. It started when he paddled a log across the moat that surrounded his habitat. He then scaled the wall and wandered off to the zoo playground. There, he acted just like a kid, taking a spin on the merry-go-round and trying out the slide. When he left in search of further fun, clever animal handlers decided to distract him with a bicycle. Sure enough, Juan stopped to examine the two-wheeler as if he were contemplating a ride. Before he could mount it, however, an officer shot him with a tranquilizer dart, thus ending Juan's excellent adventure.

"I Kick a Touchdown!"

European kickers first arrived in professional football in the 1960s. Some, like Garo Yepremian, had steep learning curves…

✻ ✻ ✻ ✻

It Beats Making Neckties

GARABED SARKO YEPREMIAN was born in 1944 to Armenian parents on the eastern Mediterranean island of Cyprus. Standing 5′7″ and prematurely bald, he was making neckties in London in 1966 while playing soccer on weekends. His brother called from Indiana: He had been watching football on TV and believed that Garo could out-kick the placekickers he saw. Garo agreed to come.

Soccer-style specialists were new to American football in 1966. Most kickers played other positions as well, and they kicked straight-on with the toe. Barred from the NCAA for having played semipro sports, Garo's only option was professional football. The Yepremian brothers wangled Garo a tryout with the NFL's Detroit Lions, and Garo boomed his kicks with accuracy. Despite English that was as limited as his football knowledge, he had a new career.

Welcome to the NFL (*crunch*)

Legends abound regarding Garo's rookie year. As Garo puts it, he didn't know a chin strap from a jockstrap, much less how to put on the uniform. During his first game, Coach Harry Gilmer advised Garo that Detroit had lost the coin toss; Garo ran to midfield to look for the coin. He also faced prejudice, as many pro players disliked both foreigners in football and placekicking specialists. On kickoffs, every opposing Neanderthal wanted to knock Garo silly. After a memorable Week 4, when he was leveled by famed Packer linebacker Ray Nitschke, Garo began wearing a face bar on his helmet.

In Week 6, Gilmer finally let Garo do some placekicking. He was 3 for 3 on extra points and 1 for 4 on field goals, a middling start. As famed Lions defensive lineman Alex Karras told, during one early losing effort, Garo kicked an extra point that didn't affect the outcome of the game and began to celebrate wildly. The Greek-speaking Karras asked Yepremian in Greek what he was celebrating. Garo answered: "I keek a touchdown!" Whether or not Karras jazzed up the story, it became part of the Garo legend. It was certainly plausible. During the Colts' return visit to Detroit later that season, Garo booted a league-record six field goals, including four in one quarter (this shared record still stands).

After two years with Detroit, Garo joined the Army. When he got out, Detroit didn't need a kicker. He played semipro football in 1968, sat out 1969, and in 1970 the Miami Dolphins invited him to try out. Garo became a two-time Pro Bowler in Miami, a fan favorite, and their 1970s kicking mainstay.

Super Bowl VII

The Dolphins' perfect 1972 season pitted them against the Washington Redskins for the NFL title. If they won, they would finish 17–0—the greatest season in NFL history. With Miami up 14–0 late in the game, Coach Don Shula sent Garo out to kick a game-icing field goal.

Redskin defensive tackle Bill Brundige blocked the kick, which bounced to Garo's right. Garo picked it up. With Brundige thundering in on him, Garo next attempted an impromptu forward pass. The ball went straight up. As it fell, Garo batted it—right to Washington defensive back Mike Bass, who dashed past Garo for a touchdown. The play went into the books as a fumble return. Miami held on to win, and Garo Yepremian's place in the history of NFL humor was forever secured.

The normally easygoing kicker was depressed after the game, but some encouragement from Coach Shula helped him overcome such a public embarrassment. Today Garo is a motivational speaker and cancer fundraiser who not only laughs at funny stories about his football days—he tells them himself!

Which is the Loudest Rock Band Ever?

This question is momentous to anyone with a shred of rock and roll DNA, and right now the answer is open to dispute.

✻ ✻ ✻ ✻

THE GUINNESS BOOK *of World Records* no longer keeps records in this category, apparently not wanting people to permanently damage to their ears. The most recent claim was made by the hardcore band Gallows in 2007, after they hooked twelve amps and twelve speaker cabinets together and reportedly pumped out 132.5 decibels of noise, which is louder than the sound you would hear standing next to a jet engine.

The band had to wear earplugs and the heavy ear covers that airline employees use on the tarmac. "You could feel the sound blasting through the amps. It was [bleeping] off the hook!"

Gallows guitarist Laurent Bernard told the music magazine *Kerrang!* "More importantly, that's louder than [previous purported record holder] Manowar. I am louder than Manowar! That's all I care about!" It's nice to have such clear goals in life.

Manowar, the bombastic U.S. metal band, claimed 129.5 decibels in 1994. This was an unofficial addendum to its *Guinness* record of loudest musical performance from 1984. Not to run down Gallows or Manowar, but the loudness record has clearly fallen into the hands of lesser lights in the past generation. Previous *Guinness*-approved record holders included The Who, Deep Purple, The Rolling Stones, and Kiss—a veritable smorgasbord of rock icons.

The Who has the added distinction of having blown up its own drum kit with a concussion bomb, on *The Smothers Brothers Comedy Hour* television show in 1967. Some cite this incident as the origin of guitarist Pete Townshend's impaired hearing, but he blames it on too much loud music through headphones in the recording studio.

Anyhow, while the answer to this question lacks certainty right now, it resonates—so to speak—with headbangers of all ages. In fact, it brings up a related question that's just as significant: What was the loudest concert you ever attended? Can you still feel your head vibrating? If so, Gallows would be proud.

The Big Sticky

Taped crusaders on the ball

✻ ✻ ✻ ✻

Sitting quietly in the corner of an otherwise loud space is, perhaps, the world's largest tape ball, at home in a Madison,

Wisconsin body shop. It now weighs at least 150 pounds and keeps growing all the time. Several years in the making, it now takes two people to shimmy it out of its corner.

Just steps away from an auto painting booth, the ball is made completely of used tape from the painting process; tape and plastic is used to block off parts of the car that shouldn't be painted, such as the windshield or door handles. The project began as a way to keep the tape out of the garbage.

But Is It the World's Largest?

In 2005, someone put a 60-pound ball of tape on eBay and dubbed it the "world's largest ball of tape"; it was scooped up for $900. In 2006, a young Canadian got minor league hockey teams to contribute used tape to his ball. He purportedly created a seven-foot-tall, 1,862-pound ball of hockey tape, but he never made the call to Guinness.

The current record holders for the world's largest ball of tape are two New Zealanders who spent two months building their version. Dubbed the "largest of its kind" when completed in 2007, it weighed in at 53 kilograms and has a circumference of 2.5 meters; in the American system, it weighs a measly 116.845 pounds and is 8.2 feet around.

The Madison body shop's ball puts the Aussies to shame weight-wise. As for circumference, the body shop's employees haven't bothered with spherical perfection, so that's more difficult to measure.

For now, the employees are elated to know their ball is bigger than Guinness's record holder, but in light of the young Canadian's poorly documented feat, they are still content with the second-best distinction and with owning the world's "probably-largest" ball of tape.

Night Flight

Up, up, and away, over the Iron Curtain

* * * *

ON SEPTEMBER 16, 1979, Peter Strelzyk, Günter Wetzel, their wives, and four children dropped from the night sky onto a field in West Germany, flying a homemade hot-air balloon. Strelzyk, an electrician, and Wetzel, a bricklayer, built the balloon's platform and burners in one of their basements. Their wives sewed together curtains, bed sheets, shower liners, and whatever other fabric was on hand to make the 75-foot-high balloon. A bid to escape communist East Germany during the days of the Berlin Wall, their famous flight was two years in the making, spanned 15 miles, and took 28 minutes to complete. Unsure whether they had reached freedom, the two families spent the next morning hiding in a barn, until they saw an Audi driving down a nearby road and realized they were in the West.

The Great Escape

Unlike sprinting out of an exploding building or jumping from a bridge onto a speeding train, this action-hero plan actually works. If someone happens to be shooting at you, you can avert the gunfire by diving underwater.

* * * *

A Shield of Water

A 2005 EPISODE OF THE Discovery Channel's *MythBusters* proved that bullets fired into the water at an angle will quickly slow to a safe speed at fewer than four feet below the surface. In fact, the bullets from some high-powered guns in this test basically disintegrated on the water's surface.

It might seem counterintuitive that speeding bullets don't penetrate water as easily as something slow, like a diving human being or a falling anchor. But it makes sense. Water has considerable mass, so when anything hits it, it pushes back. The force of the impact is equal to the change in momentum (momentum is velocity times mass) divided by the time taken to change the momentum.

In other words, the faster the object is going, the more its momentum will change when it hits, and the greater the force of impact will be. For the same reason that a car suffers more damage in a head-on collision with a wall at fifty miles per hour than at five miles per hour, a speeding bullet takes a bigger hit than something that is moving more slowly.

The initial impact slows the bullet considerably, and the drag that's created as it moves through water brings it to a stop. The impact on faster-moving bullets is even greater, so they are more likely to break apart or slow to a safe speed within the first few feet of water.

It's Not Foolproof

The worst-case scenario is if someone fires a low-powered gun at you straight down into the water. In the *MythBusters* episode, one of the tests involved firing a nine-millimeter pistol directly down into a block of underwater ballistics gel. Eight feet below the surface seemed to be the safe distance—the ballistics gel showed that the impact from the bullet wouldn't have been fatal at this depth. But if a shot from the same gun were fired at a 30-degree angle (which would be a lot more likely if you were fleeing from shooters on shore), you'd be safe at just four feet down.

The problem with this escape plan is that you have to pop up sooner or later to breathe, and the shooter on shore will be ready. But if you are a proper action hero, you can hold your breath for at least ten minutes, which is plenty of time to swim to your top-secret submarine car.

10 Things to Do if You Meet a Skunk

Skunks have it pretty rough. Their small size makes them prey for scores of large predators. They're scavengers, which means lunch is literally garbage, and many of them end up as roadkill. Read on to learn about how to avoid the path of a disgruntled skunk and what to do if you do tangle with one.

* * * *

1. **Stay Away:** The best way to avoid getting skunked is to stay away from them. Skunks only spray when they're threatened, so don't threaten them and you shouldn't have a problem.
2. **Speak Softly and Walk with a Big Stomp:** If you must approach a skunk, do so with caution. Speak in a low voice and stomp your feet. Skunks have poor vision and often spray in defense because they simply don't know what's going on.
3. **Freeze!:** Another tactic for avoiding a skunking is to stand perfectly still and wait for the skunk to go away. Passive, but effective.
4. **Run . . . or Shut Your Eyes and Hang on Tight:** Right before a skunk lets loose its spray of stinkiness, it stomps its feet and turns around, as the spray glands are located near the anus. If you see a skunk doing this little dance, run away or hang on tight, because you're about to get skunked.
5. **Flush It Out:** If you get sprayed in the face, immediately flush your face and eyes with water. The sulfur-alcohol compound that skunks emit can cause temporary blindness, which could lead to bigger problems.
6. **Take It Outside:** Now that you've been skunked, anything you come into contact with is going to smell like you do.

You smell like skunk, if you hadn't already noticed. So try to stay outside, if at all possible.

7. **Skip the V8, Air Freshener, and Lemon Juice:** No matter what Grandma said, tomato juice does not take the smell of skunk off of you, your dog, or your clothes. And unless you like "fresh morning dew" skunk, vanilla skunk, or lemony skunk, don't even bother with air fresheners or lemon juice. These products don't eliminate skunk smell, they only make it worse by coating it with another cloying scent.

8. **Mix Up a Peroxide Bath:** To get rid of the skunk smell, you must neutralize the chemicals in the spray. This home remedy seems to work well on animals or humans: Mix one quart of 3 percent hydrogen peroxide, one teaspoon mild dishwashing detergent, and 1/4 cup baking soda in a bucket. Lather, rinse, repeat.

9. **Buy Deodorizing Spray:** These special sprays are available at pet stores and some home and garden stores, too. They work well because they're specially formulated to neutralize the intense odor of skunk.

10. **Call the Public Health Department or Your Doctor...:** ...if you've been bitten. Skunks have been known to carry rabies, even though they rarely resort to biting. The same goes for your pet—get it to the vet quickly if the skunk did more than spray. Also, notify the public health department within 24 hours.

Thinking Outside The Knot

Alexander the Great was a legend in his own time, a bold young king who conquered most of the civilized world.

* * * *

IN 333 B.C., ALEXANDER, the 23-year-old Macedonian king, was a military leader to be feared. He had already secured the

Greek peninsula and announced his intention to conquer Asia, a feat no Greek had yet accomplished. His campaign eventually took him to Gordium, in the central mountains of modern-day Turkey, where he won a minor battle. Though undefeated, he still hadn't scored a decisive victory and was badly in need of an omen to show his troops that he could live up to his promise.

Conveniently, there was a famous artifact in Gordium—the Gordian Knot. Some 100 years before Alexander arrived, a poor peasant rode into the town on his oxcart and was promptly proclaimed king by the people because of a quirk of prophecy. He was so grateful, he dedicated the cart to Zeus, securing it in a temple by using a strange knot that was supposedly impossible to untie. An oracle had once foretold that whoever loosed the knot would become the king of Asia. Because Alexander was in the neighborhood and had just that goal in mind, he couldn't resist the temptation of such a potentially potent omen of future success. If he could do it, it would be a huge morale boost for his army.

Many had wrestled with this particular knot before, and Alexander found it no easy task. Apparently he, too, struggled with it as a crowd gathered around him. Irritated but not defeated, he decided to approach the problem from a different angle. He realized that if he took the prophecy literally, it said that the person who undid the knot would be king. However, the legend didn't specify that the knot had to be untied. Deciding that the sword was mightier than the pen, he simply cut the knot in half. It may have been cheating, but it certainly solved the problem.

The prophecy was fulfilled, and Alexander used it to bolster his troops as he went on to conquer the Persian Empire and some of India, taking the title "king of kings."

The Humble Fire Hydrant

A device first conceived in the 1600s saves countless lives and millions of dollars in property every year.

✻ ✻ ✻ ✻

FIRE HYDRANTS ARE one of the most ubiquitous fixtures in U.S. cities. Squat, brightly painted, and immediately recognizable, two or three of them adorn virtually every city block in the United States. New York City alone has more than 100,000 hydrants within city limits.

Fire hydrants as we know them today have been around for more than 200 years, but their predecessors first appeared in London in the 1600s. At that time, Britain's capital had an impressive municipal water system that consisted of networks of wooden pipes—essentially hollowed-out tree trunks—that snaked beneath the cobblestone streets. During large blazes, firefighters would dig through the street and cut into the pipe, allowing the hole they had dug to fill with water, creating an instant cistern that provided a supply of water for the bucket brigade. After extinguishing the blaze, they would drain the hole and plug the wooden pipe—which is the origin of the term *fireplug*—and then mark the spot for the next time a fire broke out in the vicinity.

Fanning the Flames of Invention

In 1666, a terrible fire raged through nearly three-quarters of London. As the city underwent rebuilding, the wooden pipes beneath the streets were redesigned to include predrilled plugs that rose to ground level. The following century, these crude fireplugs were improved with the addition of valves that allowed firefighters to insert portable standpipes that reached down to the mains. Many European countries use systems of a similar design today.

With the advent of metal piping, it became possible to install valve-controlled pipes that rose above street level. Frederick Graff Sr., the chief engineer of the Philadelphia Water Works department, is generally credited with designing the first hydrant of this type in 1801, as part of the city's effort to revamp its water system. A scant two years later, pumping systems became available, and Graff retro-fitted Philly's hydrants with nozzles to accommodate the new fire hoses, giving hydrants essentially the same appearance that they have today. By 1811, the city boasted 185 cast-iron hydrants along with 230 wooden ones.

Hydrants Spread Like Wildfire

Over the next 50 years, hydrants became commonplace in all major American cities. But many communities in the north faced a serious fire safety problem during the bitter winters. The mains were usually placed well below the frost line, allowing the free-flow of water year-round, but the aboveground hydrants were prone to freezing, rendering them useless. Some cities tried putting wooden casings filled with sawdust or other insulating material, such as manure, around the hydrants, but this wasn't enough to stave off the cold. Others sent out armies of workers on the worst winter nights to turn the hydrants on for a few minutes each hour and let the water flow.

The freezing problem wasn't fully solved until the 1850s, with the development of dry-barrel hydrants. These use a dual-valve system that keeps water out of the hydrant until it is needed. Firefighters turn a nut on the top of the hydrant that opens a valve where the hydrant meets the main, letting the water rise to street level. When this main valve is closed, a drainage valve automatically opens so that any water remaining in the hydrant can flow out. Very little else about the design of fire hydrants has changed since. In fact, some cities are still using hydrants that were installed in the early 1900s.

Ingenious Uses for Plastic Bottles

Okay, so you know you should recycle them. But every now and then, those plastic bottles come in handy for all kinds of projects around the house. Who knew?

⁕ ⁕ ⁕ ⁕

⁕ Tired feet love this one: Fill a two-liter soda bottle with hot water, cap it, and roll it under your feet to warm your tootsies. Or fill the bottle with cold water and freeze to cool hot, burning feet after a long day.

⁕ Keep your pantry neat with a two-liter plastic soda bottle. Cut the bottom off, turn the bottle upside down, hang it, and stuff plastic bags inside. The small opening acts as a dispenser. This works for storing string, too.

⁕ Fill clean one-liter bottles with juice and freeze them. They'll keep items cold until lunchtime; by then, the juice will be thawed enough to drink.

⁕ Speaking of freezing, fill a one-liter bottle two-thirds full of water and freeze it. Makes a great spill-proof ice pack for little boo-boos.

⁕ The top half of a one-liter soda bottle makes a great water balloon filler. Just slip the balloon over the small opening and fill.

⁕ Milk jugs aren't just for milk! Cut an empty jug at an angle to make a handy scoop. Slice the jug in half and make a funnel from the top part; use the bottom half as a disposable tray for a small paint job. A bottomless jug works great as a protector for delicate seedlings. Fill jugs with water for use as anchors or weights. Fill with sand for use as free weights in your home gym.

* Clean squeeze bottles from mustard or catsup make great summertime outdoor toys. Fill with water, hand them to the kids, and watch the fun. They are also great for bath time.
* Empty laundry detergent bottles make great portable carriers. Cut a medium-size hole opposite the handle and use for hauling whatever you need in and out of the garden, including small produce. Who knew?

Without a Clue

There's nothing puzzling about the origin of the crossword puzzle.

* * * *

THE FIRST PUBLISHED crossword puzzle was created and constructed by an editor and journalist named Arthur Wynne, who was employed by *The New York World* newspaper, a daily rag that adorned doorsteps and magazine racks in the Big Apple from 1860 until 1931. Wynne was asked by his editors to create a new puzzle for the paper that would challenge, entertain, and educate. Wynne, who was originally from Liverpool, England, designed a format similar to Magic Squares, a popular word game he played as a child.

Wynne's puzzle, which was originally dubbed Word-Cross, first appeared in the Sunday, December 21, 1913, edition of *The World*. It was diamond-shape, contained no internal black squares, and provided one free solution (the word *fun*) to get the semantic search started. The clues were not separated into across and down divisions; a numbering system was used to guide the riddle researcher. The first clue to intrigue potential puzzlers was "what bargain hunters enjoy" which, of course, is "sales." Wynne also compiled the first book of crossword puzzles, hitting the bookshelves with his publication in 1924.

In February 1922, the first British crossword was printed in *Pearson's Magazine*, a monthly publication that specialized in essays on the arts and politics and helped spawn the careers of such notable authors as H. G. Wells and George Bernard Shaw. The first crossword to appear in *The Times,* England's national newspaper, was published on February 1, 1930.

The invention of cryptic crosswords, in which each clue is a puzzle in itself, is usually credited to Edward Powys Mathers, a brilliant English scholar, translator, poet, and linguist who compiled more than 650 puzzles for the *Observer* newspaper under the pseudonym Torquemada.

Bet You Didn't Know

Colorful words tend to have colorful histories.

❋ ❋ ❋ ❋

Light-Fingered Finn

The next time you go out with friends for a drink, be careful what you order. According to some historians, Mickey Finn was a notorious Chicago bartender who plied his wares in the late 1890s. When his customers weren't looking, he would spike their drinks with chloral hydrate. After a few minutes, they would pass out and Finn would rob them blind.

F-Word Firsts

Have you ever wondered who first uttered the F-word onscreen? That dubious honor goes to Marianne Faithfull, who said it in the 1968 film *I'll Never Forget What's 'Isname.* But director Brian De Palma broke all the records in his 1983 film, *Scarface,* starring Al Pacino. The F-word was spoken exactly 226 times, or on average, four times every three minutes.

Where's the Tag Tournament?

Those of you who wrangled your way out of doing the dishes as a kid by playing Rock, Paper, Scissors with your brothers and sisters will be happy to learn that the childhood game lives on. Every year, the World RPS (that's Rock, Paper, Scissors) Society hosts their world championship, designed to pit the best players against each other. Players can also go at it with other aficionados at the Sami Tournament in Norway; the New Zealand Rock, Paper, Scissors Championships in Auckland; the Australian RPS Championships on the Gold Coast; or challenge your fists against the best at the Camden Riversharks Tournament in Camden, New Jersey. And, for the slightly more cultured, try the Sixth Annual Roshambo Winery Rock, Paper, Scissors Tournament in Sonoma, California.

Of Monks and Chefs

White chef hats date back to the 15th century, when several Greek nobles hid from the invading Turks in monasteries. The nobles dressed as monks—in tall black hats and robes—and were given cooking responsibilities. The monks grew frustrated that they couldn't tell the nobles apart from other monks, so they changed the nobles' uniforms from black to white.

The 30-Minute Scourge

To quote a popular saying, you can run but you can't hide. This is especially true if you happen to be flipping through television channels. There, the dreaded infomercial will find you.

* * * *

WITH A SIMPLE pen stroke and little fanfare, President Ronald Reagan signed off on the Cable Communications Policy Act of 1984. Among a host of other things, the act deregulated television and allowed cable stations to sell airtime to the highest bidders. Soon, the entire medium would be altered. The infomercial (information commercial) had been spawned.

Once upon a time, commercials were 15–60 seconds long. They could be an annoyance, but viewers understood them to be a necessary evil. But then the infomercial, a was born. The 30-minute marauder first invaded an overnight no-man's-land of postbroadcast test patterns and "snow." Today, the beast can be found anytime and anywhere—often in disguise.

Current infomercials are often modeled after entertainment talk shows. It's part and parcel of the genre—infomercials strive to make us believe that we absolutely need the product being pushed, and they can get personal. (*Really* personal.)

But who's to say that we don't need the "breakthrough" items being hawked? Certainly a child with an unruly mane will benefit from a hair-cutting system that employs a filthy vacuum cleaner. What about a person suffering from "shameful" acne? Might they not benefit from an undisclosed "secret liquid formula" that purportedly flogs humongous craters into a smooth expanse? Of course, this doesn't even touch on "miracle" pills for men suffering from a humiliating "gravity" problem.

Hate them or hate them more, infomercials are here to stay. In 2003, the absurdly long advertisements raked in $91 billion in overall sales. That, as they say, ain't hay. The moral: Someone is buying those Ginsu knives and Ab Rollers.

Pantyhose—Not Just for Legs

Every woman—and not just a few men—knows that pantyhose can make legs look oooh-la-la. But not every woman (and very few men) knows that there are many other things hose can do. Who knew?

* * * *

DID YOU LOSE something small and hard to find? Slide a length of pantyhose over your vacuum cleaner's hose, secure it tightly with rubber bands, and carefully vacuum where you think the lost item might be. The hose will keep the item from being sucked up into the bag.

Cut an old piece of pantyhose slightly larger than your new hairbrush. Push the bristles through the hose. When it's time to clean the brush, just pull the old pantyhose off and refresh it with a new piece.

Keep pantyhose in the craft room. When shredded, it works great for stuffing soft toys. Cut the pantyhose into small squares for use instead of cotton balls. Pantyhose wrapped and rubber-banded around a wooden stick makes an inexpensive, in-a-pinch paintbrush. Use a clean piece of pantyhose to test your sanding job for snags.

Line your houseplant's pots with pantyhose to prevent soil loss from the bottom of the pot. Cut pantyhose into strips and use them to tie seedlings to posts or to tie bundles of brush. Wider strips attached to stakes can be used as a "hammock" for melons or large produce in the garden. Or leave the legs intact, hang the hose so mice can't reach it, and store seeds in them for the winter. (This works great for storing onions and potatoes at the end of the growing season, too.)

Use recycled pantyhose as mesh bags in the laundry room. Just cut off a length of leg, drop in your delicates, and wash as usual. Use the other leg as a back-scrubber: Insert a bar of soap or two, tie a knot at both ends, and scrub-a-dub-dub. Who knew?

Why Did John Hancock Sign His Name So Big on the Declaration of Independence?

Poor John Hancock—he was the president of the Continental Congress as the United States sprang to life, and a nine-term governor of Massachusetts. Heck, he even graduated from Harvard.

* * * *

BUT WHAT'S HIS legacy? Penmanship. If he had known this would be his ignominious fate, perhaps he wouldn't have scrawled his name so largely on the Declaration of Independence or added that fancy loop to the "k." But he did write that big, and he did add that loop—so now we're forced to listen to some waiter quip, "Just give me your John Hancock," every fifth time we pay with a credit card at a restaurant.

One of the reasons Hancock's signature is so enormous is that he, as president of the Continental Congress, was the first to sign the Declaration of Independence. Hancock had plenty of real estate, and he used it. But goodness gracious, there were fifty-five other signatures that needed to be added. Leave some room for everyone else, dude.

Still, it wasn't necessarily a case of a man doing something simply because he could. Hancock felt that a big signature was important. Signing such a document did two things: First, it told American colonists and the rest of the world why the Congress felt it was necessary to break away from Great Britain. Second, by creating the Declaration of Independence, the congressional members were directly insulting England's King George III, a treasonous act that could lead to hanging. Hancock believed that a bold sweep of his feathered quill would instill confidence and courage into his fellow colonial delegates, and into everyone else who read the document.

It's been said that after signing his name, Hancock defiantly exclaimed, "There, I guess King George will be able to read that!" or "The British ministry can read that name without spectacles; let them double the reward for my head!" Sure, and George Washington never told a lie. In all likelihood, Hancock never made such a boast—there simply wasn't the audience for it. Only one other person was present when Hancock signed the Declaration of Independence: Charles Thomson, the secretary of the Continental Congress, who claimed that Hancock never uttered such words. Besides, saying something that grandiose with just one other person in the room would have been, well, weird.

The delegates voted to ratify the Declaration of Independence on the night of July 4, 1776, but they did not sign it. (Now, go out and use that piece of info to win a bar bet!) The first version was printed, copied, signed by Hancock and Thomson, and distributed to political and military leaders for their review. On July 19, the Congress ordered that the document be "fairly engrossed on parchment," a fancy way of saying officially written. On August 2, the final version was ready to be signed. Hancock signed first, putting his John, er, his name in the middle of the document below the text. As was the custom, others started signing their names below Hancock's.

Not everyone whose name is on the Declaration of Independence was present that day. Signatures were added in the coming days, weeks, months, and years. The last person to sign was Thomas McKean, in 1781. And you just know that when McKean saw what little room there was for his signature, he thought, "[Bleeping] Hancock!"

Precious Cargo: Airplane Stowaways

Some people are so desperate to flee their home country that they're willing to risk their lives. One method of escape is to stow away in the wheel well of a jet plane. The odds of pulling off such a stunt are staggering, but there appears to be little that U.S. regulators can do to prevent the practice on international flights. Despite the risks, people such as Pardeep Saini of India and Fidel Maruhi of Tahiti were willing to take the chance.

* * * *

A New Beginning

IN 1996, PARDEEP Saini, 22, and his brother Vijay, 19, were desperate to escape their native India. Details surrounding their motives are cloudy (Pardeep claims he was persecuted for alleged links with Sikh separatists), but their method of "flight" is clear. According to the elder Saini, a smuggling agent in Delhi told the brothers they could stow away in the undercarriage of a jumbo jet and wing it to England, where they could request political asylum. On paper, the idea seemed entirely workable, but such things rarely go according to plan.

In October 1996, the brothers climbed high up into the wheel well of a British Airways Boeing 747 and crossed their fingers. They were in for the ride of their lives.

In 2000, Fidel Maruhi of Tahiti set his sights on a new life in the United States. At Tahiti's Faa'a International Airport, Maruhi crawled into the wheel well of a Los Angeles-bound Air France jet and hunkered down for the forthcoming aerial onslaught.

The Death Zone

At 29,035 feet, Mt. Everest is considered a killer mountain, not for its steepness or avalanche dangers, but because of the peak's extreme height. Simply put, humans can't survive long at an altitude where oxygen density is about one-third that found at sea level. During their odyssey, the Saini brothers spent ten hours at heights approaching 39,000 feet—nearly 10,000 feet higher than the celebrated death summit.

Shortly after takeoff, Pardeep lost consciousness, sparing him the panic of gasping for air and consciously freezing in temperatures that would eventually reach –40°Fahrenheit.

On his way to Los Angeles, Maruhi's airliner reached its cruising altitude in approximately 20 minutes. During the plane's ascent, the outside air temperature quickly plunged, and oxygen in the air decreased as Maruhi's airplane cruised at 38,000 feet for nearly eight hours.

Tragedy Strikes

After ten agonizing hours, the Saini brothers' plane prepared to land at Heathrow Airport. At 2,000 feet, the pilot lowered the landing gear, and, in an instant, Vijay was sucked out of the airplane. His limp body was later recovered in Richmond, a neighborhood in southwestern London. He had frozen to death before he fell from the plane.

This left Pardeep. Surely no one could survive the oxygen deprivation and extreme cold. But he did. Baggage handlers found him suffering from extreme hypothermia. By a fateful stroke, the stowaway had defied huge odds and made it to Britain. Saini was initially denied asylum, but, upon appeal, he was later granted "compassionate leave" to remain in England.

A Shocking Discovery

When Maruhi's plane landed at LAX Airport, shocked workers spotted his body tucked deep into the wheel well. He was alive, but just barely. The ordeal had lowered the Tahitian's body temperature to an astounding 79 degrees—six degrees below that which is generally considered fatal. Maruhi was transported to a hospital where he was treated for hypothermia and frostbite. Miraculously, he survived. Maruhi remembers nothing about the trip, which is understandable because he'd blacked out just after takeoff. Despite his ordeal, U.S. authorities shipped him back to Tahiti. His dream of a life in the United States remains unfulfilled.

Random Records

What's it like to somersault the length of Paul Revere's 12-mile ride from Boston to Concord? Ask Ashrita Furman. In April 1986, he accomplished the feat—one of many that landed him in Guinness World Records. Born in 1954, this Queens, New York, native has jumped 23.11 miles on a pogo stick, raced the fastest mile hula-hooping while balancing a milk bottle on his head (11 minutes, 29 seconds), and duct taped himself to a wall in 8 minutes, 7 seconds. In fact, Furman has the record for the most records in Guinness: 225 since 1979. Furman's accomplishments may be exceptional, but he's not the only one trying to shine.

✻ ✻ ✻ ✻

✻ According to the *Book of Alternative Records*, New Zealander Mem Bourke glued a record 31,680 rhinestones to the body of her friend Alastair Galpin in 2006.

* American magician Todd Robins ate a total of 4,000 lightbulbs during his nearly 30-year career, which ran from 1980 to 2009.
* On October 25, 2008, 4,179 people from ten nations, who were joined by video feed, simultaneously performed Michael Jackson's "Thriller" routine.
* A record 552 children danced the Macarena together at Britain's Plumcroft Primary School in 2009.
* Some kids are born record-breakers. On November 30, 1998, triplets Peyton, Jackson, and Blake Coffey came into the world with a collective weight of 3 pounds, 8 ounces, making them the smallest threesome ever born. On February 9, 2003, Tomasso Cipriani of Italy became the world's largest newborn when he weighed in at 28 pounds, 4 ounces.
* But what's really the strangest record ever? Consider this: Between September 21 and October 6, 1984, Rob Gordon of Shropshire, England, sat in a bathtub filled with cold spaghetti for a total of 360 hours while working as a disc jockey at a local radio station. Try to top that one, Ashrita Furman!

Fumbling Felons

There's dumb, and then there's this.

* * * *

Helping Hands

A ROBBER IN ENGLAND broke into a local supermarket, but the police arrived quickly and apprehended him. Despite the handcuffs, however, the man somehow managed to break free before the cops could bundle him off to the station.

The devious crook had pulled off a near-miraculous escape and should have thanked his lucky stars. But there was still something that didn't feel right to him—he was still wearing the handcuffs.

The dim-witted robber went to the nearest police station, hoping that the cops would help him get the handcuffs off. After all, they were the only things marring a perfect escape. Not surprisingly, the cops didn't see it his way and quickly rearrested him.

The Flasher

Kids love those sneakers with the lights in the heels, because every time they put their foot down, the lights flash. Good for kids; bad for crooks.

A Kansas criminal found this out after he robbed a convenience store at night. Fleet of foot, he knew the area well and was confident in his ability to elude the cops that were chasing him. But to the bandit's dismay, every time he darted down another dark alley to elude a cop, more would show up right behind him to continue the pursuit. No matter how much he juked and jived, hopped fences, and cut through darkened yards, there was always an officer hot on his trail.

All good things come to an end, and ultimately, so did the crook's stamina. The cops corralled their man, who was amazed that all of his fancy footwork had gone for naught. It was then he learned how the cops had always been able to find him: They had merely followed the flashing lights in the heels of his athletic shoes, which gave them clear pursuit in the dark. Unfortunately for the crook, the next red flashing lights he saw were on the police car he rode in to the station.

❋ Chapter 11

Space Dogs, Flying Cats, and More

Bats Make Good Bombs

Small, winged, nocturnal, furry, and explosive. These critters were supposed to help end the war but succeeded only in destroying an American airplane hangar and a general's car.

❋ ❋ ❋ ❋

WORLD WAR II inspired innovation and invention as scientists and engineers from the major powers strove to develop weapons that would provide a winning edge. Many of these innovations are well-known—jet engines and rockets in Germany, for instance. However, those that did not work so well are scarcely remembered.

In the early days of the war, an American dental surgeon from Irwin, Pennsylvania, named Lytle S. Adams conceived an idea while on vacation in the American southwest. He proposed that the United States develop a method for attaching incendiary bombs to bats and releasing thousands of the flying mammals over Japan. Under Adams's logic, the bats would roost in wooden buildings and explode, causing fires that would spread out of control. On paper, Adams's idea held merit—a typical bat can carry 175 percent of its body weight, and since the Japanese populace would not detect the roosting bats, the fires could spread unchecked.

In the weeks following America's entry into the war, thousands of citizens sent ideas for new weapons to the White House, and the bat-bomb proposal was one of the very few that went into development. Approved by President Roosevelt, it eventually consumed a modest $2 million of taxpayers' money.

By March 1943 a team consisting of Dr. Adams and two university chemists had scoured the caves of the southwest in search of the perfect species for the project. Although the mastiff bat was larger and the mule-eared bat more common, the team settled on the Mexican free-tailed bat, because it could carry the requisite weight and was available in large numbers (in fact, one colony of free-tailed bats near Bandera, Texas, numbered some 20 to 30 million animals).

Months of testing followed. The creatures were tricked into hibernation with ice, then a small explosive device was surgically attached by a string. The procedure was delicate and required lifting the bats' fragile skin, which was liable to tear if done incorrectly. The prepared bats were then loaded into cardboard cartons, which were parachuted from aircraft and opened at a preset altitude. There were numerous complications, however. Many of the containers did not open or the bats did not wake up and plummeted to their deaths. Still, the bats did succeed in burning down a mock Japanese village. On the other hand, they managed to start a fire in an airplane hangar that also destroyed a visiting general's car. Perhaps for this reason in June 1943, after more than 6,000 bats had been used in tests, the army handed the project to the navy. It was renamed Project X-Ray.

The navy eventually handed off the project to the Marine Corps, which determined that the bat bombs were capable of causing tenfold the number of fires as the standard incendiary bombs being used at the time. However, when Fleet Admiral Ernest J. King learned that the bats would not be ready for deployment until mid-1945, he called off the project. Dr. Lytle

Adams was bitter about the cancellation of his novel idea. He maintained that the bat bombs could have caused widespread damage and panic without the loss of life that resulted from the use of the atomic bomb.

Laika the Astronaut Dog

The first occupied spacecraft did not carry a human being or even a monkey. Instead, scientists launched man's best friend.

* * * *

THE SPACE AGE officially began on October 4, 1957, when the Soviet Union launched humanity's first artificial satellite, *Sputnik I*. The world sat stunned, and the space race experienced its first victory. Even more astounding was the launch of *Sputnik II* on November 3. *Sputnik II* carried the first living creature into orbit, a mongrel dog from the streets of Moscow named Laika.

A Spaceworthy Dog

A three-year-old stray that weighed just 13 pounds, Laika had a calm disposition and slight stature that made her a perfect fit for the cramped *Sputnik II* capsule. In the weeks leading up to the launch, Laika was confined to increasingly smaller cages and fed a diet of a nutritional gel to prepare her for the journey.

Sputnik II was a 250-pound satellite with a simple cabin, a crude life-support system, and instruments to measure Laika's vital signs. After the success of the *Sputnik I* launch, Soviet Premier Nikita Khrushchev urged scientists to launch another satellite on November 7, 1957, to mark the anniversary of the Bolshevik Revolution. Although work was in progress for the more sophisticated satellite eventually known as *Sputnik III*, it couldn't be completed in time. Sergei Korolev, head of the Soviet space program, ordered his team to design and construct *Sputnik II*. They had less than four weeks.

The First Creature in Space

The November launch astonished the world. When the Soviets announced that Laika would not survive her historic journey, the mission also ignited a debate in the West regarding the treatment of animals. Initial reports suggested that Laika survived a week in orbit, but it was revealed many years later that she only survived for roughly the first five hours. The hastily built craft's life-support system failed, and Laika perished from excess heat. Despite her tragic end, the heroic little dog paved the way for occupied spaceflight.

Why Are Skunks So Stinky?

Oh, so you smell like a bed of roses? But seriously, skunks have earned their odiferous reputation through their marvelous ability to make other things stink to high heaven.

⁕ ⁕ ⁕ ⁕

All eleven species of skunk have stinky spray housed in their anal glands. However, as dog owners can attest, skunks aren't the only animals to have anal glands filled with terrible-smelling substances. Opossums are particularly bad stinkers; an opossum will empty its anal glands when "playing dead" to help it smell like a rotting corpse.

While no animal's anal glands are remotely fragrant, skunks' pack an especially pungent stench. This is because skunks use their spray as a defense mechanism. And they have amazing range: Skunks have strong muscles surrounding the glands, which allow them to spray sixteen feet or more on a good day.

A skunk doesn't want to stink up the place. It does everything in its power to warn predators before it douses its target with *eau de skunk*. A skunk will jump up and down, stomp its feet, hiss, and lift its tail in the air, all in the hope that the predator will realize that it's dealing with a skunk and go away. A skunk only does what it does best when it feels it has no choice. Then it releases the nauseating mix of thiols (chemicals that contain super-stinky sulfur), which makes whatever it hits undateable for the foreseeable future. Skunks have enough "stink juice" stored up for about five or six sprays; after they empty their anal glands, it takes up to ten days to replenish the supply.

Being sprayed by a skunk is an extremely unpleasant experience. Besides the smell, the spray from a skunk can cause nausea and temporary blindness. Bobcats, foxes, coyotes, and badgers usually only hunt skunk if they are really, really hungry. Only the great horned owl makes skunk a regular snack—and the fact that the great horned owl barely has a sense of smell probably has a lot to do with it.

Why Do Cats Always Land on Their Feet?

It's true: Cats have an uncanny ability to survive a fall. Maybe that's why we say they have nine lives. However, the notion that they always land on their feet isn't exactly accurate.

* * * *

EVERY ONCE IN a while, a cat does indeed go splat. And it's probably because a very sorry owner left a window or balcony door open. Cats are not afraid of heights. If they see a tasty bird or butterfly floating about outside, their predatory instincts kick in and it's jump time. This phenomenon is so common that a 1987 study of falling cats in the *Journal of the American Veterinary Medical Association* even gave it a name: High-Rise Syndrome. New York City veterinarians often use the term to describe the injuries that cats sustain after falling from the city's high-rise apartment windows.

Whether a cat lands on its feet after a fall depends on several factors, including the distance it plummets and the surface on which it lands. If a cat falls a short distance (say, fewer than one or two stories), it usually can right itself in midair. How?

For starters, cats have an amazing sense of balance and coordination. Each of their inner ears is outfitted with a vestibular apparatus, a tiny fluid-filled organ that helps them register which way is up. When the cat is falling, the fluid in the inner ear shifts, telling the cat to reorient its head until the fluid is once again equalized and level.

When a cat turns its head and forefeet, the rest of its body naturally follows. How so? Cats have super-flexible musculoskeletal systems. A cat's backbone is like a universal joint—it has thirty vertebrae (five more than a human) and no collarbone. This is why cats are so agile. With this freedom of movement, a cat can instantly bend and rotate like a pretzel to land on its feet.

If a cat falls from more than one or two stories, it likely will sustain severe or fatal injuries, even if it can right itself. Its legs and feet simply cannot absorb all the shock. That said, the study in the *Journal of the American Veterinary Medical Association* revealed something really surprising: Of the 132 high-rise cats that veterinarians examined, those that fell from above seven stories had a better chance of escaping injury.

It seems that after plummeting five stories or so, cats reach a nonfatal terminal falling velocity. At this point, they are able to relax their muscles and spread their bodies out like feline parachutes or flying squirrels, and they arch their backs just before hitting bottom, reducing the force of the impact.

Just how far can a cat fall without being killed? The longest non-fatal fall on record is forty-two floors.

How Long Would It Take a School of Piranha to Polish Off a Cow?

Are stories of this predator's ravenous appetite to be believed?

❋ ❋ ❋ ❋

First consider a few factors: How large is the cow? More important, how many fish are there, and how hungry are they? Like sharks, piranhas are drawn to blood; they're killers from the moment they are born. And it's true—a pack of piranhas can indeed strip the flesh from a much larger animal, such as a cow.

The estimated time it would take a school of piranhas to skeletonize a cow varies. Some sources claim it wouuld take less than a minute; others say up to five minutes. But marine biologists call these estimates exaggerations. The piranha has a fearsome, tooth-filled grin—but under normal circumstances, it is not considered overly aggressive.

In the United States, the legend of the piranha began with Theodore Roosevelt's 1913 trip to South America. He returned full of stories, many of which concerned the carnivorous fish. It is thought that the Brazilian tour guides who were charged with showing President Roosevelt a good time had a joke at his expense by making piranhas out to be more dangerous than they are. There was an incident in which a cow was lowered into a branch of the Rio Aripuana that was teeming with piranhas, and the outcome was every bit as grisly as legend says. But some vital facts were left out of the story. For one, the cow was sick and bleeding, which spurred the fish into a frenzy. Furthermore, the piranhas were isolated, hungry, and ornery. They saw a meal and went nuts—and we've been talking about it ever since.

Modern jungle-dwellers don't typically see piranhas as a danger. The fish usually feed on small animals—other fish, frogs, and baby caimans. It's not uncommon for a human to be bitten by a piranha, but these wounds are usually small and singular. Little flesh lost, little harm done—the fish and the human go their separate ways.

Of course, this isn't to say that piranhas lack the capacity to wreak havoc. Piranhas are known to be most vicious during the dry season. They are believed to travel in large schools for the purpose of protection, and they stimulate one another at feeding time. In light of this information, there are instances when you don't want to get anywhere near a piranha. If you have an open wound, for instance, it might be a good idea to forego that afternoon dip in the Rio Aripuana.

Freaky Facts: Platypus

With aspects of a bird, a reptile, and a mammal, the Australian platypus is one of nature's oddest animals.

- Male platypuses secrete poison from the spurs on their hind feet. The venom won't kill a human adult, but it can cause severe pain and swelling.
- The platypus' hair makes it effectively waterproof, with a dense, insulating undercoat similar to that of a Labrador retriever.
- While under water, the platypus shuts its eyes, ears, and nostrils and uses an electroreceptor system in its sensitive, ducklike bill to navigate and hunt.
- The platypus' fatty, beaverlike tail is used for energy storage, much like a camel's hump. When food runs short, the animal survives on the stored fat.
- The platypus isn't noisy, but when disturbed, it may give a call that sounds like a frog-throated puppy growl.
- Platypus eggs are rubbery or leathery, like turtle eggs, and are about the size of gumballs. Females generally lay only two or three eggs during the annual mating season, which occurs between June and October.

What Eats Sharks?

One of the most feared animals in the world, the shark has a reputation for being a people killer, ruthlessly nibbling on a leg or an arm just to see how it tastes.

* * * *

IN THE SHARK VS. people debate, guess who loses? Yup, sharks. We eat way more of them than they do of us. And we aren't the only ones partaking in their sharkliciousness.

For the most part, the big predator sharks are in a pretty cushy position ecologically. As apex predators, they get to do the eating without all that pesky struggle to keep from being eaten. They are important to the ecosystem because they keep everything below them in check so there are no detrimental population booms. For example, sharks eat sea lions, which eat mollusks. If no one ate sea lions, they'd thrive and eat all the mollusks. So if sharks are apex predators (so are humans, by the way), they aren't ever eaten, right? Wrong. Sometimes a shark gets a hankering for an extra-special treat: another shark.

Tiger sharks start eating other sharks in the womb: Embryonic tiger sharks will eat their less-developed brothers and sisters. This practice of eating fellow tiger sharks continues through adulthood. And great white sharks have been found with four- to seven-foot-long sharks in their stomachs, eaten whole.

There's also what is called a feeding frenzy. What generally happens is that an unusual prey (shipwreck survivors, for example) presents itself and attracts local sharks, which devour the unexpected meal. The sharks get so worked up while partaking, they might turn on each other.

Orcas and crocodiles have also been known to eat shark when the opportunity presents itself. Note that both orcas and crocodiles are also apex predators. So while there are no seafaring animals that live on shark alone, sharks aren't totally safe.

Finally, there's that irksome group of animals known as humans. Many people who reside in Asia regularly partake of shark fin soup, among other dishes prepared with shark ingredients. Through overfishing, humans reduced the shortfin mako's population in the Atlantic Ocean by 68 percent between 1978 and 1994.

Even with all this crazy shark-eating, it's a good bet that a sea lion or mackerel would happily trade places with the apex predator any day of the week.

Roles That Animals Have Played for Humans

Animals have been kept by humans for companionship, utility, and pleasure since prehistoric times. But the role of pets has evolved in the past few hundred years.

* * * *

Animal Stars

ANIMALS HAVE BEEN amusing people since the entertainment business began. Circuses have wowed audiences for hundreds of years, and Wild West shows have been a key attraction since the mid-1800s.

Starting in the 1950s, animals were introduced in television series, forever changing the way people related to their pets. *Fury* and *My Friend Flicka* showcased horses' relationships with their owners. Fury could kneel, limp, and knock boxes out of people's hands . . . and even smile for the camera! Mr. Ed refused to talk to anyone but his owner, but he'd count by pawing the ground the appropriate number of beats for anyone. A large black bear named Bruno costarred with Clint Howard on *Gentle Ben*; the bear slept with its trainer and drank Coca Cola as a treat.

It's been written that Suzy the dolphin, better known as Flipper, once saved TV producer Ivan Tors from drowning. Tors produced a number of shows, among them, *Gentle Ben*. That show and *Flipper* reportedly filmed near each other; supposedly, Bruno and Suzy were friendly.

But perhaps the most famous animal star to date was Lassie. A two-time Emmy winner, Lassie—whose original name was Pal—was owned by trainers Rudd and Frank Weatherwax. This honey-colored collie became a symbol of love, trust, and heroism after appearing in movies and debuting in the TV series (also called *Lassie*) in 1954. During the course of an incredible 588 episodes, Lassie was transformed from a television star into a cultural icon. The Weatherwax brothers occasionally bred Lassie, whose descendants now number in the hundreds.

Beautiful, Brave, and Smart

Some animals exhibit great feats of bravery, often saving their owners from fire, danger, or medical trauma. Dogs have been known to alert neighbors when their owners become impaired or unconscious, wake people when they smell smoke, or save the inhabitants of the house from dangerous intruders.

Guide dogs provide invaluable service, too. Educated from puppyhood, these canines undergo rigorous training in guide-dog schools; only the very talented and those with special abilities graduate to become workers for the visually impaired. Recently, several organizations have begun training guide dogs to warn diabetics before they have seizures due to dangerously low blood-sugar levels. These dogs can detect a chemical imbalance while the person is sleeping; they then wake the person so he or she can immediately inject insulin. Now that's quite a nose.

I Wanna Be Like . . .

Some people prefer coming home to a pet with eight sticky legs, one that's scaly and poisonous, or one that can swing from room to room. Hollywood celebrities have encouraged the exotic pet trend, and John D. Consumer quickly picked up the habit faster than a case of lice. Monkeys, chinchillas, kangaroos, tarantulas, rodents, eagles, pythons, iguanas, and tropical fish top the list of animals that are sold in the United States; some are legally imported, some not. Demand for exotic pets is booming, but buyers beware: The CDC reports that most imported animals arrive with minimal screening and no quarantine, making zoonotic diseases (those that jump to humans) accountable for three-quarters of all emerging infectious threats.

So, whether they take their pets to Capitol Hill (as several politicians do), bring them to their medical offices (some dentists and doctors claim that animals help relieve the stress of procedures for many patients), or stick their heads in a lion's mouth, people will always have a strong relationship to animals . . . real *and* fictional.

Why Don't Animals Need Glasses?

Humans are so quick to jump to conclusions. Just because you've never sat next to an orangutan at the optometrist's office or seen a cat adjust its contact lenses, you assume that animals don't need corrective eyewear?

* * * *

ANIMALS DO DEVELOP myopia (nearsightedness), though it seems less widespread in nature than among humans. For one thing, nearsighted animals—especially carnivores—would have an extremely difficult time hunting in the wild. As dictated by the rules of natural selection, animals carrying the myopia gene would die out and, thus, wouldn't pass on the defective gene.

For years, nearsightedness was thought to be mainly hereditary, but relatively recent studies have shown that other factors may also contribute to the development of myopia. Some researchers have suggested that myopia is rare in illiterate societies and that it increases as societies become more educated. This doesn't mean that education causes nearsightedness, but some scientists have speculated that reading and other "close work" can play a role in the development of the condition.

In accordance with this theory, a study of the Inuit in Barrow, Alaska, conducted in the 1960s found that myopia was much more common in younger people than in older generations, perhaps coinciding with the introduction of schooling and mass literacy in Inuit culture that had recently occurred. But schooling was just one component of a larger shift—from the harsh, traditional lifestyle of hunting and fishing at the edge of the world to a more modern, Western lifestyle. Some scientists believe that the increase of myopia was actually due to other changes that went along with this shift, such as the switch from eating primarily fish and seal meat to a more Western diet. This diet is heavier on processed grains, which, some experts believe, can have a bad influence on eye development.

And this brings us back to animals. Your beloved Fido subsists on ready-made kibble that's heavy on processed grains, but its ancient ancestors ate raw flesh. If this switch to processed grains might have a negative effect on the eyesight of humans, why not in animals, too?

Unfortunately, there's not much we can do for a nearsighted animal. Corrective lenses are impractical, glasses would fall off, and laser surgery is just too darn expensive. Sorry, Fido!

8 Top Dogs

Everyone who has ever loved a dog feels that their dog is the best dog ever. We're not about to get between a man (or woman) and their best friend, but here are a few dogs that truly stand out in a crowd of canines.

* * * *

1. **Coolest Dog on the Playground: Olive Oyl:** In 1998, a Russian wolfhound named Olive Oyl of Grayslake, Illinois, made the Guinness Book of World Records when she skipped rope 63 times in one minute.
2. **Smallest Dog: Tiny Tim:** Measuring three inches tall at the shoulder and four inches long from wet nose to wagging tail, Chihuahua and shih tzu mix Tiny Tim of London holds the record (as of 2004) for being the tiniest dog ever. The little guy weighs just over a pound.
3. **The Quietest Dog: The Basenji:** A yip or a yap, a whine or a woof—if you don't want a barking dog, consider a basenji. This dog was a particular favorite of ancient Egyptians. The breed is incapable of barking, instead uttering a strangely unmelodious sound called a yodel, which makes them perfect for those living in an apartment with thin walls or touchy neighbors.
4. **The Heaviest (and Longest) Dog: Zorba:** In Kazantzakis' famous novel, Zorba the Greek tackled spiritual and metaphysical quandaries; Zorba the dog apparently tackled his dinner. Zorba, an Old English mastiff, was the world's heaviest and longest dog ever recorded. Zorba weighed 343 pounds and, from nose to tail, was eight feet three inches long.

5. **The Oldest Dog: Bluey:** The oldest dog reliably documented was an Australian cattle dog named Bluey. After 29 years and 5 months of faithful service, Bluey was put to rest in 1939. We can only hope that now Bluey is chasing cows in the big cattle ranch in the sky.
6. **Most Courageous Dogs: September 11 Search and Rescue Dogs (SAR Dogs):** Okay, so any dog serving its country as a SAR dog gets the "Most Courageous Dog" distinction, but the SAR dogs that waded into the rubble in the wake of the terrorist attacks on September 11, 2001, get an extra gold star. Hundreds of SAR dogs scoured the debris and braved the chaos in the days following the attack. While German shepherds are often trained for SAR duties, any working, herding, or sporting breed can be trained to be a hero.
7. **The Dog That Might Score Low on an IQ Test: A Hound:** We couldn't bring ourselves to say dumbest, but it looks like the hound group is given this reputation most often. Hounds weren't bred for taking IQ tests, or doing much of anything except hunting and following scents, so expecting them to quickly learn how to sit or stay is a big mistake. It's just not in their nature, so have patience with your hound—they're not dumb, they're different.
8. **Dog Most Likely to Help You with Your Algebra Homework: The Border Collie:** Border collies are widely regarded as the smartest of dogs, since they have been bred to work closely with humans for centuries. Again, different dogs are better at certain tasks and are more apt to thrive in different environments. However, collies can appear hyper and less-than-brilliant if they're not given enough stimulation.

Unpleasant but True Facts About Bats

Long associated with vampires, bats have gotten a bad rap. But the truth is that bats are our friends. They keep down the insect population, pollinate flowers, and are a vital part of many ecosystems. Of course, this doesn't change the fact that they're kind of creepy.

* * * *

They Carry Rabies.

Okay, not all of them. But some of them. About .5 percent of bats have rabies. And bat bites cause about 71 percent of rabies cases in the United States. They might not, as old wives' tales have it, purposefully tangle themselves in your hair, but they do bite. It's not a bad idea to stay away from them.

They Have Razor-Sharp Teeth.

Sure these teeth are tiny, but they're still, very sharp. And scary.

Some of Them Have Strangely Long Tongues.

One species of nectar bat, the *Anoura fistulata,* has a longer tongue relative to the length of its body than any other mammal. It uses this long tongue to reach inside flowers and get pollen. When the bat is not using its tongue, the appendage curls up in the bat's rib cage like a little garden hose.

One Bat Can Eat 1,000 Insects in an Hour.

This might be less unpleasant if you consider that bats mean fewer insects flying around and bothering you.

Bloodsucking

Some species of bats suck the blood from the backs of cattle or the feet of chicken. They use their sharp toenails to slit the skin of these animals and then lap up the blood.

Why Does Australia Have So Many Poisonous Snakes?

Many people associate the cute and cuddly koala with Australia. And that's exactly the image the nation's tourism industry wants to tout: cute and cuddly. Deadly and dangerous wouldn't sell as many vacation packages, though it would be more accurate.

* * * *

AUSTRALIA IS A place that would drive the snake-phobic Indiana Jones to the brink of insanity—there are snakes, snakes, and more snakes, many of which are poisonous. Of the approximately six hundred known venomous snakes in the world, a whopping sixty-one reside in Australia, according to the University of Sydney. And the Australia Venom Research Unit reports that eight of the ten most toxic land snakes on the planet are native to the continent.

Cute and cuddly? We think not. Thirty-five percent of the snake species in Australia are poisonous. Why does this continent host so many scary slitherers? Hundreds of millions of years ago, Australia was part of the supercontinent Gondwana, which also included South America, Africa, India, New Zealand, and Antarctica. Gondwana began to break up one hundred and fifty million years ago, and Australia snapped off altogether about fifty million years ago.

The snakes that were on the terrain now known as Australia included those from the Elapidae family, a group that had many venomous varieties. Once this land mass became surrounded by water, the snakes had nowhere to go. So they developed ways to survive on this biodiverse continent, which has a rain forest, vast deserts, and the largest coral reef on the planet. As is the case with natural selection, the strongest varieties lasted; many of today's venomous serpents are descendants of the Gondwana castaways.

Australia's venomous snakes come in a variety of lengths and colors, and they reside in many of the continent's environments. The deadliest—not just on Australia, but on the entire planet—is the inland taipan. This snake, part of the ancient Elapidae family, has venom potent enough to kill one hundred humans in a single bite. Close behind on the venom chart are the eastern brown snake and mainland tiger snake, which also hail from the dreaded Elapidae clan. Both can seriously ruin a vacation.

But don't let these snakes—or the continent's other poisonous critters, such as the box jellyfish or the funnel web spider—scare you away. If you visit Australia, the main predator you'll need to beware of is the human—specifically, those who are driving cars. Auto accidents cause more deaths each year in Australia than all of its poisonous creatures combined.

Animal Rights in Hollywood

Throughout motion picture history, animals have taken it on the chin. Regular abuse was the sad reality of their Hollywood existence until a series of laws helped right the wrongs.

* * * *

A Profitable Fur Trade

DURING THE EARLY days of filmmaking, a laissez-faire attitude existed within the motion picture industry. Like any organization in its infancy, rules and regulations regarding accepted practices were made up on an as-needed basis. Not surprisingly, mistreatments and abuses occurred, particularly for Hollywood's performance animals. As a furry subgroup without a voice, animals were treated as producers and directors saw fit. With eyes firmly fixed on the financial bottom line, filmmakers treated animals as little more than props. With the exception of a few breakout performers like Rin Tin Tin and Jackie the Lion, animal treatment bordered on the appalling.

Horsing Around

Shocking examples of mistreatment to animals, both on-screen and off, were enough to give an animal lover nightmares. Horses were tripped, shocked, and run ragged. When a scene called for a horse to fall, wires were strung around its ankles or through its hooves, and a vicious yank did the job. During the making of *The Charge of the Light Brigade* (1936) and *Ben-Hur* (1959), it is believed that 31 horses were killed or euthanized after being wire-tripped.

Advocates Step In

To get a horse to plunge off a cliff, the animal was blindfolded and sent down a heavily greased metal tilt chute dangling high above a waterway. After a grisly tilt chute accident claimed a horse's life in *Jesse James* (1939), the animal-welfare group American Humane damned Hollywood for its callous and brazen mistreatment of animals. Hoping to sidestep bad press, Hollywood bigwigs made swift changes. In 1940, provisions were added to the Production Code to include the prohibition of trip wires, branding, and tilt chutes. Also, if a film featured scenes with animals, a representative from American Humane's Film & TV Unit had to be invited to the set. It was a step in the right direction.

One Step Forward . . . Two Steps Back

During the 1950s, the constitutionality of Hollywood's Production Code was challenged by the Supreme Court. Unfortunately, this led to the gradual weakening and eventual dissolution of the office that enforced the animal code, and American Humane's ability to observe film sets was revoked. Once again, Hollywood was free to treat animals as it pleased. Some moviemakers did self-police and treat their animals properly, but others were not nearly as kind.

And so, animal abuses continued for decades. *Apocalypse Now* (1979) and *Heaven's Gate* (1980) each featured incidents where animals were killed, with the latter including a scene in which a saddle was blown from a horse's back by an explosion, injuring the animal so severely that it had to be euthanized. *Heaven's Gate* also featured genuine cockfights (illegal in California and most other U.S. states) as well as intentional chicken beheadings. With such outright gore taking place virtually unchallenged, many actors and crew members clamored for change. It finally came in 1980 with an amendment to the Screen Actors Guild/Producer's Agreement. The newly implemented rules once again authorized American Humane's Film & TV Unit to oversee the treatment of animals in film and gave the organization the power to grant or deny the end credit: "No animals were harmed in the making of this film."

A Dicey Future

Today, computer-generated imagery often acts as a stand-in for stunts deemed too risky for an animal to perform. Nevertheless, the question of animal rights in Hollywood lingers. Recently, chimpanzee trainers have come under fire for abusing their simian charges during and after their film days. Pending lawsuits will attempt to sort it all out, but one thing seems certain: Because animals can't speak for themselves, they require the protective advocacy of concerned humans. It is a fact as true now as it was in moviemaking's golden era.

20 Mythical Creatures

To the best of our knowledge, none of the animals listed here are real, so please don't buy the first one you find online.

* * * *

1. **Dragon:** This legendary monster was thought to be a winged, fire-breathing snake or lizard. Oddly, most cultures that believed in dragons didn't know about dinosaurs.

2. **Griffin:** Originating in the Middle East around 2000 B.C., the griffin was said to have the body of a lion and the head of a bird (most often an eagle).

3. **Phoenix:** This ancient Egyptian bird was believed to live for 500 years or more before setting itself on fire. A new phoenix was supposed to have been born from the ashes of the previous bird.

4. **Chupacabra:** This vampiric creature, whose name means "goat sucker," has allegedly been spotted in the Americas from Maine to Chile. It has been described as a panther, dog, spined lizard, or large rodent that walks upright and smells ghastly.

5. **Yeti/Bigfoot/Sasquatch:** If these creatures existed, they would likely be the same animal, as they are all bearlike or apelike hominids that live in remote mountainous areas. (The Yeti has been reported in the Tibetan Himalayas, while Bigfoot and Sasquatch are rumored to live in the northwestern United States and Canada.)

6. **Skunk Ape:** This hairy seven-foot critter weighs in at 300 pounds, smells like a garbage-covered skunk, inhabits the Florida Everglades, and is thought to be a relative of Bigfoot.

7. **Vegetable Lamb of Tartary:** Mythical animal or vegetable? Eleventh-century travelers told tales of a Middle Eastern plant that grew sheep as fruit. Although the tales were false, the plant is real: It's the fern Cibotium barometz, which produces a tuft of woolly fiber.

8. **Rukh/Roc:** Marco Polo returned from Madagascar claiming to have seen this enormous, horned bird of prey carry off elephants and other large creatures.

9. **Jackalope:** Sometimes called a "warrior rabbit," the jackalope is a legendary critter of the American West. Described as an aggressive, antlered rabbit, it appears to be related to the German wolperdinger and Swedish skvader.

10. **Adjule:** It's reported to be a ghostly North African wolf dog but is likely just a Cape hunting dog, horned jackal, or ordinary wild dog.

11. **Andean Wolf/Hagenbeck's Wolf:** In 1927, a traveler to Buenos Aires bought a pelt belonging to what he was told was a mysterious wild dog from the Andes. This tale encouraged other crypto-enthusiasts to purchase skulls and pelts from the same market until DNA testing in 1995 revealed that the original sample came from a domestic dog.

12. **Fur-Bearing Trout (or Beaver Trout):** The rumor of a fur-covered fish dates back to Scottish visitors to the New World who regaled their countrymen with tales of "furried animals and fish" and photographs of pelt-wrapped trout. Occasional sightings of fish with saprolegnia (a fungal infection that causes a white, woolly growth) serve to perpetuate the tall tales.

13. **Gilled Antelope (or Gilled Deer):** This Cambodian deer or antelope is rumored to have gills on its neck or muzzle that enable it to breathe underwater. In reality, it is the rare Saola or Vu Quang ox, whose distinctive white facial markings only look like gills.

14. **Hodag:** In lumberjack circles, the hodag was believed to be a fetid-smelling, fanged, hairy lizard that rose from the ashes of cremated lumber oxen. In 1893, a prankster from Rhinelander, Wisconsin, led a successful "hunt" for the fearsome beast, which resulted in its capture and subsequent display at the county fair. It was later revealed that the hodag was actually a wooden statue covered in oxhide, but by then it had earned its place in Rhinelander lore.

15. **Sea Monk/Sea Bishop:** Reports and illustrations of strange fish that looked like clergymen were common from the 10th through 16th centuries, likely due to socioreli-

gious struggles and the idea that all land creatures had a nautical counterpart. Sea monks were likely angel sharks, often called "monkfish."

16. **Unicorn:** The unicorn's appearance varies by culture—in some it's a pure white horse and in others it's a bull or an ox-tailed deer—but the single horn in the middle of the forehead remains a constant. Unicorns have reportedly been seen by such luminaries as Genghis Khan and Julius Caesar, but it's likely that such reports were based on sightings of rhinoceroses, types of antelopes, or discarded narwhal horns.

17. **Cameleopard:** Thirteenth-century Romans described the cameleopard as the offspring of a camel and leopard, with a leopard's spots and horns on the top of its head. The legendary creature in question was actually a giraffe, whose modern name stems from the Arabic for "tallest" or "creature of grace."

18. **Basilisk/Cockatrice:** Pliny the Elder described this fearsome snake as having a golden crown, though others described it as having the head of a human or fowl. In fact, no one would have been able to describe it at all, because it was believed to be so terrifying that a glimpse would kill the viewer instantly.

19. **Mokele-mbembe:** Since 1776 in the Congo rainforests, there have been reports of this elephant-size creature with a long tail and a muscular neck. Its name translates as "one that stops the flow of rivers." Hopeful believers—who point out that the okapi was also thought to be mythical until the early 1900s—suggest that it may be a surviving sauropod dinosaur.

20. **Lake Ainslie Monster:** Lake Ainslie in Cape Breton, Nova Scotia, is frequently said to be home to a sea monster, usually described as having a snakelike head and long

neck (similar in appearance to the Loch Ness Monster). Recently, these sightings have been attributed to "eel balls," a large group of eels that knot themselves together in clumps as large as six feet in diameter.

Cockroaches: Nuke-Proof, to a Point

We've all heard that cockroaches would be the only creatures to survive a nuclear war. But unless being exceptionally gross is a prerequisite for withstanding such an event, are cockroaches really that resilient?

* * * *

COCKROACHES ARE INDEED that resilient. For one thing, they've spent millions of years surviving every calamity the earth could throw at them. Fossil records indicate that the cockroach is at least three hundred million years old. That means cockroaches survived unscathed whatever event wiped out the dinosaurs, be it an ice age or a giant meteor's collision with Earth.

The cockroach's chief advantage—at least where nuclear annihilation is concerned—is the amount of radiation it can safely absorb. During the Cold War, a number of researchers performed tests on how much radiation various organisms could withstand before dying. Humans, as you might imagine, tapped out fairly early. Five hundred Radiation Absorbed Doses (or rads, the accepted measurement for radiation exposure) are fatal to humans. Cockroaches, on the other hand, scored exceptionally well, withstanding up to 6,400 rads.

Such hardiness doesn't mean that cockroaches will be the sole rulers of the planet if nuclear war breaks out. The parasitoid wasp can take more than 100,000 rads and still sting the heck out of you. Some forms of bacteria can shrug off more than one million rads and keep doing whatever it is that bacteria do.

Clearly, the cockroach would have neighbors.

Not all cockroaches would survive, anyway—definitely not the ones that lived within two miles of the blast's ground zero. Regardless of the amount of radiation a creature could withstand, the intense heat from the detonation would liquefy it. Still, the entire cockroach race wouldn't be living at or near ground zero—so, yes, at least some would likely survive.

9 Top Cats

Some people are cat people. They take pictures of their cats, tell stories about their cats, and feed their cats designer food. For a cat lover, even the most unremarkable cat is special, but the following cats have been singled out for extra-noteworthy achievements or distinctions.

✻ ✻ ✻ ✻

1. **Cat Most in Need of a Babysitter: Bluebell:** Bluebell, a Persian cat from South Africa, gave birth to 14 kittens in one litter. She holds the record for having the most kittens at once, with all of her offspring surviving—rare for a litter so large.
2. **Most Aloof Cat: Big Boy:** When Hurricane George hit Gulfport, Mississippi, in 1998, Big Boy was blown up into a big oak tree. In 2001, Big Boy's owner claimed the cat never left the tree. The feline eats, sleeps, and eliminates in the tree and climbs from branch to branch for exercise.
3. **Big Mama: Dusty:** In 1952, a seemingly ordinary tabby cat gave new meaning to the term "maternal instinct." Texas-born Dusty set the record for birthing more kittens than any other cat in history. Dusty had more than 420 kittens before her last litter at age 18.
4. **Oldest Cat: Cream Puff:** More than 37 years old at the time of her death, Cream Puff, another Texan, is recognized

as the oldest cat to have ever lived. In human years, she was about 165 years old when she died.

5. **Best-Dressed Cat: The Birman:** The Birman cat breed originally came from Burma (now Myanmar) where these longhairs were bred as companions for priests. A Birman cat can be identified by its white "gloves." All Birmans have four white paws, which give them that oh-so-aristocratic look.

6. **Most Ruthless Killer: Towser:** In Scotland, a tortoiseshell tabby named Towser was reported to have slain 28,899 mice throughout her 21 years—an average of about four mice per day. Her bloodlust finally satiated, Towser died in 1987. (The mice of Scotland are rumored to celebrate her passing as a national holiday.)

7. **Most Itty-Bitty Kitty in the Whole World: Tinker Toy:** Though this Blue Point Himalayan died in 1997, this cat still holds the record for being the smallest cat ever. Tinker Toy was just 2.75 inches tall and 7.5 inches long and weighed about one pound eight ounces.

8. **Cat Most in Need of a Diet: Himmy:** According to the Guinness Book of World Records, the heaviest cat in recorded history was an Australian kitty named Himmy that reportedly weighed more than 46 pounds in 1986. If the data is accurate, Himmy's waistline measured about 33 inches. Guinness has removed this category from their record roster, so as not to encourage people to overfeed their animals.

9. **First Cat: The Eocene Kitty:** Fossils from the Eocene period show that cats roamed the earth more than 50 million years ago. Sure, they looked a little different, but these remains show that today's domestic cats have a family tree that goes way, way back.

7 Animals that Can Be Heard for Long Distances

Animals send out messages for very specific reasons, such as to signal danger or for mating rituals. Some of these calls, like the ones that follow, are so loud they can travel through water or bounce off trees for miles to get to their recipient.

⁕ ⁕ ⁕ ⁕

1. **Blue Whale:** The call of the mighty blue whale is the loudest on Earth, registering a whopping 188 decibels. (The average rock concert only reaches about 100 decibels.) Male blue whales use their deafening, rumbling call to attract mates hundreds of miles away.
2. **Howler Monkey:** Found in the rain forests of the Americas, this monkey grows to about four feet tall and has a howl that can travel more than two miles.
3. **Elephant:** When an elephant stomps its feet, the vibrations created can travel 20 miles through the ground. They receive messages through their feet, too. Research on African and Indian elephants has identified a message for warning, another for greeting, and another for announcing, "Let's go." These sounds register from 80 to 90 decibels, which is louder than most humans can yell.
4. **North American Bullfrog:** The name comes from the loud, deep bellow that male frogs emit. This call can be heard up to a half mile away, making them seem bigger and more ferocious than they really are. To create this resonating sound used for his mating call, the male frog pumps air back and forth between his lungs and mouth, and across his vocal cords.
5. **Hyena:** If you happen to hear the call of a "laughing" or spotted hyena, we recommend you leave the building.

Hyenas make the staccato, high-pitched series of hee-hee-hee sounds (called "giggles" by zoologists) when they're being threatened, chased, or attacked. This disturbing "laugh" can be heard up to eight miles away.

6. **African Lion:** Perhaps the most recognizable animal call, the roar of a lion is used by males to chase off rivals and exhibit dominance. Female lions roar to protect their cubs and attract the attention of males. Lions have reportedly been heard roaring a whopping five miles away.
7. **Northern Elephant Seal Bull:** Along the coastline of California live strange-looking elephant seals, with huge snouts and big, floppy bodies. When it's time to mate, the males, or "bulls," let out a call similar to an elephant's trumpet. This call, which can be heard for several miles, lets other males—and all the females nearby—know who's in control of the area.

How Can It Rain Cats and Dogs?

It can't. Many of us are familiar with strange-but-true stories that describe fish, frogs, or bugs raining from the sky. Indeed, waterspouts and odd, windy weather patterns can suck up small animals, carry them a few miles, and drop them from the sky. But nowhere on record are there any confirmed reports of feline or canine precipitation.

* * * *

IT'S A FIGURE of speech, and its origins are unknown. However, that hasn't prevented etymologists from speculating. One unlikely theory claims that in days of yore, dogs and cats that were sleeping in the straw of thatched roofs would sometimes slip off the roof and fall to the ground during a rainstorm. We know—weak.

Almost as unlikely is the belief that the phrase was cobbled together from superstitions and mythology. Some cultures have associated cats with rain, and the Norse god Odin often was portrayed as being surrounded by dogs and wolves, which were associated with wind. (Anybody who's had an aging dog around the house can vouch for it being an occasional source of ill wind, but that's hardly the stuff of legend.) The components seem right with this one, but it's hard to imagine someone stitching everything together to coin a catchy phrase.

A couple of simpler theories seem rather more plausible. Some folks think that "cats and dogs" stems from the Greek word *catadupe* or the archaic French *catadoupe*, both meaning "waterfall." Others point to the Latin *cata doxas* ("contrary to experience").

The most believable explanation, however, is the least pleasant. The earliest uses of the term occur in English literature of the seventeenth and eighteenth centuries. Around that time in London, dead animals, including cats and dogs, were thrown out with the trash. Rains would sweep up the carcasses and wash them through the streets. Jonathan Swift used the phrase "rain cats and dogs" in his book *A Complete Collection of Polite and Ingenious Conversation* in 1738. Twenty-eight years earlier, Swift had published a poem, "A Description of a City Shower," that included the lines: "Drown'd Puppies, stinking Sprats, all drench'd in Mud/Dead Cats and Turnip-Tops come tumbling down the Flood."

Hardly a love sonnet, but perhaps it answers our question.

How to Get Your Dog in the Movies

Every pet owner thinks his or her dog is the cutest, the smartest, and the best trained, but there's more to it than just being able to sit and roll over on command.

* * * *

Q: What skills does a dog need to get a role in a movie?

A: The dog needs a lot of specific training. To be used in a movie, your dog needs to be comfortable obeying hand commands in unfamiliar locations, under the glare of hot lights and surrounded by lots of strangers. In some cases, owners cannot be on set with their animals, so your dog must be able to take commands from strangers. If you feel confident that your pooch possesses all the skills to become a movie star, you must ask yourself: Are you ready to be the owner of a celebrity dog?

Q: So the owner needs special skills as well?

A: Show business can be just as competitive for canine actors as it is for human ones. As the owner, are you able to attend an audition or turn up on the set with your dog at a moment's notice? Movie schedules can be long with plenty of downtime spent waiting for your dog's scene to shoot. This might be all right if you're retired and have a lot of spare time but not so convenient if you have a full-time job.

Q: How much can a dog earn?

A: Depending on Fido's skills and talents, payment can range from $100 to $500 per day.

Q: Where do I start?

A: First, you should check out the American Humane organization (www.americanhumane.org), which protects animals on sets. You should know something about this watchdog group because it will be watching out for your dog's welfare. Once you've done your homework, take some photos and video clips of your dog and contact an animal-talent agent. It's also a good idea to contact local film schools to see if any student filmmakers need your dog's services. They may not be able to pay much, but it's a good way to get Fido some on-set experience and make sure that acting really is for him. Some dogs prefer to curl up with their owners while they watch a movie on TV rather than actually be part of the cast.

Got Goose, Will Travel

Geese may look like docile creatures, but these quackers can get territorial and downright cranky when provoked. Here's a look at some surprising uses for the common goose.

✻ ✻ ✻ ✻

Beware the Roused Goose

An incensed goose is a frightful thing, with a nearly five-foot-wide wingspan; a serrated beak capable of biting a hole in a metal bucket; and a loud, piercing war-honk. If attacked by a goose, one should never break eye contact with it—this will be perceived as a sign of weakness and will only encourage the aggressor. It's best to back away slowly, and try to position yourself so that there is a fence between you and the angry goose; although they are marvelous flyers, geese hate to go over a fence.

Guardians of Western Civilization and Whiskey

For most of recorded history, man has kept geese. We know that the ancient Egyptians practiced goose husbandry, as did the Romans, who valued the animal so highly that they regularly offered the bird as a sacrifice to the god Juno. These noble beasts had uses far beyond that of supplicating the gods, however; in 390 B.C., the holy geese in the Roman temple roused the sleeping guards in time to fend off attacking Gallic hordes. Indeed, geese are still used as watch animals throughout the world. For instance, at the Ballantine's whiskey-aging facility in Glasgow, Scotland, geese guard 240 million liters of alcohol. Farmers have also learned that geese are capable of looking after a flock of smaller waterfowl and will protect them from foxes and other predators.

A far more quotidian employment for the goose, however, is that of gardener; they will eat weeds and regrowth without disturbing crops. It takes only two geese to keep an acre of row crops free of weeds. Best of all, a goose gardener works for just a bit more than what it can eat on the job and leaves behind nothing but webprints and fertilizer.

Monster on the Chesapeake

Chesapeake Bay, a 200-mile intrusion of the Atlantic Ocean into Virginia and Maryland, is 12 miles wide at its mouth, allowing plenty of room for strange saltwater creatures to slither on in.

* * * *

ENCOUNTERS WITH GIANT, serpentine beasts up and down the Eastern seaboard were reported during the 1800s, but sightings of Chessie, a huge, snakelike creature with a football-shape head and flippers began to escalate in the 1960s. Former CIA employee Donald Kyker and some neighbors saw not one, but four unidentified water creatures swimming near shore in 1978.

Then in 1980, the creature was spotted just off Love Point, sparking a media frenzy. Two years later, Maryland resident Robert Frew was entertaining dinner guests with his wife, Karen, when the whole party noticed a giant water creature about 200 yards from shore swimming toward a group of people frolicking nearby in the surf. They watched the creature, which they estimated to be about 30 feet in length, as it dove underneath the unsuspecting humans, emerged on the other side, and swam away.

Frew recorded several minutes of the creature's antics, and the Smithsonian Museum of Natural History reviewed his film. Although they could not identify the animal, they did concede that it was "animate," or living.

The Chessie Challenge

Some believe Chessie is a manatee, but they usually swim in much warmer waters and are only about ten feet long. Also, the fact that Chessie is often seen with several "humps" breaking the water behind its head leads other investigators to conclude that it could be either a giant sea snake or a large seal.

One Maryland resident has compiled a list of 78 different sightings over the years. And a tour boat operator offers sea-monster tours in hopes of repeating the events of 1980 when 25 passengers on several charter boats all spotted Chessie cavorting in the waves.

A Tall Tale from the Animal Kingdom

The notion that elephants are scared of mice has been especially persistent, reaching at least as far back as the first century AD. It seems, however, that the elephant has gotten a bad rap.

* * * *

The Theory of Unreasonable Fear

NO, IT DOESN'T seem logical that such an impressive beast would cower before such a diminutive one, but that might be the appeal of the myth. Unreasonable fear is a human trait, and applying human traits to animals is one way we form connections with them. After all, humans fear harmless insects and, yes, mice: The image of a shrieking housewife standing on a kitchen chair is as ubiquitous as that of an elephant rearing back on its hind legs at the sight of a rodent.

The difference is this: Many a housewife has taken refuge on many a kitchen chair, yet few elephants have been spooked by mice. In fact, upon being presented with a mouse face to face, most elephants don't react at all. If they have a fear, it's of stepping on the little creature and getting creamed mouse all over the bottom of their feet.

We May Never Know How the Rumor Started

The origins of the myth are lost to history. Pliny the Elder, an ancient Roman writer and philosopher, mentioned the elephant's fear of mice about 2,000 years ago in his *Natural History*. There may have once been an elephant that had the misfortune of letting a mouse catch it off guard, and that elephant may have overreacted in front of the wrong crowd (elephants are skittish creatures and can be easily spooked). Once it happened, there was no taking it back—an amused crowd never forgets.

One theory is that elephants are afraid of mice because they think the rodents might crawl up their trunks. It's more likely that the elephant smells or hears something unfamiliar and reacts as though it's in danger. The idea that a mouse might climb up an elephant's trunk, and that the elephant would be afraid of this happening, is another example of humans projecting human traits on animals. We imagine how we would feel with a mouse crawling around on us, trying to find a way inside us, and shudder with the unpleasantness of the scenario.

There may have been an original incident with an elephant and a mouse, or there may not have. No one can say for sure. But we can say this with relative certainty: Outside of the movies, elephants are not especially afraid of mice. The thought of public speaking, on the other hand, makes them break into a cold sweat.

Birds No Longer Watched

What became of some of the world's most interesting birds? In short, people either slaughtered them or slaughtered their habitats. Sadly, you will never see one, unless DNA can someday provide a miracle.

✻ ✻ ✻ ✻

Q: What were passenger pigeons?

A: They were about the size and shape of a mourning dove, but mostly blue-gray with orange on the neck. They once represented 25 to 40 percent of all birds in North America. Their flocks darkened the skies for hours or days in passing—which sounds like exaggeration but is not. Passenger pigeons were all but wiped out in the wild by 1900, and the last known bird died in a zoo in 1914. The passenger pigeon is one of history's most incredible examples of transforming plenty to zero by sheer ignorance.

Q: Were dodos actually stupid?

A: It's fairer to call them naïve. The dodo lived only on Mauritius in the Indian Ocean and had no natural predators before European settlers arrived. Had the Dutch hunters not slaughtered them, the settlers' dogs and pigs would have. The last one died by 1681. Dodos were great big flightless things weighing about 50 pounds, mostly gray with nine-inch bills.

Q: What in the world was a Great Auk?

A: It was a distant relative of the penguin that lived in the North Atlantic, mostly in modern Canada, Greenland, and Iceland. This ungainly looking creature stood nearly a yard tall and weighed ten pounds at most; though flightless like a penguin, it swam as fluidly as most birds fly. It was black-backed, black-beaked, and white-fronted, and the last live sighting was off Newfoundland in the mid-1800s. Humans hunted the Great Auk to extinction for food, eggs, and its soft, downy feathering.

Q: Did the United States really once have a native parrot species?

A: Yes. The Carolina parakeet, or Carolina conure, was a lovely small parrot with shading that ranged from spring green to aqua to yellow to orange, going from tail to head. A combination of hunting (for the feathers), disappearing forest habitat, and disease wiped them out by about 1918. Mexico's thick-billed parrot used to range into the extreme southwestern United States too; it isn't extinct, but it now stays south of the border in its core habitats.

Q: Are there any ivory-billed woodpeckers left?

A: We hope so, and for many years we didn't dare hope. Also called the "Good God Bird" because that's what people exclaimed when they saw one, the ivory-bill is (or was) one of the largest woodpeckers north of Mexico. Twenty inches tall with a wingspan of nearly a yard, this blue-black, white, and red bird looks so much like the more common pileated woodpecker that it's hard to confirm sightings. Timber companies cut down most of its habitat—the swamps and forests of the lower Mississippi in Arkansas and Louisiana. Since 2000, some intriguing sightings, nest discoveries, and distinctive tapping sounds have led scientists to believe that a few survive.

My Cat Is Smarter Than Your Dog

Until either a dog or a cat develops a cure for cancer, we can't settle this ongoing debate. The concept of intelligence is too nebulous, and dogs and cats are too different from one another. Furthermore, most people—whether they own a cat or a dog—are convinced that their little wookums is the most amazing pet in the world, and no one can tell them otherwise.

* * * *

THE BIGGEST OBSTACLE in crowning an ultimate pet genius is that cats and dogs have contrasting goals. Dogs evolved as pack animals—their ancestors hunted in groups—so they are highly social. Dogs are hardwired to pick up signals and understand commands, and they are driven to please their pack leaders. These days, those pack leaders tend to be humans, which helps explain why canines are so easily trained.

Cats, on the other hand, evolved to hunt alone; consequently, they're motivated to take care of themselves. Most felines exhibit remarkable intelligence when it comes to self-preservation and self-reliance. They're extremely skilled at mapping out their surroundings: They can travel long distances, escape from tight spaces, and pull off spectacular leaping, balancing, and landing maneuvers.

If you define intelligence broadly—as the mastery of complex skills—there's a good case to be made for both dogs and cats. But if you define intelligence the way we do in school—as the ability to absorb information and then utilize this data when tested—dogs appear to be at the head of the class.

In experiments in which animals are rewarded for figuring out complex tasks (like hitting levers or navigating a maze), dogs invariably outperform cats. Dogs have learned to do things that cats can't come close to doing, like distinguishing photos of different dog breeds and even human faces. The evidence suggests that dogs also possess greater language abilities. Some dogs understand well more than two hundred words or signals; this is roughly the equivalent of a two-year-old human's vocabulary. Cats, meanwhile, seem to top out at approximately fifty words, which is about the same comprehension that an eighteen-month-old human displays.

But the evidence might be misleading—dogs are more innately driven to perform in order to earn praise and treats. It's difficult to motivate cats to do anything in experiments because they're so independent. And language is a social ability, so it's more suited to a pack animal like a dog.

The counter argument is that cats could master the same skills as dogs, but they're smart enough not to bother. Why bust your butt if you can lounge around while someone feeds and shelters you? Doesn't this mean that cats are smarter than their owners, too? Most cat owners have long suspected as much.

Toro!

A bullfight brings a certain image to mind: a magnificently attired matador waving his crimson cloak at a snorting, stampeding bull. Most observers would turn red-faced upon learning that the color of the cape does not cause the animal to charge.

⁕ ⁕ ⁕ ⁕

IF YOU QUESTIONED the average person about what transforms a bull from a passive, pasture-loving bovine into a rip-roaring lethal ton of bolting beef, the answer you'd receive would probably revolve around the rotating red cape brandished by the sartorially splendid matador. If it could offer a retort, the bull would tell you that it isn't the color of the cloak that causes it to snort and stomp; rather, it's the matador's tormenting and provoking mannerisms that raises its ire. The constant furling and unfurling of the red cape by a skilled matador unleashes an aggressive streak in the bull, which has been specially bred and trained to be belligerent and hostile. The exaggerated movements cause the animal to charge at full speed with the intent of doing harm to the manipulator of the muleta. To further anger (and weaken) the bull, a horseman called a picador stabs the animal in the neck and shoulders repeatedly with a sword.

Seeing Red

Cape color has never been a factor, because bulls are colorblind. They are charging at the movement of the matador and his cape, which they perceive to be gray in color. The traditional bullfighter's cape is crimson for two reasons: Red is a color that can be easily seen by the onlookers who enjoy watching this type of spectacle, and it also camouflages the blood that is inevitably spilled by the slowly butchered bull.

Cows Are in Herds, and Fish Are in Schools. What About Other Animals?

Ever heard of a sloth of bears? How about an unkindness of ravens?

* * * *

THERE'S AN ALMOST endless array of collective nouns for animals, people, and objects. Many of them—and many of the most interesting—come from *The Book of St. Albans*published in 1486. The book includes a long list of collective nouns for groups of animals; most were meant to be used by hunters in the field.

Some group names describe attributes we ascribe to the animals. For example, a parliament of owls reflects the notion that owls are wise. An ambush describes a group of tigers, reflecting their predatory tactics. What do you call a group of apes, those regal mammals known for their superior intellect? A shrewdness, of course. Some terms stem from physical characteristics, like a prickle of porcupines or a tower of giraffes. Others are just plain poetic: a murder of crows.

Some collective nouns are used for multiple species. You can have a pod of dolphins, whales, or seals. You can have a flock of birds or, get this, a flock of camels. Some animals are described by more than one collective noun. You can have a storytelling of ravens or an unkindness of ravens.

The names can be highly specific. A group of geese on the ground is a gaggle; flying, it's a skein; flying in a V formation, you've got yourself a wedge.

Collective nouns don't just apply to animals. They can describe objects and people, too. There's a flood of tears, a quiver of arrows, a range of mountains. And if one isn't enough, you can call upon a host of angels, a slate of candidates, or a sentence of judges.

Why not create your own collective descriptions? Gather a lead of pencils and a tree of papers, and get to work.

Strange Doings Beneath the Sea

Most sea creatures are quite comfortable with habitats and relationships that human land-dwellers find rather odd. They're flexible about how they look, where they live, and even what gender they claim as their own.

* * * *

No-Brainers

SEA SQUIRTS—SO NAMED because they squirt water at whatever annoys them—are small, blobby creatures that appear in all oceans and seas. Many are short and fat, while others are elongated. Sea squirts can grow to the size of an egg, though most are much smaller. Some live alone and some form colorful colonies that look like flowers blooming on the ocean floor. Although usually found in shallow water, sea squirts also turn up as deep as 28,000 feet.

Sea squirts are categorized as chordates, the same phylum that humans belong to. That's because the larval stage has a notochord (a flexible skeletal rod) and a simple nervous system. With a head, mouth, sucker, and tail, the young sea squirt looks and moves like a tadpole. But this adolescent goes through some major changes as it grows up—more than a human teenager.

Attaching itself to a piling, a seashell, a sandy bottom, gravel, algae, or even the back of a big crab, the youngster absorbs its own tail and nervous system. The mature sea squirt is a spineless, sedentary, immobile glob. Tufts University science philosopher Daniel C. Dennett put it this way: "When it finds its spot and takes root, it doesn't need its brain anymore, so it eats it! (It's rather like getting tenure.)"

Sex Shifters

Worldwide, the oceans' coral reefs harbor about 1,500 species of fish, including some with adaptable sexual identities. Wrasses, parrotfish, and other reef fish start out female and eventually become male. However, other types of reef fish change sex according to the needs of the group. If there aren't enough males or females, the problem is easily taken care of.

Gobies that live in Japan's coral reefs can change back and forth as need dictates. If the dominant male dies or leaves, a female will become male, changing gender in about four days. If a larger male shows up, the gobie that changed simply switches back to female. Many fish that change sex do so quickly. A particular variety of sea bass found in reefs from North Carolina to Florida and in the northern Gulf of Mexico are both female when they meet for mating. One switches to male, they mate, then both switch sex and they mate again. This toggling between sexes is accompanied by color changes; the female is blue, and the male is orange with a white stripe.

Dual Sexuality

The belted sandfish (a coastal sea bass) is a hermaphrodite, with active male and female organs. It can theoretically self-fertilize, meaning that a single individual can release eggs, then shift to its male self (in about 30 seconds) and release sperm. More often, two fish take turns fertilizing each other's eggs. Hermaphroditic sea slugs are underwater snails without shells. The Navanax inermis variety, found off the coast of California and Mexico and in the Gulf of Mexico, have male sex organs on one end and female sex organs on the other. They sometimes mate in chains of three or more, with suitable ends attached. The slugs in the middle of the line act as male and female simultaneously.

The Perfect Couple?

Seahorses, those bony little fish that swim upright, live in sea grasses, mangrove roots, corals, and muddy bottoms in both tropical and temperate oceans and lagoons. They keep the sex they were born with, and seahorse couples tend to remain monogamous throughout a mating season. Couples perform a little dance when they meet, joining tails, swimming around together, and circling each other. It's the male seahorse that gets pregnant. After he opens a special pouch in his body, the female aligns with the opening and lays her eggs. The male fertilizes the eggs, his pouch swells, and two weeks later he gives birth to as many as 1,500 live offspring.

Male seahorses sometimes experience false pregnancies; the pouch swells but no eggs or babies are present. Males can even die of postpartum complications such as infections caused by dead, unborn ponies.

Partners Forever

Far down in the ocean, between 3,000 and 10,000 feet, is a cold, dark world of sharp-toothed hunters. There, many fish use built-in lights to confuse pursuers, to signal a mate, or to bait a trap. Among these deep-sea hunters are anglerfish that grow a "fishing rod" with deceptive "bait" dangling from it. They move about slowly, waving their glowing lures to attract potential meals toward their big toothy mouths. However, only the female anglerfish grows a lure—the male doesn't need one. He's also born without a digestive system, because he isn't going to need that either.

When a young male anglerfish is just a few inches long, he searches out a (much larger) female and sinks his teeth into her. His jaws begin to grow into her skin, and after a few weeks he is unable to let go. The male's eyes get smaller and eventually disappear. Most of his internal organs also disappear. His blood vessels connect to those of the female, so he gets nutrition from whatever she eats. The male grows a little larger, but the gain is all in testes.

Finally, he's the sex object he was destined to be—a producer of sperm and little else. The female gains a mate that's literally attached to her forever. Sometimes she doesn't settle for just one but drags several males along through life.

The Sea Creature with 1,000 Stomachs

The longest of all ocean-dwellers, the praya, assigns sex to small entities that are also its body parts. A praya is a "colonial" animal called a siphonophore, made up of many individuals called polyps. Each polyp is adapted for a special duty: some breed, some swim, and some are just stomachs.

In a praya, the various kinds of polyps are strung together into a well-coordinated monster that moves through the water like a snake on a roller coaster. It roams the ocean vertically, from near the surface to depths of 1,500 feet. Though only as thick as a finger, a praya can grow to be 130 feet long. A mere six-foot siphonophore can have more than 100 stomach polyps, and a large praya might have 1,000 stomachs.

Rin Tin Tin to the Rescue

Audiences have always been captivated by furry creatures, going back to the days of small circuses and vaudeville, where trained animal acts were always part of the show.

* * * *

IN THE SILENT era of Hollywood, perhaps no animal was as famous or as talented as a dog named Rin Tin Tin. Like his heroic on-screen counterparts, this lovable German shepherd is the center of much Hollywood folklore. Some stories claim he saved a studio from the brink of bankruptcy; others claim he generated more fan mail than human stars. Most of these stories are exaggerated; however, each story reveals the dog's special relationship to the moviegoing public and suggests the magnitude of his stardom. Myth and legend swirl around the famous canine, much like they do around American folk heroes and Hollywood icons.

The German Shepherd that Saved Warner Bros.

"Rinty," as Rin Tin Tin was nicknamed, took an indirect path to movie stardom. Born during World War I at a bombed-out kennel in Lorraine, France, the puppy was amongst a small group of surviving dogs rescued by U.S. Corporal Lee Duncan, who adopted Rin Tin Tin and his sister Nenette (named after French puppets) and took them on an arduous ocean voyage to his home in America. Sadly, Nenette died from canine distemper shortly after her arrival in the States, but Rin Tin Tin forged on. From the start, Rinty's pluck was apparent.

Duncan became fascinated with teaching the young pup tricks. And he knew that he was dealing with a truly special animal; Rin Tin Tin had an innate ability to learn and an athletic prowess that permitted him to leap nearly 12 feet high. The pair frequented the dog show circuit and eventually approached Hollywood studios in search of a contract. The latter proved fruitless until one day in 1922, when the twosome happened upon a Warner Bros. film crew working with a wolf. Take after take had to be reshot as the unruly animal continued to misbehave. As the story goes, Duncan bragged that his dog could do the scene in just one take. The crew initially refused the offer but eventually relented, and Rin Tin Tin was given a shot. One successful scene later, the breakout sensation of *The Man from Hell's River* (1922) was on his way.

Such good fortune had come none too soon for the studio. Warner Bros. was trying to establish itself as a major studio and needed cash flow, a higher profile, and new strategies to make that happen. *The Man from Hell's River* helped give them a success to tout and money to pay off debts incurred during a turbulent time. Subsequent films featuring Rin Tin Tin kept the gravy train rolling and put the studio firmly in the black. In a cutthroat town where countless human actors had failed, Rinty, the German shepherd, prevailed. It was an outcome that Hollywood itself would be hard-pressed to dream up.

Rin Tin Tin starred in 27 successful movies, a feat that earned him the nickname "the mortgage lifter." At the height of his popularity, the canine reportedly pulled down $6,000 per week and attracted 40,000 fan letters per month, though these figures vary based on the source. All suggest he was an animal star of unprecedented magnitude. Fanzines claimed that, like any coddled star, Rinty received his share of perks—a tenderloin steak was his for the barking each day, a delicacy allegedly prepared by his own private chef.

In making a successful transition from silent films to "talkies," the dog won the hearts of millions. Rin Tin Tin's last project was a serial titled *Lightning Warrior* (1931). Later that year, the famous dog retired. Sadly, his "out to pasture" years were short-lived. In 1932, Rin Tin Tin passed away. Legend has it that he died in the adoring arms of his Beverly Hills neighbor, blonde bombshell Jean Harlow. Though Rinty was originally buried in California, Duncan had his remains reinterred in his native

France, in *Cimetie're Des Chiens* (a cemetery for dogs) in Paris, as a gesture of honor.

Rin Tin Tin's offspring effectively kept the dog's name alive for years to come, and while successful, they lacked the charisma and special abilities of the original. There were also many imitators at other studios, but they were little more than understudies. There was only one *original* Rin Tin Tin.

How Did the Stork Get the Job of Delivering Babies?

It's been the salvation of uptight parents everywhere. Inquisitive little children all eventually arrive at the same brutal question: "Where do babies come from?"

* * * *

THE MOST EXPEDIENT response? "The stork brings them." But stork folklore goes way beyond lazy parenting. The large, long-necked bird has been associated with maternity and fertility in many cultures for thousands of years.

In Greek mythology, the story of the stork that we know gets turned on its head. Gerana, queen of the Pygmies, angers the goddess Hera and is changed into a stork. She then tries to steal her own child away but is constantly foiled by other family members. Early Christians believed the stork to be a symbol of marital chastity. In Norse mythology, the stork represents a commitment to family, based on the (erroneous) belief that storks are monogamous. The Hebrew word for stork means "kind mother."

The myth of the stork delivering babies appears to have taken hold in Europe, in places such as Holland, Germany, and Poland, where the migratory birds arrived for breeding about nine months after midsummer—during springtime, just as all the babies who were conceived during the lusty fun of midsummer festivals were being born. A stork nesting on the roof at the same time that a new baby arrived was seen as much more than a coincidence, so storks came to be associated with the welcome addition of little blessings in a home. In the hyper-prudish Victorian era, this stork/baby association provided a handy way to avoid the embarrassment of discussing childbirth with curious youngsters.

Although the stork remains a common symbol of birth, children today who ask where babies come from are likely to receive a much more progressive response, such as, "Go ask your mother."

Deadliest Creepy Crawlies

Insect stings kill between 40 and 100 Americans every year. But in other parts of the world, bugs kill many times that number. Here are some of the most fearsome.

- **Locusts:** Individually, locusts are just weird-looking grasshoppers. But when they swarm, look out! Locusts can rapidly devastate huge regions of farmland, leading to mass starvation.
- **Mosquitoes:** Believe it or not, mosquitoes are responsible for more deaths than any other creature. The reason? They spread a variety of potentially deadly diseases including malaria, which kills an estimated two million annually.
- **Africanized Bees:** Big honeybees with a bad attitude, these bees attack with little provocation. Worse, when one bee stings, it releases a chemical that provokes the entire nest.
- **Fleas:** Little more than a nuisance today, fleas are one of history's greatest mass killers. Fleas are thought to have wiped out one-third of the population of Europe through the transmission of bubonic plague in the 14th century.

* **Brazilian Wandering Spider** The most venomous spider in the world, this deadly nocturnal hunter likes to hide in banana bunches during the day. One bite, and you're done.
* **Black Widow** Another deadly arachnid, the black widow's venom is more powerful than that of the cobra or the coral snake.
* **Tsetse Fly** This harmless-looking bug is responsible for spreading sleeping sickness throughout Africa, resulting in hundreds of deaths each year.

Order in the Court? Animals on Trial

During the Middle Ages, people believed that animals were legally responsible for their crimes and misdeeds. But punishment was not administered without fair trial.

* * * *

THE YEAR WAS 1386. In the French city of Falaise, a child was killed and partially devoured by a sow and her six piglets. Locals refused to let such a heinous crime go unpunished. However, rather than killing the sow, they brought her to trial. The pig was dressed in men's clothing, tried for murder, convicted, and hanged from the gallows in the public square.

Porkers weren't the only animals to face trial during medieval times. Bees, snakes, horses, and bulls were also charged with murder. Foxes were charged with theft. Rats were charged with damaging barley.

In the early 1700s, Franciscan friars in Brazil brought "white ants" (probably termites) to trial because "the said ants did feloniously burrow beneath the foundation of the monastery and undermine the cellars . . . threatening its total ruin."

History

The first record of animal trials exists in Athens. More than 2,000 years ago, the Athenians instituted a special court to try murderous objects (such as stones and beams) as well as animals that caused human deaths. They believed that in order to protect moral equilibrium and to prevent the wrath of the Furies, these murders had to be avenged.

Animal trials peaked in the Middle Ages, ranging from the 9th century to as late as the 18th century. During this time, people believed that animals committed crimes against humans and that, like humans, animals were morally and legally responsible for their actions. As a result, animals received the same punishment as humans, ranging from a knock on the head to excommunication or death.

Legal Rights

Animals accused of crimes in Europe's Middle Ages received the same rights under the law as humans, which included a fair trial. Domestic animals were often tried in civil courts and punished individually. Animals that existed in groups (such as weevils, eels, horseflies, locusts, caterpillars, and worms) were usually tried in ecclesiastical courts. They weren't stretched on the rack to extract confessions, nor were they hanged with individual nooses. Instead, they received a group malediction or anathema.

The accused animals were also entitled to legal representation. When the weevils in the French village of St. Julien were accused of threatening the vineyards in 1587, Pierre Rembaud argued in their defense. The innocent weevils should not be blamed, said Rembaud. Rather, the villagers should recognize God's wrath and don sackcloth. The court ruled in favor of the weevils and gave them their own parcel of land.

As for the six little piglets in Falaise? They also must have had good counsel—they were acquitted on the grounds of their youth and their mother's poor example.

Capital Punishment

Murder wasn't the only crime to carry a death sentence. Often, animals accused of witchcraft or other heinous crimes received similar punishment. In 1474, a cock was burned at the stake in Basel, Switzerland, for the crime of laying an egg. As was widely understood, this could result in the birth of a basilisk, a monster that could wreak havoc in a person's home.

Pigs were often brought to the gallows for infanticide (a perennial problem since 900-pound sows often ran free). A mother pig smothering her infants was most likely an accident, but in those times people saw it as a sign of evil thanks to the Biblical account of the demon-possessed herd at Gadarenes.

Animals had slim hopes for survival when accused of severe crimes. However, there is the amazing account of a jenny that was saved when the parish priest and the citizens signed a certificate that proclaimed her innocence. It stated that they had known the "she-ass" for four years and that "she had always shown herself to be virtuous and well-behaved both at home and abroad and had never given occasion of scandal to anyone."

Contemporary Courtrooms

Although animals are not tried as humans in the United States, they are not immune to the gavel. In April 2007, a 300-pound donkey named Buddy entered the courtroom at the North Dallas Government Center in Texas.

While technically it was Buddy's owner who was on trial, the donkey was accused of the "crime." His owner's neighbor had been complaining about Buddy's braying and foul odor.

When the defense attorney asked Buddy if he was the said donkey, Buddy twitched his ears and remained silent. For the next few minutes, he was calm and polite—hardly the obnoxious beast that had been described in the accusations.

While the jury pondered, the neighbors reached an agreement. The day ended peacefully. Buddy had his day in court.

Charming Charlatans

The only thing mystifying about snake charmers is their cunning ability to bamboozle. Here's what's really happening when a cobra "dances" to the music.

* * *

STROLLING PAST A bazaar in India, you spot a man sitting cross-legged in front of a basket. After he raises a flute and begins to play, a venomous cobra starts to sway, rhythmically "dancing" to the music. You look on, mesmerized. At times the serpent is mere inches from the snake charmer's face; indeed, the man even fearlessly "kisses" it. The man and his music seem to have a calming effect on the surly serpent.

In truth, it doesn't require much nerve to charm a cobra—and here are the dynamics behind the demonstration. Contrary to folklore, the much-feared cobra is not aggressive. In fact, it will try to scare off potential predators rather than fight them. A cobra accomplishes this by standing erect and flaring its hood. When it sets its sights on a potential threat, it will sway its body along with the motions of the intruder. Sound familiar?

A snake charmer knows this behavior well—and capitalizes on it. To get a cobra to rise from a basket and stand erect, the performer simply lifts the lid. Startled by the sudden light, the serpent emerges. But before the cobra will "dance" along with the flute music, it must first be conditioned to regard the instrument as an enemy (the performer will often tap the snake with it during "training" sessions). Once accomplished, the snake will follow the flute's every move.

✳ Chapter 12

Unprofessional Sports

Why Is a Football Shaped that Way?

Would you rather call it a bladder? Because that's what footballs were made of before mass-produced rubber or leather balls became the norm.

✳ ✳ ✳ ✳

THE ORIGINS OF the ball and the game can be traced to the ancient Greeks, who played something called harpaston. As in football, players scored by kicking, passing, or running over the opposition's goal line. The ball in harpaston was often made of a pig's bladder. This is because pigs' bladders were easy to find, roundish in shape, relatively simple to inflate and seal, and fairly durable. (If you think playing ball with an internal organ is gross, consider what the pig's bladder replaced: a human head.)

Harpaston evolved into European rugby, which evolved into American football. By the time the first "official" football game was played at Rutgers University in New Jersey in the fall of 1869, the ball had evolved, too. To make the ball more durable and consistently shaped, it was covered with a protective layer that was usually made of leather.

Still, the extra protection didn't help the pig's bladder stay permanently inflated, and there was a continuous need to reinflate the ball. Whenever play was stopped, the referee unlocked the ball—yes, there was a little lock on it to help keep it inflated—and a player would pump it up.

Footballs back then were meant to be round, but the sphere was imperfect for a couple reasons. First, the bladder lent itself more to an oval shape; even the most perfectly stitched leather covering couldn't force the bladder to remain circular. Second, as a game wore on, players got tired and were less enthused about reinflating the ball. As a result, the ball would flatten out and take on more of an oblong shape. The ball was easier to grip in that shape, and the form slowly gained popularity, particularly after the forward pass was introduced in 1906.

Through a series of rule changes relating to its shape, the football became slimmer and ultimately developed its current look. And although it's been many decades since pigs' bladders were relieved of their duties, the football's nickname—a "pigskin"—lives on.

Wacky Sports Injuries

Freak falls and weird war wounds.

❋ ❋ ❋ ❋

- **Ryan Klesko**—In 2004, this San Diego Padre was in the middle of pregame stretches when he jumped up for the singing of the national anthem and pulled an oblique/rib-cage muscle, which sidelined him for more than a week.
- **Freddie Fitzsimmons**—In 1927, New York Giants pitcher "Fat Freddie" Fitzsimmons was napping in a rocking chair when his pitching hand got caught under the chair and was crushed by his substantial girth. Surprisingly, he only missed three weeks of the season.
- **Clarence "Climax" Blethen**—Blethen wore false teeth, but he believed he looked more intimidating without them.

During a 1923 game, the Red Sox pitcher had the teeth in his back pocket when he slid into second base. The chompers bit his backside and he had to be taken out of the game.

* **Chris Hanson**—During a publicity stunt for the Jacksonville Jaguars in 2003, a tree stump and ax were placed in the locker room to remind players to "keep chopping wood," or give it their all. Punter Chris Hanson took a swing and missed the stump, sinking the ax into his non-kicking foot. He missed the remainder of the season.

* **Lionel Simmons**—As a rookie for the Sacramento Kings, Simmons devoted hours to playing his Nintendo Game Boy. In fact, he spent so much time playing the video game system that he missed a series of games during the 1991 season due to tendonitis in his right wrist.

* **Jaromir Jagr**—During a 2006 playoff game, New York Ranger Jagr threw a punch at an opposing player. Jagr missed, his fist slicing through the air so hard that he dislocated his shoulder. After the Rangers were eliminated from the playoffs, Jagr underwent surgery and continued his therapy during the next season.

* **Paulo Diogo**—After assisting on a goal in a 2004 match, newlywed soccer player Diogo celebrated by jumping up on a perimeter fence, accidentally catching his wedding ring on the wire. When he jumped down he tore off his finger. To make matters worse, the referee issued him a violation for excessive celebration.

* **Clint Barmes**—Rookie shortstop Barmes was sidelined from the Colorado Rockies lineup for nearly three months in 2005 after he broke his collarbone when he fell carrying a slab of deer meat.

* **Darren Barnard**—In the late 1990s, professional British soccer player Barnard was sidelined for five months with knee ligament damage after he slipped in a puddle of his

puppy's pee on the kitchen floor. The incident earned him the unfortunate nickname "Whiz Kid."

* **Marty Cordova**—A fan of the bronzed look, Cordova was a frequent user of tanning beds. However, he once fell asleep while catching some rays, resulting in major burns to his face and body that forced him to miss several games with the Baltimore Orioles.
* **Gus Frerotte**—In 1997, Washington Redskins quarterback Frerotte had to be taken to the hospital and treated for a concussion after he spiked the football and slammed his head into a foam-covered concrete wall while celebrating a touchdown.
* **Jamie Ainscough**—A rough and ready rugby player from Australia, Ainscough's arm became infected in 2002, and doctors feared they might need to amputate. But after closer inspection, physicians found the source of the infection—the tooth of a rugby opponent had become lodged under his skin, unbeknownst to Ainscough who had continued to play for weeks after the injury.
* **Sammy Sosa**—In May 2004, Sosa sneezed so hard that he injured his back, sidelining the Chicago Cubs all-star outfielder and precipitating one of the worst hitting slumps of his career.

Baseball's Worst Trades

High hopes; bad deals. There's an inherent risk in all trades, and today's GMs know that any swap they make has the potential to backfire—and earn its spot among the worst deals of all time.

⁕ ⁕ ⁕ ⁕

Listed below are some of the most one-sided swaps in baseball history. In each, future superstars (noted in bold) with great years ahead of them were dealt for players beyond their prime or on the fast track to mediocrity. General managers and owners had to live with the ramifications—and reminders from the public—for decades. All you have to do is read about them.

The Date: January 1, 1894

The Trade: Brooklyn Grooms send **OF Wee Willie Keeler** and **1B Dan Brouthers** to Baltimore Orioles for 3B Billy Shindle and OF George Treadway.

The Fallout: Keeler (.371) and Brouthers (.347) shined for the Orioles in 1894, with Wee Willie eventually batting above .300 for 13 straight seasons after the trade—including .424 with the Orioles in 1897. Shindle, while productive, never went over .300 for the Grooms. Treadway had one good year but was gone from the majors in three.

The Date: December 15, 1900

The Trade: Cincinnati Reds send **P Christy Mathewson** to New York Giants for P Amos Rusie.

The Fallout: Mathewson shares the NL record for most career wins (tied with Grover Cleveland Alexander at 373)—all but one of which were with the Giants, whom he pitched to five pennants. Rusie had won 246 but went just 0–1 with the Reds before retiring.

The Date: April 12, 1916

The Trade: Boston Red Sox send **OF Tris Speaker** to Cleveland Indians for P Sam Jones, 3B Fred Thomas, and $55,000.

The Fallout: This was the first of many disastrous deals made by Boston. Speaker, a .336 hitter and top-notch fielder with the Sox over nine years, lasted another 13 stellar seasons (11 with the Indians) and retired with a lifetime .345 batting average and 3,514 hits. Jones helped the Sox to the 1918 World Series title but was a mediocre 64–59 with Boston overall. Thomas, a

.225 lifetime hitter, spent just one forgettable year with the Sox.

The Date: January 3, 1920

The Trade: Boston Red Sox send **OF Babe Ruth** to New York Yankees for $425,000 in cash and loans.

The Fallout: The granddaddy of bad swaps shifted baseball's power base from Boston to New York, where Babe hit 659 homers and rewrote the record books while leading the Yanks to seven pennants and four World Series championships in 15 years. The Sox, World Series victors four times from 1912 through 1918, wouldn't win it all again until 2004. Then-Sox owner Harry Frazee is still reviled in New England for this one.

The Date: May 6, 1930

The Trade: Boston Red Sox send **P Red Ruffing** to New York Yankees for OF/1B Cedric Durst and $50,000.

The Fallout: The last in a string of stars sent from cash-poor Boston to New York in one-sided deals, Ruffing rebounded from a 39–96 mark with woeful Sox clubs to a Cooperstown-worthy 231–124 slate with the Yanks—plus a 7–2 mark in ten World Series starts. Durst hit .245 with one homer for Boston during the rest of 1930, his last year in the majors.

The Date: June 15, 1964

The Trade: Chicago Cubs send **OF Lou Brock,** P Jack Spring, and P Paul Toth to St. Louis Cardinals for P Ernie Broglio, P Bobby Shantz, and OF Doug Clemens.

The Fallout: It was essentially a Brock-for-Broglio swap. The Cubs hoped Broglio (having racked up an 18–8 record in 1963) would anchor their staff; instead, he went 7–19 over parts of three seasons. Brock, a .251 hitter with Chicago, caught fire with the Cards, batting .348 with 33 steals the rest of '64 to help St. Louis win the World Series title. He retired in 1979 with

3,023 hits, a then-record 938 stolen bases, and a .391 average in 21 World Series games.

The Date: December 9, 1965

The Trade: Cincinnati Reds send **OF Frank Robinson** to Baltimore Orioles for P Milt Pappas, P Jack Baldschun, and OF Dick Simpson.

The Fallout: After ten stellar seasons (including 324 home runs), Robinson was deemed "an old 30" by Reds owner Bill DeWitt, architect of the swap. Robinson went on to win the Triple Crown and AL MVP Award in 1966 and led the Orioles to four pennants and two World Series titles through 1971, eventually hitting 586 career homers. Pappas won 209 games in his career but went just 30–29 with the Reds. Throw-ins Baldschun and Simpson gave Cincinnati one win and five homers, respectively.

The Date: April 21, 1966

The Trade: Philadelphia Phillies send **P Ferguson Jenkins,** OF Adolfo Phillips, and OF John Herrnstein to Chicago Cubs for P Larry Jackson and P Bob Buhl.

The Fallout: Jenkins, who went 2–1 in two years with the Phillies, won at least 20 in each of the next six years for Chicago en route to 284 lifetime victories. Phillips (mediocre) and Herrnstein (awful) were soon gone from the scene, while the aging Jackson and Buhl—once very good pitchers—were a combined 47–53 with the Phils.

The Date: December 10, 1971

The Trade: New York Mets send **P Nolan Ryan**, P Don Rose, OF Leroy Stanton, and C Frank Estrada to California Angels for SS Jim Fregosi.

The Fallout: The Mets saw Fregosi, a six-time All-Star shortstop with the Angels, as the long-term key to their third-base woes; instead, he failed to adapt to his new position and hit just .232 in

1972 before he was sent packing in '73. Young fireballer Ryan, 29–38 with New York, was an instant sensation with the Angels, winning 19 with 329 strikeouts in '72 on the way to staggering life time marks of 324 wins, seven no-hitters, and a record 5,714 Ks. Rose, Stanton, and Estrada? Mere footnotes to history.

The Date: February 25, 1972

The Trade: St. Louis Cardinals send **P Steve Carlton** to Philadelphia Phillies for P Rick Wise.

The Fallout: Wise was a dependable pitcher before and after this trade, going 32–28 in 1972–73 and winning 188 games in his career, but he was no Steve Carlton. "Lefty" went an incredible 27–10 for the last-place Phillies of '72 and, all told, won 241 games and four Cy Youngs for the club en route to 329 lifetime victories.

The Date: January 27, 1982

The Trade: Philadelphia Phillies trade **2B Ryne Sandberg** and SS Larry Bowa to Chicago Cubs for SS Ivan DeJesus.

The Fallout: Sandberg played just 13 games for Philadelphia but became a legend with the Cubs, hitting 282 homers and earning nine Gold Gloves over 15 years. The aging Bowa, a perennial All-Star with the Phils, was Ryne's double-play partner for three years with the Cubs, the same amount of time DeJesus—a .250 hitter with zero power and a so-so glove—lasted in Philadelphia.

The Date: August 31, 1990

The Trade: Boston Red Sox send **1B Jeff Bagwell** to Houston Astros for P Larry Andersen.

The Fallout: It wasn't as bad as Ruth for cash, but it was close. The Sox, seeking a reliever for the stretch drive, got 22 solid innings from Andersen and won the 1990 AL East. He then left as a free agent—while Boston-born Bagwell debuted as

'91 Rookie of the Year with Houston and eventually smashed 449 homers and 488 doubles as the greatest hitter in Astros history.

I believe the sale of Babe Ruth will ultimately strengthen the team.

—Red Sox owner Harry Frazee, January 1920

Disorder in the Court!

Hear Ye! Hear Ye! Kangaroo court is now in session.

* * * *

The sight of George "Boomer" Scott in a black robe and white wig in the Milwaukee clubhouse in the 1970s could only mean one thing: kangaroo court was in session. Frank Robinson, who later became Major League Baseball's "director of discipline," trained for that job wearing a (hopefully unused) mophead for a wig in Baltimore's kangaroo court in the 1960s. Such courts in baseball date back to the late 19th century, but the purpose today remains the same: Promote camaraderie, punish stupidity.

The kangaroo court was a fun way for a team to "punish" a player for doing something stupid either on or off the field. It allowed teammates to become more aware of things they were doing wrong, in a way that promoted camaraderie. The "judge" was usually someone with seniority—or an especially good sense of humor. (Don Baylor and Steve Reed excelled in the role.) The fines imposed were collected and used for a party or given to charity at the end of the season.

Mishaps on the field—throwing to the wrong base, multiple whiffs, missing signs, forgetting the number of outs—ranked high on the list of offenses. Off the field there were even more ways to get on the court's bad side: making out in public, wearing a hideous outfit (and in the heyday of the kangaroo court in the 1970s, there were lots of questionable clothing decisions), or fraternizing with the "enemy." No one, from batboys to team owners, was safe from the court's watchful eye and imposition of justice.

Kangaroo courts have become rare these days, though they do happen occasionally. Players today are more of a collection of independents, and they tend to be friendlier with the opposition, eliminating one of the major infractions. Some players are just too touchy for the ribbing that goes along with the fines. Many young players aren't even aware of the lore of the kangaroo court. But for those who took part, the court was about more than pointing out mistakes—it was a way to bring their team together (and have a lot of fun in the process!).

Strangest Homers

Not every home run is a cut-and-dried event; some are messy and some are just weird.

✻ ✻ ✻ ✻

SOMETIMES IT IS obvious that a home run has been hit the second the ball leaves the bat. Other times fans anxiously wait to see whether the ball sneaks over the wall or collides with the foul pole before they are able to celebrate a home run. And then there are those times when the ball takes an entirely different route and does something downright strange.

In the Doghouse

One of the oddest homers in history took place at American League Park, which was the home of the Washington Senators from 1904 to 1910. In this ballpark there was a doghouse near the outfield flagpole. The groundskeeper stored the flag inside the doghouse between games. One afternoon the doghouse door was left open during a game, and a member of the Senators hit the ball inside of it. Philadelphia A's center fielder Socks Seybold crawled inside to retrieve the ball and got stuck, allowing the batter to circle the bases for an "inside-the-doghouse" home run.

Not a Stolen Base—A Stolen Ball!

That wasn't the first time an open door figured into a homer. On May 3, 1899, the Louisville Colonels were enjoying a comfortable lead over the home-team Pittsburgh Pirates until the Bucs staged a rally in the bottom of the ninth. Jack McCarthy had already homered when teammate Tom McCreery drove a ball to the right-field fence. A Pirates employee opened the right-field gate, picked up the ball, and ran off with it. McCreery circled the bases, and despite protests from the Louisville players, the umpire allowed the play to stand. The outcome was a 7–6 Pittsburgh victory. However, later that season at a league meeting, the game was thrown out and was replayed. It did not count in the final National League standings.

Feeling Heat from the Warm-Ups

In 1911, American League President Ban Johnson attempted to speed up games by eliminating warm-up pitches between innings. On June 27, at the Huntington Avenue Baseball Grounds, the Athletics' Stuffy McInnis capitalized on the rule change when he noticed Boston pitcher Ed Karger tossing warm-ups. McInnis hurried to the plate and drilled a ball that center fielder Tris Speaker refused to chase because he thought it was hit during warm-ups. McInnis rounded the bases for a home run. Umpire Ben Egan had no choice but to allow the homer to count because Johnson was sitting in the stands.

Oddities at Ebbets

In 1940 at Ebbets Field, Lonnie Frey of the Cincinnati Reds hit a ball to right field. It bounced off the screen and landed on top of the wall that extended between the scoreboard and foul pole. The ball bounced up and down but never fell back to the field of play, allowing Frey to complete an inside-the-park home run. Brooklyn's Pee Wee Reese duplicated the feat in the final game of the 1950 season.

Another home run that occurred at Ebbets Field seemed to defy the laws of gravity. George Cutshaw, who played for the Dodgers from 1912 to 1917, hit a line drive to the left-field wall. Apparently the ball had a lot of topspin on it; when it hit the wall, the ball rolled up and over the fence for a home run.

It Is! Or Is It?

One of the most memorable homers of George Brett's career was disallowed for a brief period of time. Brett's two-out, two-run, ninth-inning homer off Yankees relief ace Goose Gossage was a majestic drive. The game took place on July 24, 1983, at Yankee Stadium, and the blast gave the Royals a 5–4 lead over the Yanks. But after Brett circled the bases, New York manager Billy Martin came out of the dugout and asked the home plate umpire to examine Brett's bat. It was determined that the pine tar on Brett's bat exceeded the legal limit of 18 inches. Brett was called out, and the home run was nullified. The umpire was later overruled by American League President Lee MacPhail; the home run counted, and the game was picked up at that point and finished three weeks later despite Martin's furious protests. The Royals held on to their lead to claim the victory.

Just a Hop, Skip, and a Jump

Sometimes the opposition actually helps the hitter. In May 1993, Cleveland's Carlos Martinez hit a fly ball to right field that bounced off Jose Canseco's head and over the fence for a home run. The gift round-tripper gave the Tribe a 7–6 win over the Rangers.

While that was embarrassing for Canseco, it was at least less frustrating than what happened to Dick Cordell during a minor-league game on August 9, 1952. In the seventh inning of a scoreless game between Denver and Omaha in the Western League, Cordell ran down a long drive off the bat of Denver's Bill Pinckard. Cordell caught the ball before crashing into the left-field wall. The ball was jarred from his glove on impact, ricocheted off the wall, then bounced off his head and over the fence. After a lengthy discussion, the umpires ruled that Pinckard's drive was indeed a homer. It turned out to be the only run of the game.

Only the Best of Intentions

For Jim Bottomley, one home run wasn't worth all of the grief. Bottomley, who spent 16 seasons in the majors and led the NL with 31 homers in 1928, was once sued after one of his home runs hit a spectator in the face. The suit stated that Bottomley "swung on that ball deliberately and with the intention of creating a situation commonly known as a home run."

During questioning at a deposition, an attorney suggested that a skilled contact hitter could place the ball to whichever part of the field he determined. He then asked Bottomley, "Did you deliberately intend to hit anyone when you batted that ball?"

"No sir," replied Bottomley. "There is no malice in any of my home runs."

The Absolute Worst Teams in Baseball History

The teams that are found underneath the bottom of the barrel.

* * * *

MANY ARGUMENTS HAVE raged about the best baseball team ever. But the worst? Many qualify for the dubious honor of worst team for a single season. Each was terrible in its own way in its own time. Only teams that completed a season schedule are considered here (eliminating several fly-by-night clubs from the 19th century).

1899 Cleveland Spiders

Record: 20–134 (84 games out)

Managers: Lave Cross (8–30); Joe Quinn (12–104)

Nightmare Season: Just three years earlier, the Cleveland

Spiders had played in consecutive Temple Cups (the championship series of the day) and boasted Cy Young in his prime as well as a cast of top hitters. With no rules prohibiting multiple team ownership, the best players from Cleveland were shifted to St. Louis in 1899. The result for Cleveland was the most losses in big-league history. The Spiders finished the year going 1–40. They were such a bad draw (fewer than 150 fans per home game!) that ownership forced them to play 112 road games.

1916 Philadelphia Athletics

Record: 36–117 (54 1/2 games out)

Manager: Connie Mack

Nightmare Season: Two years removed from consecutive pennants, Connie Mack's Athletics compiled the worst winning percentage (.235) of the 20th century. Competition from the Federal League, anger at being swept in the 1914 World Series, and dwindling finances led Mack to sell off his best players, leading to 109 losses in 1915 and seven straight last-place finishes. And Mack let his pitchers take the punishment: Despite being the only team in the American League with an ERA above 3.00 in 1916, the A's led the league with 94 complete games.

1935 Boston Braves

Record: 38–115 (61 1/2 games out)

Manager: Bill McKechnie

Nightmare Season: The Red Sox and Braves staged a heated contest for the dubious title of worst team in Boston throughout the 1920s and '30s. With each racking up five 100-loss seasons, neither team was good. But the Braves earned the prize with their doozy of a showing in 1935. After three consecutive .500 or better seasons, the Braves dropped off a cliff. Opponents batted .303, and Boston's 4.93 ERA was the highest in the NL since the offensive explosion of 1930. Aged Babe Ruth was on this team; he quit the game in May.

1941 Philadelphia Phillies

Record: 43–111 (57 games out)

Manager: Doc Prothro

Nightmare Season: The Phillies lost 100 games six times in a seven-season span, with 1941 being the lowest point. Unlike in 1930, when the pitching staff racked up an unfathomable 6.71 ERA at the tiny Baker Bowl in a year the team batted .315, the '41 Phillies at Shibe Park couldn't hit, either. They were last in runs scored and allowed, hit for the lowest batting average yet had the highest batting average against, needed more relief appearances than any other team, and, not surprisingly, had the NL's lowest attendance. The Phils hired a new manager (Hans Lobert) and lost 109 times in '42.

1952 Pittsburgh Pirates

Record: 42–112 (54½ games out)

Manager: Billy Meyer

Nightmare Season: The Pirates have the distinction of being the worst team between America's entry into World War II and base-ball's expansion era. And what a bad team it was. The Bucs allowed more walks, hits, and home runs than any other NL team and gave up 134 more runs than anyone else as well. However, thanks to a shortened fence in left field, future Hall of Famer Ralph Kiner tied for the home run crown—his seventh straight year winning or sharing it. When Kiner asked for a salary increase, owner Branch Rickey famously responded, "We finished last with you, we can finish last without you." After trading him to the Cubs, that's just what the Pirates did . . . for the next three seasons.

1962 New York Mets

Record: 40–120 (60½ games out)

Manager: Casey Stengel

Nightmare Season: The 1962 Mets are the stars of this dreadful, gloomy list. Oh, they were bad. No one had piled up more losses since 1899. They were outscored by 331 runs. This expansion team lost the first nine games they ever played and had three losing streaks in double digits. But with players like "Marvelous Marv" Throneberry, the Mets lost with panache. New York was so starved for National League baseball that fans ate it up at the Polo Grounds and gleefully followed the team to Shea Stadium. More than 40 years later, books are still published about Casey Stengel's fun, flea-bitten crew.

1962 Chicago Cubs

Record: 59–103 (42½ games out)

Managers: College of Coaches: El Tappe (4–16); Lou Klein (12–18); Charlie Metro (43–69)

Nightmare Season: It's hard to believe that two teams from the same league in the same year could have made it onto this list, but the 1962 Cubs are worthy thanks to the College of Coaches. Owner P. K. Wrigley's bright idea put the sputtering franchise in the hands of the overmatched coaching staff on a rotating basis. Playing their tenth decade as a franchise, the Cubs lost 100 games for the first time, finished seven games behind expansion Houston, and were the only team with a .500 record against the moribund Mets. Three times, the Cubs drew crowds of under 1,000 to Wrigley Field the last week of the season.

1969 San Diego Padres

Record: 52–110 (41 games out)

Manager: Preston Gomez

Nightmare Season: If you think the Mets are the only epically bad expansion team, think again. The Padres suffered on the field and at the gate. Just 512,970 fans showed up for their first year of existence. (The Expos, playing at a tiny, frigid facility, drew twice that number that year with the same record.) Even playing at a very pitcher-friendly park, San Diego pitchers struck out the fewest batters in the bigs, while the .225-hitting offense fanned the most times in the NL. The Padres beat out the Expos and the 1977 Blue Jays as the worst first-year team not named the Mets, and they finished last in each of their first six seasons, averaging 101 losses.

1988 Baltimore Orioles

Record: 54–107 (34½ games out)

Managers: Cal Ripken, Sr. (0–6); Frank Robinson (54–101)

Nightmare Season: While there were many, many, many terrible seasons between 1970 and 2002, the staggering beginning and finish of the 1988 Orioles edges out the likes of the 1979 A's, the 1991 Indians, and the 1998 Marlins. The Baltimore Orioles began the season with three Ripkens and a record 21 consecutive losses. Cal the elder was fired six games in. (Cal Jr. and Billy stayed on.) The team managed a 51–69 mark for Frank Robinson between May and mid-September before a dismal 3–17 finish.

2003 Detroit Tigers

Record: 43–119 (47 games out)

Manager: Alan Trammell

Nightmare Season: The Tigers would have made the list anyway for their brutal 53–109 season in 1996, but let's not get greedy. The 2003 Tigers were terrible on a truly epic scale. Detroit had the most losses in American League history. They had the third most ever in the major leagues—pitcher Mike Maroth was the first 20-game loser since 1980—and the lowest AL batting average since the 1988 Orioles. Three years later, though, the Tigers took control of the AL Central, making it all the way to the World Series. Revenge is certainly sweet.

Wingnuts on the Field of Play

It was always a laugh when any of these characters were around.

* * * *

PEOPLE COMPLAIN THAT today's game, with its stratospheric payrolls and second-by-second analysis, has lost the sense of good fun and genuine silliness it once had. No one did more to keep things humorous than this gaggle of goofies.

Yogi Berra

This popular Hall of Fame catcher had a unique way with words. His marvelous, sometimes uproarious, nearly Zen (but not quite) "Berra-isms" are inextricably intertwined with baseball lore, and they often show off a mind with a keen understanding of the game. His malapropisms made people scratch their heads, but they also coaxed a chuckle.

Dizzy Dean

No ballplayer ever had a more accurate nickname. Horrible English ("there is a lot of people in the United States who say 'isn't,' and they ain't eating") and a genuine boyish love of the game characterized Dizzy's personality. After a short but highly successful pitching career in the 1930s, he was a broadcaster for more than 20 years.

Arlie Latham

A song written about Latham called him the "freshest man on Earth," and the fans of the late 1800s loved him for his roaring enthusiasm. Latham would lead the fans in cheers and heckle the opposition without mercy. Then, just to show them all how much fun he was having, he'd somersault his way out to his position. In the off-season, fans would turn out around the country for his stage act.

Bill Lee

"The Ace from Outer Space," Bill Lee marched to the beat of a different drummer—or two or three. Saying outrageous things was as natural to him as breathing. The first time he saw Fenway Park's Green Monster, he asked, "Do they leave it up during games?" After Game 4 of the 1975 World Series, a reporter asked his impression of the Series so far. Lee answered, "Tied."

Rabbit Maranville

It may be that no one had more fun playing baseball than the Rabbit—the eternal puckish clown, the one with the funniest faces, the loudest pratfalls, and the highest consumption of goldfish. As a defensive player in the 1910s, '20s, and '30s, this longtime Brave was a superstar. As an on-the-field entertainer and practical joker, he was a genius. The fans couldn't take their eyes off him.

Germany Schaefer

Schaefer's famous steal of first base—from second—was not a joke. He was trying to get a run home from third by drawing a throw from the catcher. When the catcher didn't bite (and the umpire ruled there was nothing illegal about the act), Schaefer set out from first to re-steal second. The catcher threw; the run scored. It worked.

Casey Stengel

"Stengelese" was the way this colorful manager dealt with the pestering questions of the press. "Best thing wrong with Jack Fisher is nothing," said Case. But Stengel was a beloved, fun-loving player long before he became a manager. Traded from Brooklyn to Pittsburgh in 1918, he returned to Ebbets Field by tipping his cap to the crowd and having a bird fly out.

Rube Waddell

A colorful oddball, Waddell possessed a childlike sense of life and baseball—an endearing trait attached to a sensational left arm. He loved fire trucks and fishing trips, he wrestled alligators, and his roommate had it written in his contract that Rube was forbidden to eat crackers in bed.

The Sin Bin: Hockey's House of Humility

This is the time-out seat of professional sports, where hotheaded hockey players go "to feel shame." It is the penalty box, an off-ice office of purgatory for on-ice transgressors.

❋ ❋ ❋ ❋

FOR THE FIRST 50 years of the National Hockey League's existence, every league arena had only one penalty box, which meant that players who engaged in a lively tussle on the ice served their penance together, with only an obviously nervous league official sitting between them to act as a buffer. Quite often, the combatants would continue their fisted arguments off the ice and inside their temporary, cramped quarters.

On one occasion, this led to the infamous "pickling" of New York Rangers' forward Bob Dill. On December 17, 1944, Dill and Montreal Canadiens fireball Maurice "The Rocket" Richard engaged in a raucous set-to that banished them both to the shower stall of shame. Inside the box, the obviously dazed and confused Dill attacked The Rocket again and received another sound thumping for his lack of common sense.

It wasn't until midway through the 1963–1964 season that the league introduced a rule requiring every rink to have separate penalty benches. A particularly vicious confrontation between Toronto Maple Leaf Bob Pulford and Montreal Canadien Terry Harper on October 30, 1963, precipitated by Harper's questioning of Pulford's sexual preference, spearheaded the NHL's decision to arrive at a sensible solution.

The undisputed king of the sin bin was Dave "Tiger" Williams, who logged nearly 4,000 minutes sitting on his punitive throne during his 15-year career in the NHL. Having spent his formative years with the Toronto Maple Leafs, Williams had a personal affinity for the Maple Leaf Gardens' penalty box, which he described as "a gross place to go. The guys in there are bleeding...and no one's cleaned the place since 1938."

Williams may hold the career mark for sin bin occupancy, but the rap sheet for a single-season sentence belongs to Dave "The Hammer" Schultz. During the 1974–1975 campaign, the Philadelphia Flyers enforcer cooled his carcass in the hotel of humility for 472 minutes, nearly 8 full games. He was so at home in the house, he actually recorded a single titled "The Penalty Box," which became something of a cult hit in and around the City of Brotherly Love.

Philadelphia's post of punition was also the scene of one of hockey's most hilarious highlights. During a game between the Flyers and Maple Leafs in 2001, Toronto tough guy Tie Domi was sent to the box. Upon his arrival in the cage, he was verbally accosted by a leather-lunged Philly fan named Chris Falcone, who wisely used the glass partition to shield himself from Domi. Known as "The Albanian Assassin," Domi responded to the goading by spraying his heckler with water. The broad-shouldered Falcone lunged toward Domi, fell over the glass, and landed in a heap at Domi's feet, which resulted in a comic wrestling match between lug head and lunatic.

Spitballers and Greasers

Toss me that Vaseline, and keep your eye on the ball.

* * * *

PITCHERS BEGAN SEEKING an advantage over hitters almost as soon as Jim Creighton developed the wrist snap in the 1850s. The rules were much simpler then, leaving more room for creative interpretation. So in 1868, when 16-year-old Bobby Matthews of the Lord Baltimores spat on the ball and fired it with the underhand stiff wrist the rules then required, the ball danced, and the batters went crazy.

Doctoring the Ball

The spitball has had dozens of names, from "country sinker" to the "aqueous toss" and "humidity dispenser" or, more directly, "the wet one." But the spitball belongs in a larger class of pitches in which the ball is altered in one way or another to break or twist when it heads toward the batter. These pitches haven't always been fair, but they've usually done the trick.

In the 1890s, Clark Griffith, who amassed more than 200 wins in his pitching career, would bang the ball against his spikes, cutting it and leaving it subject to off-balance aerodynamic forces—thereby baffling hitters. In later years, pitchers would alter their gloves to leave a hole through which they could scrape the ball on a doctored ring. Or they would have a friendly teammate wear a belt with a sharp buckle and tear it against the ball as they warmed up before an inning.

How the Spitball Got Its Name

In the early 1900s, Ed Walsh of the Chicago White Sox learned how to moisten a ball just right on the tips of his fingers so it would slide off and be harder to hit. Before long, the spitter was the pitch of choice for dozens of hurlers. Historians John Thorn and John Holway have said, "The dead-ball era could be called the doctored ball era."

Walsh's spitball was especially devastating because he could make it break four different ways: down and in, straight down, down and out, and up (which he threw underhand). So that the batters wouldn't know what to expect, Walsh put his hand to his face on every pitch, but he threw the spitter only about half the time. He and fellow spitball artist Jack Chesbro became the only two pitchers to win 40 games in one season in the 20th century.

Cleaning Up and Playing Dirty

After years of wild pitches (culminating in the death of Ray Chapman, who was hit in the temple by a pitched ball), baseball decided to clean up its act. Since the 1890s, it had been "illegal" for pitchers to damage a ball to alter pitches, but that rule was rarely enforced. The spitter was officially banned before the 1920 season (with stricter punishments for rule-breakers), although 17 pitchers were grandfathered in and allowed to throw it until their careers ended. The new rule outlawed spit, sandpaper, resin, talcum powder, and other "foreign substances" that produced trick pitches. So hurlers had to find better ways to cheat.

Some did and later admitted it; some have denied all wrongdoing. Hall of Famer Whitey Ford of the Yankees has been accused of using every trick he could muster, from gouging the ball with a ring, to covering one side of the ball with mud, to creating a special invisible gunk that he slathered on his fingers between innings.

Lew Burdette of the Braves in the 1950s always said that having the hitters think he had a spitter was just as good as actually throwing one. He'd wipe his hands on his pants and in his hair and then spit between his teeth.

But the all-time artist of loading the ball was Gaylord Perry, who used his wiles (and a lot of Vaseline) to win Cy Young Awards in both leagues. Perry's gyrations between each pitch were phenomenal. He'd grab here, scratch there, flick here, wipe there.

No one could possibly know what was coming. In all his years of cheating, Perry was caught just once.

The Dirt Behind the Dribble

There are numerous rules on how to properly dribble a basketball, but bouncing the ball with such force that it bounds over the head of the ball handler is not illegal.

* * * *

ALTHOUGH IT MIGHT fun-up the standard NBA game to see players drumming dribbles with the exaggerated effort of the Harlem Globetrotters, it wouldn't do much to move the game along. And contrary to popular belief, there is no restriction on how high a player may bounce the ball, provided the ball does not come to rest in the player's hand.

Anyone who has dribbled a basketball can attest to the fact it takes a heave of some heft to give the globe enough momentum to lift itself even to eye-level height. Yet, the myth about dribbling does have some connection to reality. When Dr. James Naismith first drafted the rules for the game that eventually became known as basketball, the dribble wasn't an accepted method of moving the ball. In the game's infancy, the ball was advanced from teammate to teammate through passing. When a player was trapped by a defender, it was common practice for the ball carrier to slap the sphere over the head of his rival, cut around the befuddled opponent, reacquire possession of the ball, and then pass it up court. This innovation was known as the overhead dribble, and it was an accepted way to maneuver the ball until the early part of the 20th century. The art of "putting the ball on the floor" and bouncing it was used first as a defensive weapon to evade opposing players.

By the way, there is absolutely no credence to wry comments made by courtside pundits that the "above the head" rule was introduced because every dribble that former NBA point guard Muggsy Bogues took seemed to bounce beyond the upper reaches of his diminutive 5′3″ frame.

Fantasy Tennis Star

No one can accuse Sports Illustrated of not having a sense of humor. For laughs, it invented an attractive, camera-ready tennis star to rival Anna Kournikova. Her name was Simonya Popova.

❋ ❋ ❋ ❋

Sports Satire

A SEPTEMBER 2002 ISSUE OF *Sports Illustrated* told of an unstoppable 17-year-old tennis force named Simonya Popova, a Russian from Uzbekistan and a media dream: 6′1″, brilliant at the game, fluent in English, candid, busty, and blonde. She came from an appealing late-Soviet proletarian background and had a father who was often quoted in Russian-nuanced English. But she wouldn't be competing in the U.S. Open—her father forbade it until she turned 18.

The magazine verged on rhapsody as it compared Popova to Ashley Harkleroad, Daniela Hantuchová, Elena Dementieva, and Jelena Dokic. Editors claimed that, unlike Popova, all of these women were public-relations disappointments to both the Women's Tennis Association (WTA) and sports marketing because they chose to resist media intrusions to concentrate on playing good tennis. As a result, U.S. tennis boiled down to Venus and Serena Williams, trailed by a pack of hopefuls and won't-quite-get-theres. The gushing article concluded with this line: "If only she existed."

Just Kidding!

Popova *was* too good to be true. The biography was fiction, and her confident gaze simply showcased someone's digital artistry.

Some people got it. Many didn't, including the media. They bombarded the WTA with calls: Who was Popova and why wasn't she in the Open? The article emphasized what many thought—the WTA was desperate for the next young tennis beauty. WTA spokesperson Chris DeMaria called the story "misleading and irritating" and "disrespectful to the great players we have." Complaining that some people didn't read to the end of articles, he said, "We're a hot sport right now and we've never had to rely on good looks."

Sports Illustrated claimed it was all in grand fun. It hardly needed to add that it was indulging in puckish social commentary on the sexualization of women's tennis.

The Phantom Punch

When Sonny Liston hit the canvas less than two minutes into his second heavyweight-belt bout with Muhammad Ali, pundits immediately accused the former champ of taking a dive. Did the lumbering Liston really fake a fall?

* * * *

EVEN WITHOUT THE controversial conclusion to the widely publicized Sonny Liston–Muhammad Ali rematch on May 25, 1965, there was enough ink and intrigue to fill a John LeCarre spy novel. The bout against Ali—who had just joined the Nation of Islam and changed his name from Cassius Clay—was held in a 6,000-seat arena in Lewiston, Maine, after numerous states refused to sanction the fight because of militant behavior associated with the Muslim movement.

Robert Goulet, the velvet-voiced crooner entrusted with singing the national anthem, forgot the words to the song, and the third man in the ring, Jersey Joe Walcott, was a former heavyweight champion but a novice referee. One minute and 42 seconds into the fight, Ali threw a quick uppercut that seemed to connect with nothing but air. Liston tumbled to the tarmac, though no one seemed sure whether it was the breeze from the blow or the blow itself that put him there. Liston was ultimately counted out by the ringside timer, not the in-ring referee.

Since it was a largely invisible swing (dubbed the "phantom punch" by sports scribes) that floored Liston, he was accused of cashing it in just to cash in. Evidence proves otherwise. Film footage of the bout shows Liston caught flush with a quick, pistonlike "anchor" punch that Ali claimed was designed to be a surprise. Liston actually got back up and was trading body blows with the Louisville Lip when the referee stepped in, stopped the fight, and informed Liston that his bid to become the first boxer to regain the heavyweight title was over.

Good Ol' Boys Make Good

The 1930s and 1940s saw a new breed of driver tearing up the back roads of the American South.

* * * *

YOUNG, HIGHLY SKILLED, and full of brass, these road rebels spent their nights outwitting and outrunning federal agents as they hauled 60-gallon payloads of illegal moonshine liquor from the mountains to their eager customers in the cities below. In this dangerous game, speed and control made all the difference. The bootleggers spent as much time tinkering under their hoods as they did prowling the roads. A typical bootleg car might be a Ford Coupe with a 454 Cadillac engine, heavy-duty suspension, and any number of other modifications meant to keep the driver and his illicit cargo ahead of John Law.

And They're Off!

With all that testosterone and horsepower bundled together, it was inevitable that these wild hares would compete to see who had the fastest car and the steeliest nerve. A dozen or more of them would get together on weekends in an open field and spend the afternoon testing each other's skills, often passing a hat among the growing number of spectators who came to watch. Promoters saw the potential in these events, and before long organized races were being held all across the South. As often as not, though, the promoters lit out with the receipts halfway through the race, and the drivers saw nothing for their efforts.

Seeking to bring both legitimacy and profitability to the sport, driver and race promoter William "Bill" Henry Getty France Sr. organized a meeting of his colleagues at the Ebony Bar in Daytona Beach, Florida, on December 14, 1947. Four days of haggling and backslapping led to the formation of the National Association for Stock Car Auto Racing (NASCAR), with France named as its first president. The group held its inaugural race in 1948, on the well-known half-sand, half-asphalt track at Daytona. Over the next two decades, the upstart organization built a name for itself on the strength of its daring and charismatic drivers. Junior Johnson, Red Byron, Curtis Turner, Lee Petty, and the Flock Brothers—Bob, Fonty, and Tim—held regular jobs (some still running moonshine) and raced the circuit in their spare time. And these legendary pioneers were some colorful characters: For example, Tim Flock occasionally raced with a pet monkey named Jocko Flocko, who sported a crash helmet and was strapped into the passenger seat.

Dawn of a New Era

During these early years, NASCAR was viewed as a distinctly Southern enterprise. In the early 1970s, however, Bill France Jr. took control of the organization from his father, and things began to change. The younger France negotiated network television deals that brought the racetrack into the living rooms of Middle America. In 1979, CBS presented the first flag-to-flag coverage of a NASCAR event, and it was a doozy. Race leaders Cale Yarborough and Donnie Allison entered a bumping duel on the last lap that ended with both cars crashing on the third turn. As Richard Petty moved up from third to take the checkered flag, a fight broke out between Yarborough and Allison's brother Bobby. America was hooked.

France also expanded the sport's sponsorship beyond automakers and parts manufacturers. Tobacco giant R. J. Reynolds bought its way in, as did countless other purveyors of everyday household items, including Tide, Lowe's Hardware, Kellogg's Cereal, the Cartoon Network, Nextel, and Coca-Cola. Today, NASCAR vehicles and their beloved drivers are virtually moving billboards. Plastered with the logos of their sponsors as they speed around the track, Jeff Gordon, Dale Earnhardt Jr., Tony Stewart, Bobby Labonte, and their fellow daredevils draw the eyes of some 75 million regular fans and support a multibillion-dollar industry that outearns professional baseball, basketball, and hockey *combined.*

Playing Chicken

Part cheerleader, part contortionist, and part comedian, the mascot was inspired by the "fowl" play of a college student.

* * * *

THE COLORFUL AND enthusiastic mascots that provide comic relief and intermission entertainment at sporting events have become as essential to the games as snacks and sodas. Most of these individuals are talented athletes in their own rights, as dexterity in gymnastics, dance, and occasional stunts are vital ingredients of their performances.

The first mascot to gain notoriety for his amusing antics was a farcical fowl known as the San Diego Chicken. In 1974, a San Diego State University student named Ted Giannoulas was hired by a local radio station to dress like a chicken and hand out Easter eggs at the San Diego Zoo. Giannoulas's shtick was so popular among the children that he decided to attend San Diego Padres baseball games and act as the team's "unofficial" mascot. The Chicken attracted worldwide acclaim—in fact, the mascot was named one of the 100 most powerful people in sports for the 20th century by *Sporting News Magazine*. Soon, most professional sports teams were using mascots to entertain, amuse, and fire up the fans.

The first official sideline mascot wasn't a gaily garbed goliath or dapperly draped demon, but it was definitely a beast. In 1889, Yale University enlisted an English bulldog named Handsome Dan to slobber along the sidelines and support the college football team, which became known as the Bulldogs.

In Hockey, Why Is Scoring Three Goals Called a Hat Trick?

The sports world is full of weird, wonderful jargon. Third base in baseball is called "the hot corner." In football, a deep pass thrown up for grabs is known as a "Hail Mary." In hockey, scoring three goals is labeled a "hat trick." The first term makes sense, the second one kind of makes sense, and the third one makes absolutely no sense.

⁕ ⁕ ⁕ ⁕

FURTHERMORE, THROUGHOUT MOST of hockey's history, players didn't wear helmets, let alone hats. What gives?

Etymologists agree that the term "hat trick" originated in cricket, a British game that few Americans care about or even want to understand. Evidently, back in the mid-1850s, when a cricket "bowler" captured three consecutive "wickets," he earned a free hat. Though we have no idea what that means, (and still don't want to) it still raises the question: How did the term "hat trick" infiltrate the lexicon of hockey, another game that, admittedly, a fair amount of Americans also don't understand?

As legend has it, hockey's use of "hat trick" originated in the early 1940s with a Toronto haberdasher (someone who sells hats) named Sammy Taft. The story goes that a Chicago Blackhawks forward named Alex Kaleta visited Taft's shop one day in search of a new fedora, only to find that Taft's hats were too pricey for his meager professional athlete's salary (my, how times have changed).

Taft, feeling generous, offered to give Kaleta the hat for free if he could score three goals in that evening's game against the Toronto Maple Leafs. Kaleta did, and the hat was his. Taft, sensing a potential marketing boon, made a standing offer to any player who could score three goals in a Maple Leafs home game. Sometimes he even threw the prize hat onto the ice after the third goal. The hat-tossing became a fad, and soon other fans—apparently far wealthier than poor Kaleta—were tossing their hats onto the ice.

For many hockey fans, any excuse to throw something onto the ice is seen as cause for celebration. Besides hats, octopuses have been tossed onto the rink by fans in Detroit, and in Florida the lovely tradition of throwing rats onto the ice began after a Panthers player scored two goals in a game after killing a rat in the locker room. It's a far cry from the heart-warming tradition started by Sammy Taft. Then again, the whole concept of hockey in Florida is pretty weird in itself.

Decoding the Scoring System in Tennis

The British are an odd bunch. They call trucks "lorries," drugstores "chemists," and telephones "blowers." They put meat in pies and celebrate a holiday called Boxing Day that has nothing to do with boxing. So it shouldn't be surprising that tennis, one of Britain's national pastimes, has such a bizarre scoring system.

* * * *

FOR THOSE WHO haven't been to the tennis club lately, here's a refresher on how scoring works. The first player to four points is the winner of the match, but points are not counted by one, two, five, six, or any other logical number—they go by fifteen for the first two points of the game, then ten for the third point. The sequence, then, is: 0–15–30–40. Except it's not zero—it's called "love." So: love–15–30–40. To confuse matters further, if both players are tied at forty, it's not a tie—it's called "deuce." Say what? Just trying to figure out this scoring system makes one long for a gin gimlet and a cold compress.

Gin gimlets, in fact, may have been the order of the day when modern tennis was invented. According to most tennis historians, it dates back to the early 1870s, when the delightfully named Major Walter Clopton Wingfield devised a lawn game for the entertainment of party guests on his English country estate. Wingfield (whose bust graces the Wimbledon Tennis Museum) based his game on an older form of tennis that long had been popular in France and England, called "real tennis."

Unfortunately, the origin of tennis' odd scoring system is as obfuscated as the system itself. A number of historians argue that Wingfield, being something of a pompous ass, borrowed the terms for his new game from the older French version, even though they made no sense once adapted into English. Hence, *l'oeuf* (meaning "egg") turned into "love." And a *deux le jeu* ("to two the game") became "deuce."

Furthermore, Wingfield opted to borrow the counting system from earlier versions of tennis—in French, scoring mimicked the quarter-hours of the clock: 15–30–45. For some unknown reason (possibly too many gin gimlets), 45 became 40, and we have the scoring system that we know and love (no pun intended) today.

There are plenty of other theories about where the scoring system originated, including "love" coming from the Flemish *lof* (meaning "honor") and "deuce" originating in ancient card games. Others argue that scoring by fifteen was based on the value of the *sou*, a medieval French coin. However, in the absence of definitive evidence, we prefer to attribute the ludicrous scoring system to tipsy Brits.

PETA Would Doubtless Disapprove of This Sport

Imagine a game in which the "ball" is the fresh carcass of a goat—decapitated, dehoofed, and soaked overnight in a bath of cold water to make it stiff. Such a game really exists, and it's called buzkashi.

⁕ ⁕ ⁕ ⁕

BUZKASHI IS THE national sport of Afghanistan. In addition to the interesting choice of a ball, the players in this rough-and-tumble game are mounted on horseback and wear traditional Uzbek garb: turbans, robes, and scarves around their waists.

There's no complicated playbook, only a minimally regimented strategy that requires—encourages—no-holds-barred violence. The referees carry rifles, in case things really get out of hand.

The field has no set boundaries; spectators are in constant danger of being trampled. The objective is to gain possession of the goat and carry it to a designated goal. And the winning players cook and eat the carcass.

Buzkashi translates to "goat pulling" and likely evolved from ordinary herding. It originated with nomadic Turkic peoples who moved west from China and Mongolia from the 10th to 15th centuries. Today, it's played mainly in Afghanistan, but you can also find folks yanking the ol' carcass in northwestern China and in the Muslim republics north of Afghanistan.

The game has two basic forms: modern and traditional. The modern involves teams of ten to 12 riders. In the traditional form, it's every man for himself. Both require a combination of strength and expert horsemanship.

The competitions often are sponsored by *khans* ("traditional elites") who gain or lose status based on the success of the events. And in this case, success is defined by how little or how much mayhem erupts. Biting, hair-pulling, grabbing another rider's reins, and using weapons is prohibited in buzkashi. Anything else goes.

The Golf Wardrobe: A Trail of Fashion Casualties

In most of the major sports, athletes don't have much choice when it comes to what they wear. Basketball, football, baseball, and hockey teams all have uniforms. But other athletes aren't so lucky (and neither are their fans). Golfers, for example, are allowed to choose their own garb, leading to a parade of "uniforms" that look as if they were stitched together by a band of deranged clowns.

* * * *

WHY BIG-TIME GOLFERS wear such hideous clothes is a source of bewilderment. Some apologists blame it on the Scots. Golf, after all, was supposedly invented by shepherds in Scotland back in the twelfth century, and it almost goes without saying that a sport born in a country where man-skirts are considered fashionable is doomed from the start. We'd like to point out that we are no longer in twelfth-century Scotland—let's move on, people.

But history may indeed play a role in golf's repeated fashion disasters. Kings and queens were reputed to have hit the links in the sixteenth and seventeenth centuries, and by the late nineteenth century, golf was a popular pastime amongst the nobility of England and Scotland.

The nobility, however, wasn't exactly known for its athletic prowess. The other "sports"

many of these noblemen participated in were activities like steeplechase (which has its own awful fashion), and so most early golfers had absolutely no idea what types of clothes would be appropriate for an athletic endeavor. Early golfers simply took to the links wearing the fashionable attire of the day—attire that, unfortunately, included breeches and ruffled cravats (these were like neckties, and equally useless).

The tradition of wearing stuffy, silly attire continued into the twentieth century (as did the tradition of wealthy, paunchy white guys playing the sport), with awful sweaters and polyester pants replacing the ruffled cravats and knee-length knickers.

Yet, remarkably, modern golfers take umbrage at the stereotype that duffers have no sense of fashion. According to one golf wag, the knock on golfers for being the world's worst-dressed athletes is unfair because nowadays almost everybody wears Dockers and polo shirts. (We'll pause while that gem sinks in.)

To be fair, the dreadful golf fashions of the 1970s and 1980s have given way to a more benign blandness that is at least less offensive, if not remotely what anybody would call "stylish." Of course, all fashion is less offensive than it was in the 1970s and 1980s, so perhaps golf fashion is proportionally no better.

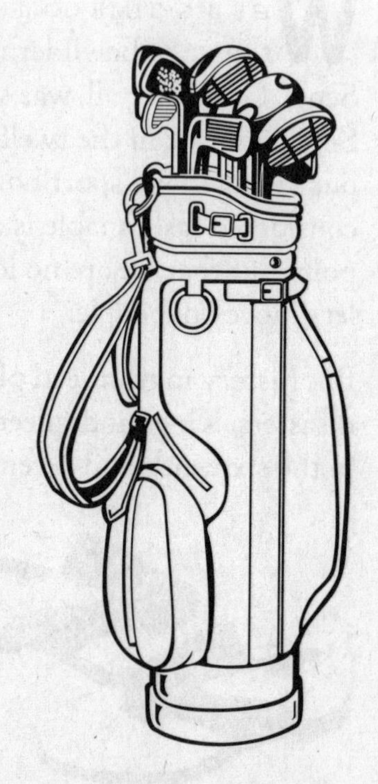

"Golf," Mark Twain once complained, "is a good walk spoiled." We love Mark Twain, but we have to clarify his words and note that spoiling a good walk is the least of golf's transgressions.

How Come Nobody Else Calls It Soccer?

Millions of kids across the United States grow up playing a game that their parents hardly know, a game that virtually everyone else in the world calls football. It's soccer to us, of course, and although Americans might be ridiculed for calling it this, the corruption is actually British in origin.

* * * *

SOCCER—FOOTBALL, AS THE Brits and billions of others insist—has an ancient history. Evidence of games resembling soccer has been found in cultures that date to the third century BC. The Greeks had a version that they called *episkyro*. The Romans brought their version of the sport along when they colonized what is now England and Ireland. Over the next millennium, the game evolved into a freewheeling, roughneck competition—matches often involved kicking, shoving, and punching.

In England and Ireland, the sport was referred to as football; local and regional rules varied widely. Two different games—football and rugby—slowly emerged from this disorganized mess. The Football Association was formed in 1863 to standardize the rules of football and to separate it from rugby. The term "soccer" most likely is derived from the association's work.

During the late nineteenth century, the Brits developed the habit of shortening words and adding "-er." (We suffer this quirk to this day in expressions like "preggers." A red card to the Brits on this one.) One popular theory holds that given the trend, it was natural that those playing "Assoc." football were playing "assoccers" or "soccer." The term died out in England, but was revived in the United States in the early twentieth century to separate the imported sport with the round white ball from the American sport with the oblong brown ball.

Soccer has long struggled to catch on as a major spectator sport in the United States. For most Americans, there just isn't enough scoring or action. In fact, many Yanks have their own word for soccer: boring.

Biggest Soccer Riots of All Time

Despite its lack of popularity in the United States, soccer is by far the most popular sport worldwide. Only slightly less popular, though, is the less skilled sport of "soccer rioting"—also known as hooliganism. These riots are no laughing matter. Here are some of the biggest riots in soccer history.

* * * *

Lima, Peru: May 24, 1964

This one was a doozy. During an Olympic qualifying match between Peru and Argentina, frenzied Peruvian fans grew irate when referees disallowed a goal for the home team. The resulting riot left 300 people dead and 500 injured.

Calcutta, India: August 16, 1980

Tensions were already high in post-partition India when an official's call sparked rioting during a soccer match in Calcutta. The result: 16 dead, 100 injured.

Brussels, Belgium: May 29, 1985

Nobody does soccer riots like the British. In fact, the British are so good they can cause riots in other countries. Take the case of the "Heysel Disaster"—a match in Brussels, Belgium between the British team Liverpool and the Italian club Juventus. The game hadn't even begun when a crowd of drunk and disorderly Liverpool supporters charged toward a group of Juventus fans. The mass stampede caused a stadium wall to collapse, resulting in 39 deaths and a five-year ban on all British soccer teams in Europe.

Zagreb, Croatia: May 13, 1990

In a grim harbinger of the ethnic violence that would ensnare the region over the next few years, Serbs and Croats fought each other before, during, and after a match between the Dinamo Zagreb and the Red Star Belgrade soccer teams, leaving hundreds wounded and throwing the city into a state of chaos.

Orkney, South Africa: January 13, 1991

Fights broke out in the grandstand during a game between the Kaizer Chiefs and Orlando Pirates after a disputed goal. In the ensuing rush of panicked fans trying to flee the fights, more than 40 people were killed and another 50 were injured. Ironically, most of the deaths were a result of being crushed against riot-control fencing. Fans of these two teams would combine for another riot in 2001, in which 43 people were killed.

Accra, Ghana: May 9, 2001

Unruly fans throwing bottles and chairs onto the field during a Ghanaian soccer match were bad enough, but to make it worse, police responded by firing tear gas into the jammed grandstands. The resulting panic killed more than 100 people.

Moscow, Russia: June 9, 2002

When Russia lost to Japan in the 2002 World Cup, Russian fans decided to express their disappointment by setting fire to Moscow. The ensuing riots left one dead and more than two dozen injured, including a group of Japanese tourists.

Basel, Switzerland: May 13, 2006

The Swiss might be politically neutral, but they're hotheads about their football. Never was this more apparent than when FC Basel lost their chance at the Swiss League title when FC Zurich scored a late goal. The resulting riot—which included fans storming the field—resulted in more than 100 injuries and became known as the "Disgrace of Basel."

Manchester, England: May 2008

Observers knew there was going to be trouble when hooligans began fighting the day *before* the 2008 UEFA Cup Final. But the rioters kicked it up a notch on game day, attacking police officers and lighting things on fire in a sad display that became known as the "Battle of Piccadilly." The impetus? The failure of a large television screen erected to give fans without tickets a view of the game.

Disasters of the Olympics

After years of training, even Olympic-caliber athletes are vulnerable to last-minute injuries that dash their hopes. Athletes are sidelined by everything from the common pulled muscle or cold to more unexpected ailments. For instance, in 1912 Sweden's cyclist Carl Landsberg was hit by a motor wagon during a road race and was dragged down the road. The performance of runners Pekka Vasala (Finland) and Silvio Leonard (Cuba) suffered in 1968 and 1976 when Vasala got Montezuma's Revenge and Leonard cut his foot on a cologne bottle. Perhaps the most memorable Olympic disaster was when Janos Baranyai of Hungary dislocated his elbow while lifting 148 kg during the 2008 Beijing Olympics. Who knew the Olympics could be so dangerous?

* * * *

- The U.S. track-and-field team for the 1900 Paris games was weakened because the French unexpectedly held events on the Sabbath. Several universities forbade their collegiate athletes to compete.
- Runner Harvey Cohn was almost swept overboard, and six athletes required medical treatment, when the SS *Barbarossa* was hit by a large wave enroute to Athens in 1906. Several favored U.S. athletes did poorly or dropped out because of their "ocean adventure."

* Francisco Lazaro of Portugal collapsed during the 1912 marathon and died the next day from sunstroke.
* After losing his opening round at the Berlin 1936 Olympics, Thomas Hamilton-Brown, a lightweight boxer from South Africa, drowned his sorrows with food. But the competitors' scores had accidentally been switched. Sadly, the damage was done—a five-pound weight gain kept Hamilton-Brown from the final round.
* Shortly after arriving in London for the 1948 Olympics, Czech gymnast Eliska Misakova was hospitalized. She died of infantile paralysis the day her team competed and won the gold. At the award ceremony, the Czech flag was bordered in black.
* During the 1960 cycling road race in Rome, Dane Knut Jensen suffered sunstroke, fractured his skull in a fall, and died.
* In 1960, Wym Essajas, Suriname's sole athlete, misunderstood the schedule and missed his 800-meter race. Suriname couldn't send another athlete to the Olympics until 1972.
* Australian skier Ross Milne died during a practice run for the men's downhill at Innsbruck in 1964 after smashing into a tree.
* Mexico City's altitude of 7,347 feet slowed the times of endurance events in the 1968 games. Three men running the 10,000-meter were unable to finish while others fell unconscious at the finish line.
* The Munich Massacre of 1972 resulted in the deaths of eleven Israeli athletes, five Palestinian terrorists, and one German policeman after the kidnapping of the athletes.
* In 1972, U.S. runners Eddie Hart, Rey Robinson, and Robert Taylor, supplied with an outdated schedule, rushed to the 100-meter semifinals at the last minute. Hart and

Robinson, both winners in the quarterfinals, missed their heats. Taylor ran and won the silver medal.

* Sixteen-year-old swimmer Rick DeMont took two Marex pills for an asthma attack the day before his 400-meter freestyle race. His gold medal was revoked when he failed the drug test. The 1972 team physicians never checked to see whether his prescription contained banned substances. The same thing happened to Romanian gymnast Andreea Raducan in 2000. She was stripped of her gold medal for the all-around competition when she tested positive for the banned substance pseudoephedrine—an ingredient in the cold medicine provided by team doctors.
* In 1996, two people were killed and 111 were injured when American Eric Robert Rudolph detonated a bomb at the Atlanta Olympics.
* During the 2010 Winter Olympics in Vancouver, Georgian luger Nodar Kumaritashvili died during a training run on the luge track, losing control of his sled in a tight turn and crashing headlong into a steel support pole. Kumaritashvili's death prompted officials to alter the luge course in an attempt to make it less dangerous. Sadly, the modifications came too late for Nodar.

Beefed Up

You're probably familiar with the terms "juiced," "roid-raged," "hyped," and "pumped"—all used to describe the hyperbolic effects of anabolic steroids. For better or for worse, steroids have invaded the worlds of professional and amateur sports, and even show business.

* * * *

Better Living Through Chemistry

ANABOLIC STEROIDS (ALSO called anabolic-androgenic steroids or AAS) are a specific class of hormones that are related to the male hormone testosterone. Steroids have been used for thousands of years in traditional medicine to promote healing in diseases such as cancer and AIDS. French neurologist Charles-Édouard Brown-Séquard was one of the first physicians to report its healing properties after injecting himself with an extract of guinea pig testicles in 1889.

In 1935, two German scientists applied for the first steroid-use patent and were offered the 1939 Nobel Prize for Chemistry, but they were forced to decline the honor by the Nazi government.

Interest in steroids continued during World War II. Third Reich scientists experimented on concentration camp inmates to treat symptoms of chronic wasting as well as to test its effects on heightened aggression in German soldiers. Even Adolf Hitler was injected with steroids to treat his endless list of maladies.

Giving Athletes a Helping Hand

The first reference to steroid use for performance enhancement in sports dates back to a 1938 *Strength and Health* magazine letter to the editor, inquiring how steroids could improve performance in weightlifting and bodybuilding. During the 1940s, the Soviet Union and a number of Eastern Bloc countries built aggressive steroid programs designed to improve the performance of Olympic and amateur weight lifters. The program was so successful that U.S. Olympic team physicians worked with American chemists to design Dianabol, which they administered to U.S. athletes.

Since their development, steroids have gradually crept into the world of professional and amateur sports. Their use has become commonplace in baseball, football, cycling, track—even golf and cricket. In the 2006 Monitor the Future survey, steroid use was measured in eighth-, tenth-, and twelfth-grade students; a little more than 2 percent of male high school seniors admitted to using steroids during the previous year, largely because of their steroid-using role models in professional sports.

Bigger, Faster, Stronger—Kinda

Steroids have a number of performance enhancement perks for athletes such as promoting cell growth, protein synthesis from amino acids, increasing appetite, bone strengthening, and the stimulation of bone marrow and production of red blood cells.

Of course, there are a few "minor" side effects to contend with as well: shrinking testicles, reduced sperm count, infertility, acne, high blood pressure, blood clotting, liver damage, headaches, aching joints, nausea, vomiting, diarrhea, loss of sleep, severe mood swings, paranoia, panic attacks, depression, male pattern baldness, the cessation of menstruation in women, and an increased risk of prostrate cancer—small compromises in the name of athletic achievement, right?

While many countries have banned the sale of anabolic steroids for non-medical applications, they are still legal in Mexico and Thailand. In the United States, steroids are classified as a Schedule III controlled substance, which makes their possession a federal crime, punishable by prison time. But that hasn't deterred athletes from looking for that extra edge. And there are thousands of black-market vendors willing to sell more than 50 different varieties of steroids.

Largely produced in countries where they are legal, steroids are smuggled across international borders. Their existence has spawned a new industry for creating counterfeit drugs that are often diluted, cut with fillers, or made from vegetable oil or toxic substances. They are sold through the mail, the Internet, in gyms, and at competitions. Many of these drugs are submedical or veterinary grade steroids.

Impact on Sports and Entertainment

Since invading the world of amateur and professional sports, steroid use has become a point of contention, gathering supporters both for and against their use. Arnold Schwarzenegger, the famous bodybuilder, actor, and politician, freely admits to using anabolic steroids while they were still legal. "Steroids were helpful to me in maintaining muscle size while on a strict diet in preparation for a contest," says Schwarzenegger, who held the Mr. Olympia bodybuilding title for seven years. "I did not use them for muscle growth, but rather for muscle maintenance when cutting up."

Lyle Alzado, the colorful, record-setting defensive tackle for the Los Angeles Raiders, Cleveland Browns, and Denver Broncos admitted to taking steroids to stay competitive but acknowledged their risks. "Ninety percent of the athletes I know are on the stuff. We're not born to be 300 lbs. or jump 30 ft. But all the time I was taking steroids, I knew they were making me play better," he said. "I became very violent on the field and off it. I did things only crazy people do. Now look at me. My hair's gone, I wobble when I walk and have to hold on to someone for support and I have trouble remembering things. My last wish? That no one else ever dies this way."

Recently, a few show business celebrities have come under scrutiny for their involvement with steroids and other banned substances. In 2008, 61-year-old *Rambo* star Sylvester Stallone paid $10,600 to settle a criminal drug possession charge for smuggling 48 vials of Human Growth Hormone (HGH) into the country. HGH is popularly used for its anti-aging benefits. "Everyone over 40 years old would be wise to investigate it (HGH and testosterone use) because it increases the quality of your life," says Stallone.

"If you're an actor in Hollywood and you're over 40, you are doing HGH. Period," said one Hollywood cosmetic surgeon. "Why wouldn't you? It makes your skin look better, your hair, your fingernails. Everything."

The MVS (Most Valuable Sod)

A soft mound, a swampy outfield, a puddle surrounding first base. These are just some of the tricks groundskeepers have used to help their team walk away with a victory.

❋ ❋ ❋ ❋

"A GOOD GROUNDSKEEPER CAN be as valuable as a .300 hitter," owner Bill Veeck once said. He would have known. Veeck's head groundskeeper was Emil Bossard, who spawned a legacy that has helped keep the "home" in home field advantage.

Bossard had a big job in Cleveland from the 1930s to the 1950s as caretaker for League Park and Municipal Stadium, the two fields used by the Indians at the time. He built the mound tall when fireballer Bob Feller pitched, and he kept the grass on the left side thick, as player-manager Lou Boudreau requested. When slugging clubs like the Yankees came in, the outfield grass was left especially long and wet to turn their doubles into singles. And that's not all. Years later, Roger Bossard confessed that his grandfather used to move the portable fence back 10 to 15 feet against the Yankees to diminish their power. Interestingly enough, Cleveland was the only American League team to win multiple pennants during the Yankees' run from 1941 to '64.

The Bossards branched out. Harold and Marshall took over in Cleveland for their father. Brother Gene went to Chicago's Comiskey Park, where the club won its only pennant in 88 years with Veeck and Bossard having a hand in things. Gene watered down the field—earning Comiskey the nickname of "Camp Swampy"—and kept baseballs in a freezer to cut down on home runs by slugging opponents. Grandson Roger maintains Comiskey's successor (U.S. Cellular Field) and claims to be one of the last groundskeepers to know the special maneuvers used by previous generations. "I won't let the tricks die," he told the Sports Turf Managers Association. ESPN listed the

Bossard family as number seven on the all-time list of baseball cheaters.

Other groundskeepers took to drowning the field to help ensure victory. During the 1962 best-of-three playoff between the San Francisco Giants and Los Angeles Dodgers, Candlestick Park groundskeeper Matty Schwab stepped in to give his players an edge. At the behest of Giants manager Alvin Dark, the ground was soaked around first base to slow Dodgers speedster Maury Wills. While umpire Jocko Conlan made Schwab work to dry out the right side of the infield, the left side went untouched and remained a sponge to slow down grounders. The Giants won the game (Wills never reached base), and shortstop Joe Pagan fielded eight chances flawlessly. The Giants went on to win the pennant. Schwab, who'd been lured from the Dodgers to the Giants back when the teams were in New York, received a full World Series share. Dark would forever retain the nickname "Swamp Fox."

But Schwab was not alone in his tactics. The Tigers grounds crew regularly drenched the area around home plate so Ty Cobb's bunts would stay fair. The Indians watered down third base to protect Al Rosen, who broke his nose nine times while fielding ground balls. Kansas City groundskeeper George Toma, among others, wet the mound and then let it bake in the sun when the opposing pitcher was Catfish Hunter, who preferred a soft mound. If a team wants to use every possible advantage, they might as well start with the foundation beneath their feet.

✳ Chapter 13

Heirlooms, Treasures, and Arty-Facts

Treasures Lost And Found

Millions of dollars worth of treasure was pillaged, buried, sunken—you name it—during the riotous days of the Civil War.

✳ ✳ ✳ ✳

TREASURES OF COIN, bullion, and other valuables had already been traveling along America's waters, rails, and roads before the Civil War began. Once that conflict erupted, these stores of wealth became targets for pirates, outlaws, and soldiers to intercept either for personal gain or to hit the enemy in the pocketbook—or both. The paths this loot took on the high seas, on the tracks, and in the backwoods of the country became more and more treacherous as the war developed.

The Sources of Treasure

As early as January 1861, the Union's secretary of the Navy sent word to his commanders stationed along the water route from New York to California to "be vigilant and if necessary be prompt to use all the means at your command for the protection of the California steamers and their treasure." Commodore Cornelius Vanderbilt also recommended adding guns on passenger ships because "our steamers may be seized or robbed on their voyage." Treasury Secretary Salmon Chase and others also begged for strong cannons to be mounted on merchant boats to protect them from Confederate bandits. One group of merchants petitioning for such assistance estimated that $40 million in gold traveled from San Francisco to New York every year.

The Confederate navy simply didn't match up to that of the Union, so the South hired private raiders to attack Northern merchant ships in an effort to stop general trade and the transport of items that assisted the Union war effort. In the first half of the war, Confederate-sponsored raiders captured 40 Northern ships. The South eventually commissioned about 200 warships that had been built in England. Many carried British sailors driven by the promise of reward. One daring raider, the Tallahassee, had a field day sinking ships off the coast of New Jersey and New York. It sank six ships in six hours before moving northward to attack coastal and transatlantic trading vessels.

Weather Woes

Other Union merchant ships were subject to damage from the weather. The SS Republic, a twin-paddlewheel steamer reportedly carrying $400,000 in coins, sailed from New York to New Orleans in 1865 and was pounded not by a Confederate attack but by a powerful hurricane. The Republic disappeared beneath the waves many miles off the coast of Georgia.

After more than a decade of failing deep-sea hunts and haggling with the government over proper rights and authorization, two modern undersea explorers, John Morris and Greg Stemm of Odyssey Marine Exploration, hit the jackpot in 2003 when they discovered more than 50,000 coins and 13,000 artifacts totaling $75 million. Such excavations and treasure hunts are expensive, though, leaving much underwater Civil War loot out there waiting to be discovered.

Bury the Loot

Money and assets on land during the war were also precious and subject to looting. Notorious Confederate General John

Hunt Morgan pillaged Union towns in Kentucky and throughout the Ohio River Valley. This harsh Southerner rode with his raiders up to wealthy homes and threatened to burn them down unless the owner could pay the ransom. With a command of more than 2,000 troops, Morgan ravaged towns, stole from businesses, and even took collection plates from churches.

After Morgan's force conducted a tour of robberies in central Kentucky, which included a hit on a bank for $80,000, Union cavalry came upon the raiders. The cavalry dispersed Morgan's troops, but their plunder was never found—was it buried in the Kentucky hills? The total wealth Morgan looted on his infamous raids will never be known, but it is likely that much of it was dispersed and hidden throughout the area along the trails and roads his forces traveled. One estimate is that Morgan accumulated nearly $1 million in gold and silver bullion.

The Treasury's Depleting Treasure

Perhaps the largest stash that traveled through Confederate hands was the Confederate Treasury after Union soldiers encroached on the Southern capital of Richmond, Virginia. Nearing the point of surrender, President Jefferson Davis ordered the area evacuated and assigned the Treasury to Captain William Parker. Parker and his soldiers loaded the sum, which totaled upward of $1 million, onto boxcars and sent it as far as Danville, Virginia, the new Confederate capital for the next eight days. Already on the run and trying to evade capture, several Southern leaders chose to distance themselves from the loot in favor of their own safety.

After the money bounced from town to town, it landed in Washington, Georgia. Here, Confederate troops charged with protecting the wealth feared for their safety and demanded payment on the spot. The military escort dwindled, as did the Treasury itself. As the loot traveled from farmhouse to farmhouse, it shrunk to $288,022.90 by the time it reached

President Davis and what was left of the acting government. Some believe Davis took a large chunk of the money himself and buried it in several locations before he was captured.

On May 14, 1865, two Virginia bankers arrived in Washington with a federal order to commandeer the money, and Clark gave it up. The party carrying the money back to Richmond pulled over for the night 12 miles outside of the city only to be robbed. The outlaws, understanding that both law enforcement and Confederate soldiers were on their trail, buried their take before Confederates shot and killed them. Some believe the loot was buried on the south bank of the Savannah River.

Secret Society

According to legend, documented history, and some modern-day discoveries, rebels buried much more wealth than this. The mysterious Knights of the Golden Circle likely left behind millions in coin and precious metals. This secret society, explains historian and Knights of the Golden Circle detective Bob Brewer, grew out of anti-Union, proslavery sentiment and had several chapters by 1855. Famous politicians and rank-and-file composed the secret membership society, helping foster the ideas behind nullification and secession. Through a complex system of Masonic codes, secret signals, handshakes, and other rites, the organization, it is be lieved, buried much of the South's wealth in the hopes of financing a later uprising to reassert and complete the goals of the temporarily defeated South.

Some of this loot has been found. Brewer discovered a pint jar with coins and gold pieces—worth about $28,000 by today's values—at a location 65 miles from Oklahoma City called Buzzard's Roost. Treasure hunters report that four caches of booty have been unearthed over the years near this location, totaling more than $1 million. Those who study the Knights predict a lot more is out there. The organization likely developed some type of grid system marked with tree and rock carvings directing fellow Knights to the wealth. While little or no known record of the alleged postwar scheme fully defines its scope, it is believed that information to locate the money and the group's cause has been handed down by word of mouth, from father to son, in the hopes that one day the South will rise again.

Curious Displays and Museums

From a museum that "digs up the dirt," to another that "plants" people six feet into it, these showcases feature anything but the expected as they celebrate the wacky and wild.

* * * *

Burlingame Museum of Pez

WHO CAN RESIST the quirky little candies that are ejected from a whimsical dispenser? Certainly not Gary Doss. In 1995, he opened the Burlingame Museum of Pez in California to showcase his collection of more than 500 Pez dispensers. Favorites include Mickey Mouse, Tweety Bird, and Bugs Bunny, but offerings such as Uncle Sam, a wounded soldier, and a stewardess are also on display.

The name "Pez" comes from *pfefferminz,* the German word for peppermint. The candy was originally introduced in Austria in 1927 to aid smokers trying to kick the habit. The first dispensers, utilitarian in design, appeared around 1950. By 1952, cartoon heads and fruit-flavored candies had made the scene. The popularity of the dispensers has turned them into highly collectible items. Doss's collection features a Mr. Potato Head version that, due to its small parts, was deemed a hazard to children and was yanked from store shelves in 1973. Today, the dispenser is valued at around $5,000.

Leila's Hair Museum

If you walk into Leila's Hair Museum in Independence, Missouri, expecting to find curlers, dryers, and similar hair-styling implements, you'll be sorely disappointed. If, however, you go ape for jewelry, wreaths, chains, broaches, hat pins, postcards, and a host of other intriguing things made *from* hair, you'll feel like a veritable Lady Godiva at this shrine to follicles. With more than 2,000 pieces of hair jewelry and 150 hair wreaths at which to gawk, a visitor could grow out of their haircut if they stayed to scrutinize each one.

Created by Leila Cohoon in 1990, the museum was originally a place to store her burgeoning hair-wreath collection. Cohoon continually increases her hair collection by visiting garage sales, antique dealers, auctions, and other sources. She even has a hair wreath donated by comedian Phyllis Diller, who is noted for her kooky hairdos.

Museum of Dirt

Here's a dirty little secret: Boston's Museum of Dirt isn't just about dirt. Oh sure, you can find the brown, gritty, and cruddy stuff there, but the museum also contains such filthy offshoots as a ball of lint, crumpled leaves, and rocks. (In case you're wondering, the ball of lint came to the museum compliments of humorist Dave Barry, and the crumpled leaves are straight from Martha Stewart's yard.)

Glenn Johanson started his collection when he dug up some dirt as a souvenir from the Liberace Museum in Las Vegas. With this offbeat memento acting as impetus, the soil-buff dug up more dirt and things simply got filthier from there.

Much of the museum's dirt collection has been donated by people from around the world. But visitors will also find granules scraped from Gianna Versace's front step, as well as dirt lifted from Eartha Kitt's star on the Hollywood Walk of Fame. Some morbid pieces are also on display, such as a rock retrieved from the grounds of the Alfred P. Murrah Federal Building in Oklahoma City after it was blown up in 1995, and pieces of thatching taken from the estate of O. J. Simpson, after Nicole Brown Simpson and Ron Goldman were murdered.

Hoover Historical Center

The Hoover Historical Center, aka the Vacuum Cleaner Museum, in North Canton, Ohio, is one of the few places where the statement "this thing really sucks" means something good. The name Hoover is synonymous with vacuum cleaners, so it comes as no surprise that, in 1978, the boyhood home of the company's founder, William H. "Boss" Hoover (who popularized the machine but did not invent it), became a showcase for great "suckers" of the past.

Visitors to the dirt-removal museum will find antique cleaning devices and manual machines dating from the late 1800s. The 1908 portable "Model 0" (the first electric Hoover vacuum, which was invented by janitor James Murray Spangler in 1907) is featured at the museum beside such thoroughly odd pieces as the 1910 Kotten suction cleaner. This piece of ingenuity required the user to stand on a platform and rock side-to-side to create suction. Not too practical by today's standards.

Another interesting piece found at the Hoover Historical Center is a 1923 Hoover Model 541. Made of super-sleek, diecast aluminum, it's easy to imagine a 1920s-era flapper pushing the device to and fro as she sashayed her dirt away.

National Museum of Funeral History

We all have to go sometime, but if Robert L. Waltrip has his way, it won't be until we've visited his National Museum of Funeral History in Houston, Texas. With more than 20,000 square feet, this museum is packed to the rafters with caskets, coffins, hearses, and other items related to dying.

Opened in 1992 as a way to honor "one of our most important cultural rituals," the scope of the operation is impressive, as are some of its more unusual exhibits. A real crowd pleaser is a 1916 Packard funeral bus. This bizarre vehicle seems to ask, "Why deal with long funeral processions when this baby can do it all?" Designed to carry a coffin, pallbearers, and up to 20 mourners, the bus was retired after it tipped over during a funeral in San Francisco in the 1950s, ejecting mourners and the deceased onto the street.

Another interesting item is the "casket for three." Built for a heartbroken couple bent on suicide after their baby passed away, it ended up at the museum after they dashed their plan.

A playful sign in the museum reads, "Any day above ground is a good one." True, but after checking out the museum's ultra-cool fantasy coffins, designed to resemble objects such as a car, an airplane, or a fish, six-feet under doesn't seem so bad.

Museum of Questionable Medical Devices

Can you imagine zapping a case of arthritis with a "Cosmos Bag" (a 1920s-era cloth bag stuffed with low-grade radioactive ore) or curing a case of constipation with a painfully large "Recto Rotor," the anal answer to a plumber's mechanical snake? Imagine no more.

In the early 1980s, founder Bob McCoy opened the Museum of Questionable Medical Devices in St. Anthony, Minnesota. But since 2002, the operation has been located at the Science Museum of Minnesota in St. Paul. Visitors gawk at a collection of the most dubious medical instruments ever devised by humankind, such as a prostate gland warmer, a "phrenology" machine (designed to measure human head bumps to determine personality), a foot-operated breast enlarger, "electropathy" machines (to electrically treat diseases and conditions), radium cures, a violet-ray generator (to cure everything from heart disease to writer's cramp), and hundreds of others. There's even a Contemporary Quackery section that proves that such nuttiness lives on.

Museum of Bad Art

Some would consider a velvet Elvis or *Dogs Playing Poker* too atrocious for an art collection, but the Museum of Bad Art, located in Dedham, Massachusetts, is dedicated to the acquisition and exhibition of such rarely revered works.

The museum began where only such an institution could: in the garbage. That's where, in 1992, Boston antiques dealer Scott Wilson spotted the serendipitous work that's now cherished as the museum's *Mona Lisa*, a subtly disturbing painting he's dubbed *Lucy in the Field with Flowers*, which depicts a gray-haired matriarch sitting among a field of lilies that undulate beneath a fluorescent yellow-green sky. The old lady's lawn chair hovers nearby.

Since then the collection has grown to about 250 pieces that are, as their motto states, simply "too bad to be ignored." Twenty-five of those are on display at any given time on the lower floor of the Dedham Community Theater, just outside the men's room.

The *Belle* Shipwreck

Mired in gook for more than three centuries, a 17th-century shipwreck provides a unique window into Texas's past.

* * * *

IN 1995, A team of researchers led by the Texas Historical Commission found a sunken vessel in the mud of Matagorda Bay. This in itself was not unusual. The area had given up scores of wrecked ships in the past, each a slowly rotting testament to the ravages of the sea. But this one was different. It dated back to 1687 and was remarkably intact. Better still, it was one of four vessels sailed by French explorer René-Robert Cavelier Sieur de La Salle. Although the adventurer's goal to start a coast colony never materialized, the archaeological bonanza left in the ship's wake proved to be worth its weight in gold.

Stroke of Luck

While a tale of accidental discovery would surely add to the drama, the *Belle* was not found by chance. The Texas Historical Commission knew that the ship rested somewhere in

Matagorda Bay; a Spanish map drawn up in 1689 hinted at its location. But it wasn't until 1995 and the advent of geographical positioning systems that the ship was finally unearthed—quite literally—from the gooey seabed. Ironically, it was this muck encasing the *Belle* that had saved it from bacterial attack, thus preserving the vessel and its priceless artifacts.

Treasure Trove

Pottery, glass beads, bronze cannon, hawk bells, and other items were recovered alongside organic materials such as bone, wood, rope, and cloth. Even the skeletal remains of an unlucky crew member came up with the eclectic treasure trove. In all, over one million pieces were recovered in what is considered one of the most important shipwrecks ever discovered in North America. A vast number of *Belle* artifacts are displayed at the Texas State History Museum in Austin.

Quirky Collections

Judging from the number of online museums and collectors' guides, it seems that many people are trying to make a name for themselves with one-of-a-kind collections. American Idol contestant Brandon Green even proudly displayed his toenail and fingernail collection for a national television audience.

Navel Fluff

In 1984, Graham Barker of Perth, Australia, started collecting his navel lint. Since then, he has seldom missed a day's "harvest" and collects an average of 3.03 milligrams each day; he currently has about 2 1/2 jars of of the stuff. He was rewarded for his efforts in 2000 when *Guinness World Records* declared his navel lint collection the world's largest. Barker also collects his own beard clippings, bakery bags, fast-food tray inserts, and ski-lift tickets.

Adhesive Bandages

Marz Waggener of Long Beach, California, has been collecting (unused) adhesive bandages since February 1994, after convincing her mother that she needed a box of Incredible Crash Dummies bandages for a Girl Scout badge. Fourteen years later, her collection had reached 3,300 different bandages, including a 1979 Superman bandage and some rare samples from the 1950s.

Traffic Signs and Signals

Stephen Salcedo asks visitors to his Web site to refrain from calling the cops on him—all 500-plus traffic signs and signals in his collection were obtained legally. Salcedo started his collection in 1986 at age five. The Fort Wayne, Indiana, collector has always been attracted to the graphic design aspect of road signs and has a special fondness for older ones (pre-1960). The "treasure" of his collection is the street sign that stood on the corner near his childhood home in Merrillville, Indiana.

Police and Prison Restraints

If the handcuff-collecting world has a celebrity, it is Stan Willis of Cincinnati. Since 1969, Willis has been collecting police and prison restraints and has built his reputation selling rare cuffs to other collectors. In 2003, Guinness World Records recognized his collection as the largest—it now contains nearly 1,400 items. He also collects police and fire department-related items, such as badges, lanterns, and helmets.

Mustard

Barry Levenson began his mustard collection in October 1986, with 12 jars he bought to soothe his grief when the Boston Red Sox lost to the New York Mets in the World Series. He vowed to assemble the largest collection of prepared mustard in the world. In April 1992, he opened the Mount Horeb Mustard Museum in Mount Horeb, Wisconsin. He now displays nearly 5,000 mustards from all 50 states and more than 60 countries, as well as historic mustard memorabilia. He is currently working on getting widespread recognition of the first Saturday in August as World Mustard Day, which is already celebrated at the museum.

Toothpaste

Dentist Val Kolpakov began collecting toothpaste and dental artifacts in March 2002 as a way to advertise his new dental practice. His hobby became a mission, however, and he now displays more than 1,400 tubes in his Saginaw, Michigan, office. Although the star of his collection is a rare silver Georgian tooth powder box from 1801, he is especially interested in vintage toothpastes, tubes from around the world, toothpastes from TV and movie sets, and flavors other than the traditional mint. Dr. Kolpakov boasts a variety of liquor-flavored pastes (including bourbon, Scotch, wine varietals, and champagne) as well as curry, lavender, and pumpkin pudding. Kolpakov believes his to be the largest collection in the world.

Barf Bags

Although Steve Silberberg of Massachusetts has never been out of the United States, his 2,000-plus barf bags come from around the world. Silberberg began collecting "happy sacks"—as they're known in piloting circles—in 1981 and now has a wide range of bus, car, train, and helicopter bags as well. Although not the largest collection in the world (he guesses it might be the tenth largest), he does have the largest collection of non-transportation bags, including novelty bags not intended for use, as well as political and movie sickness bags. The treasures of his collection include those given away on the Disneyland Star Tours ride and one from the Space Shuttle.

Mechanical Memorabilia

Marvin Yagoda has been a pharmacist for 50 years, but he is best known for his hobby of collecting mechanical novelties, vintage oddities, strange curiosities, and wonders, as he calls them. He started his collection of vintage coin-operated games and toys in the early 1960s and displayed some pieces in a local food court. After the food court closed in 1988, he opened Marvin's Marvelous Mechanical Museum in 1990—a 5,500-square-foot building in a strip mall in Farmington Hills, Michigan. Yagoda has packed the 40-foot ceilings with such things as fortune-tellers-in-a-box, nickelodeons, a carousel, more than 50 model airplanes, antique electric fans, animatronic dummies, vintage arcade games (from the early 1900s through present day), prize machines, one of the infamous P. T. Barnum Cardiff Giant statues, the electric chair from Sing Sing Prison, and much, much more. More than a thousand electrical outlets power the machines, which are all in working condition. Even the walls are not overlooked, sporting his collection of magic posters and 20-foot-long carnival canvases. Admission to his museum is free, but be sure to come with a few rolls of quarters.

Diamonds Are Forever

Most diamonds have had a peaceful existence. Still, there are others awash in blood or are even rumored to be cursed. Read on for some gem-studded legends.

* * * *

The Good, the Bad, and the Pretty

THE CULLINAN DIAMOND, discovered in 1905 and weighing in at 1.3 pounds, is still the largest rough-cut gem-quality diamond that has ever been found. Some gemologists hypothesize that it may be part of a much larger diamond.

Although it is rumored that the **Koh-i-noor Diamond** was originally found 5,000 years ago, historians definitely know that it was a British prize of war in the 1849 bloody conquest of the Punjab. It later ended up in Britain's Crown Jewel collection. Still, many of the diamond's previous owners suffered terribly, and stories of lootings of kingdoms, torture, and the like prevail. In one battle, 20,000 people died—not for the Koh-i-noor itself, but because of the power, pride, and rapaciousness of its owner, symbolized by the gem.

The **Orlov Diamond** is another gem with an allegedly bloody history. At one point, the diamond was stolen from a Hindu statue by a soldier, who then offered it to an Armenian merchant. When a price could not be reached, the soldier sold it to someone else. The Armenian killed the soldier and the purchaser; later, when he and his brothers could not agree on the division of spoils, he killed them too.

As the story goes, Russian Count Grigoryevich Orlov later bought the diamond in 1775 and gave it to Czarina Catherine the Great in order to win her favor. His ploy didn't work—though she did keep the diamond.

Perhaps it is the **Hope Diamond** that carries the most baggage. However, most of the curses surrounding the fabled diamond—insanity, disease, violent death, and suicide—appear to be stories invented in 1910 by the renowned jeweler Cartier to pique a potential buyer. Since then, any curse seems to have dissipated.

Pawned, You Say?

The famous **Sancy Diamond** spent much of its early history in pawnshops that catered to European royal families. Its owner, a mercenary named Nicholas de Sancy, first used it as collateral to finance several military campaigns on behalf of monarchs. Later, the exiled queen of Charles I used the gem for a loan to support her lifestyle, but she blew her bankroll and failed to redeem it. The diamond was sold again, this time to the French King Louis XIV. Later, in the aftermath of the French Revolution, it was again pawned to finance military operations.

The **Regent Diamond** was part of the French Crown Jewels collection, pawned by Napoleon Bonaparte to pay for the cavalry and supply horses he needed to win the pivotal Battle of Marengo. That win opened the door to his historic empire-building military initiatives.

All About Harry

Harry Winston was not your average storefront jeweler. In fact, his name still comes up when some of the world's most famous diamonds are discussed, including the **Hope, Idol's Eye, Jonkers,** and the **Lesotho.** Here are some other Winston-related facts:

* When Winston donated the Hope Diamond to the Smithsonian Institution in 1958, he trusted its delivery to the U.S. Postal Service.

* Winston's firm also cut the diamond that became known as the **Taylor-Burton Diamond** when actor Richard Burton gave it to actress Elizabeth Taylor.

* In 1974, Winston negotiated a deal with DeBeers Consolidated Mining for $24.5 million—the largest single diamond purchase in the world. As negotiations were being finalized, he casually asked for and received a "deal sweetener" of a 181-carat rough diamond. When cut, this gem provided several stones, the largest of which was 45.31 carats. Appropriately, Harry named that one the **Deal Sweetener.**

Like Lumpy Money from Heaven

Money may not grow on trees, but it does occasionally fall from the sky. Welcome to the strange but lucrative market in space rocks (and the objects they impact).

* * * *

Heads Up!

EVERY DAY THE Earth is bombarded by meteorites. Most originate in our solar system's asteroid belt, some are pieces of passing comets, and others originate from the Moon or Mars. On a clear dark night it's possible to observe as many as a few per hour; during a meteor shower the count may rise to as many as 100 per hour.

Despite the large numbers of meteorites, most of them burn up upon hitting the Earth's atmosphere. Of those that survive, most fall into an ocean; the meteorites that fall on land are typically the size of a pea or smaller.

Aside from the few kilograms of moon rocks retrieved from various *Apollo* and *Luna* space missions, meteorites are the only extraterrestrial artifacts on the planet. Not only does this make them valuable to scientists seeking knowledge about other planets, but it also makes them valuable to collectors seeking rarities and oddities. As such, finding a bona fide meteorite is a rare and potentially lucrative event.

Watch the Skies . . . for Cash!

In 2007, a small meteorite from Siberia sold for $122,750 at an auction in New York City. In 1972, a cow in Valera, Venezuela, was hit by a falling meteorite. Tiny fragments of the stone that hit the beast are worth more than $1,000. (Unfortunately, the cow was eaten—perhaps its carcass could have fetched a good deal more. "Space steaks," anyone?)

Looking to get married? As of 2009, you can purchase a custom wedding ring made from fragments of the Gibeon Meteorite that hit Africa more than 30,000 years ago (after spending 4 billion years hurtling through space) for as little as $195. If you'd rather save some money, a dime-size sliver of Mars goes for about $100 online.

Lucrative Destruction

Interestingly, the most sought-after meteorite fragments are those that hit other objects. These meteorites are called "hammers," and the smaller the object they hit the better. Even more valuable than hammer meteorites, however, are the objects themselves.

While many people would order their meteorite off the Internet, one woman had hers delivered in a more direct manner. On the evening of December 10, 1984, Carutha Barnard of Claxon, Georgia (who was not a meteorite collector at the time), received a surprise package when a small stone slammed into the back of her mailbox and thudded into the ground below. Though the box itself was knocked from its post, the outgoing mail flag intrepidly remained in the upright position. In October 2007, the mailbox, without the accompanying meteorite, sold for a whopping $83,000 at auction.

The Most Valuable Chevy on Earth

On October 9, 1992, thousands of people across the Mid-Atlantic States witnessed one of the most spectacular meteor showers on record. Glowing brighter than a full moon, and breaking into more than 70 fragments as it arched through the heavens, the meteorite was recorded by more than a dozen video cameras.

In Peekskill, New York, Michelle Knapp was startled by a loud noise outside her home and emerged to find the rear end of her red 1980 Chevy Malibu badly damaged. The police were soon on the scene and filed a criminal report, not realizing that the perpetrator was still on the premises. The smell of leaking gasoline, however, brought the fire department, which discovered the fallen meteorite fragment in the Malibu's gas tank. Knapp sold the car and rock to collector R. A. Langheinrich for an estimated $30,000. Since then, the car has traveled the world as a display piece to places such as New York, Paris, Tokyo, and Munich.

What Makes Something "Art"?

If you want to see a name-calling, hair-pulling intellectual fight (and who doesn't?), just yell this question in a crowded coffee shop. After centuries of debate and goatee-stroking, it's still a hot-button issue.

* * * *

BEFORE THE FOURTEENTH century, the Western world grouped painting, sculpture, and architecture with decorative crafts such as pottery, weaving, and the like. During the

Renaissance, Michelangelo and the gang elevated the artist to the level of the poet—a genius who was touched by divine inspiration. Now, with God as a collaborator, art had to be beautiful, which meant that artists had to recreate reality in a way that transcended earthly experience.

In the nineteenth and twentieth centuries, artists rejected these standards of beauty; they claimed that art didn't need to fit set requirements. This idea is now widely accepted, though people still disagree over what is and isn't art.

A common modern view is that art is anything that is created for its own aesthetic value—beautiful or not—rather than to serve some other function. So, according to this theory, defining art comes down to the creator's intention. If you build a chair to have something to sit on, the chair isn't a piece of art. But if you build an identical chair to express yourself, that chair *is* a piece of art. Marcel Duchamp famously demonstrated this in 1917, when he turned a urinal upside down and called it *Fountain*. He was only interested in the object's aesthetic value. And just as simply as that: art.

This may seem arbitrary, but to the creator, there is a difference. If you build something for a specific purpose, you measure success by how well your creation serves that function. If you make pure art, your accomplishment is exclusively determined by how the creation makes you feel. Artists say that they follow their hearts, their muses, or God, depending on their beliefs. A craftsperson also follows a creative spirit, but his or her desire for artistic fulfillment is secondary to the obligation to make something that is functional.

Many objects involve both kinds of creativity. For example, a big-budget filmmaker follows his or her muse but generally bends to studio demands to try to make the movie profitable. (For instance, the movie might be trimmed to ninety minutes.) Unless the director has full creative control, the primary function of the film is to get people to buy tickets. There's nothing wrong with making money from your art, but purists say that financial concerns should never influence the true artist.

By a purist's definition, a book illustration isn't art, since its function is to support the text and please the client—even if the text is a work of art. The counter view is that the illustration *is* art, since the illustrator follows his or her creative instincts to create it; the illustrator is as much an artistic collaborator as the writer.

Obviously, it gets pretty murky. But until someone invents a handheld art detector, the question of what makes something art will continue to spark spirited arguments in coffee shops the world over.

Oscar Who?

Every year, film buffs and celebrity oglers around the world tune in to watch the Academy of Motion Picture Arts and Sciences hand out its Academy Awards of Merit. But why are they referred to as "Oscars"?

* * * *

THE EPIC FILMS *Ben-Hur* (1959), *Titanic* (1997), and *The Lord of the Rings: The Return of the King* (2003) each won 11 of them. The legendary Katharine Hepburn set the standard for professional acting, having won a record four of them in the category of Best Actress. Renowned British actor Peter O'Toole was nominated for eight over the course of his career, only to go home empty-handed every time.

"Them," of course, refers to the Academy Awards of Merit, the illustrious prizes handed out annually by the Academy of Motion Picture Arts and Sciences for excellence in the fields of movie acting, writing, directing, producing, and technology.

The eight-and-a-half-pound gold-plated statuette that symbolizes the epitome of film industry success is formally called the

Academy Award of Merit, or the Academy Award, for short. So, how is it that the award became known as "Oscar"?

Credit for originating the name is generally given to former Academy executive director Margaret Herrick. In 1931, as a young librarian with the Academy, Herrick cheerfully remarked that the statuette's art-deco figure reminded her of her Uncle Oscar. The name stuck, and in 1939 the Academy began using both "the Academy Award" and "the Oscar" as the official tag of its prized prize.

The Washington Monument: George in Roman Drag?

We cannot tell a lie: The Washington Monument definitely wasn't built in a day. In fact, it wasn't built in a century. From concept to completion, the narcoleptic project spanned 102 years. You might mention that the next time someone claims that today's Congress is inefficient!

* * * *

George's Lifetime

THE REVOLUTIONARY WAR concluded in 1783. The colonies were now the United States of America, and General George Washington was their greatest war hero. A veteran proposed a statue of the great man on horseback (as was typical for statues of generals). Congress approved the project but initiated no planning or building. Washington passed away in 1799 without seeing a single stone laid. In any case, Washington's passing stirred public opinion: There should be at least one memorial or sepulcher somewhere in the capital city. Of course, Congress got right on that.

Cornerstone

Just kidding. It wasn't until 1833 that someone finally got around to founding a Washington Monument Society, but this time the plan bore fruit. In 1836, architect Robert Mills's design won the Society's approval. The Monument was to be a 600-foot, flat-topped obelisk with a circular colonnade at the base. Atop the colonnade would stand a heroic statue of Washington in a Roman toga, driving . . . a chariot. Thirty other Revolutionary heroes would also get statues. The work was finally under way.

Well, not really. The project stalled until 1848 when—amid great fanfare—workers finally laid the obelisk's cornerstone. Progress continued until 1854, when funds ran low. Then a bigoted political party found a creative way to compound the existing fiasco.

Knowing Nothing

In the mid-19th century, a nativist party and quasi-secret society known as the Know-Nothings became a player in U.S. politics. (Their name, while intellectually honest, wasn't meant as such. It came from their response to queries: "I know nothing.") The first huge wave of Catholic immigrants had arrived in the United States, many fleeing the Irish potato famines for the land of religious freedom (and some lunch). As conspiracy theorists and haters of Catholicism and immigration (especially Catholic immigration), the Know-Nothings considered the immigrants a papal plot to Catholicize the United States.

We wouldn't care, except in 1854 the Know-Nothings gained control of the Washington Monument Society. Annoyed, Congress chopped the funding. The Know-Nothings knew little about masonry; what little construction they achieved had to be ripped out later due to crummy workmanship. In 1858, the Know-Nothings finally did something beneficial for the Monument: They quit the Society. By 1860, the monument was just an ugly vertical rectangular stub about 152 feet high—barely one-fourth complete and going nowhere.

Civil War

In 1861, of course, the Confederate States of America seceded from the Union. This, at least, was a legitimate reason to suspend work. During the Civil War, the sacred site served various hallowed purposes: cattle grazing, beef slaughtering, troop encampments, and drill training. The Confederacy surrendered in 1865, so the construction immediately resumed.

At Last!

Ha! Surely you're wise to this by now. Congress goofed around until 1876, when the Centennial reminded everyone that it was about time to finish memorializing the first president and commander in chief of the Continental Army. By now, Robert Mills was no longer living, so the new guard edited his design. There would be no colonnade, no George Ben-Hur; also, to allay the concerns of the Corps of Engineers about how much weight the ground could sustain, the obelisk would taper off at 555 feet.

The change in color you see on the modern Monument is where construction resumed—a slightly different type of marble was used to finish the project where the Know-Nothings left off. Construction was completed in 1885, and the government opened the interior to the public in October 1888.

George Washington would have turned 156 that February.

The Washington Monument consists of 36,491 marble, granite, and sandstone blocks and weighs approximately 90,854 tons.

At the time of its completion, the Monument was the world's tallest structure, a title it held only until 1889, when the Eiffel Tower was completed in Paris. Today, it holds a place in the record books as the world's tallest freestanding stone structure not created from a single block of stone.

You'll have to climb 897 steps to reach the top of the Washington Monument. That's a lot of steps! Fortunately, you can also make the trip by elevator if you so choose. If you take the elevator, it will take you 70 seconds to reach the top. We can't tell you how long it will take if you choose to climb the stairs!

Who Chose the Seven Wonders of the World?

Humans love their lists—to do, grocery, pros and cons—so it makes sense that we would obsessively list cool stuff to see. There is a "wonder list" for every kind of wonder imaginable. The most famous is the Seven Wonders of the Ancient World, or "Six You'll Have to Take Our Word for, Plus One You Can Actually See."

* * * *

ONE OF THE earliest references to "wonders" is in the writings of Greek historian Herodotus from the fifth century B.C. Herodotus wrote extensively about some of the impressive wonders he had seen and heard about. However, the concept of the Seven Wonders didn't really catch on until the second century B.C.

For the next fifteen hundred or so years, six of the seven were a lock, often appearing on compiled lists, with the seventh spot being a rotating roster of hopefuls. By the time the list became the accepted seven of today (around the Renaissance), the Lighthouse of Alexandria had taken up the seventh spot.

No one person actually got to pick the seven; it was more of a generally accepted concept based on the frequency with which certain wonders landed on different lists. Furthermore, by the time the Middle Ages rolled around, most of the wonders couldn't be seen in their full glory because of damage or destruction, so the selections were based primarily on reputation.

The Seven Wonders of the Ancient World were the Pyramids of Giza in Egypt, the Hanging Gardens of Babylon in Iraq, the Statue of Zeus at Olympia in Greece, the Mausoleum of Maussollos at Halicarnassus in Turkey, the Colossus of Rhodes, the Temple of Artemis at Ephesus, and the Lighthouse of Alexandria in Egypt.

The Pyramids of Giza are the Energizer Bunny of the wonders. They were the oldest when the lists began, and they are the only wonder still standing. The Colossus of Rhodes was big in size (107 feet high) but not big on longevity—the statue stood for only fifty-four years, but was impressive enough to stay in the minds of list-makers.

In 2001, the New 7 Wonders Foundation was established by a Swiss businessman. The foundation's intention was to create a new list of seven wonders of the world based on an online vote. (Notably, users could vote more than once.) In 2007, the wonders chosen were Chichén Itzá in Mexico, Christ the Redeemer in Brazil, the Colosseum in Rome, the Great Wall of China, Machu Picchu in Peru, Petra in Jordan, and the Taj Mahal in India. The Pyramids of Giza were given honorary finalist status after Egypt protested that these great historical structures shouldn't have to compete against such young whippersnappers.

So make your own lists, and bury them in a time capsule. Maybe one day that cool treehouse you built with your dad will finally get its due.

Central Park's Carousel

For most New Yorkers, the city is all about moving—quickly—from one place to another. But some residents, especially kids, would much rather go in circles.

⁕ ⁕ ⁕ ⁕

ASK LOCAL GROWNUPS about Manhattan landmarks, and you'll hear about the Statue of Liberty, the Empire State Building, and the clock at Grand Central Station. Ask a New York kid about the most famous place in the city, and you're likely to hear about the carousel in Central Park, one of New York's iconic and best-loved sites, and one of the largest merry-go-rounds in the country. Some 250,000 children and grownups head to the park annually to whirl around on painted horses while a calliope grinds out classic merry-go-round music to the delight of all.

Going 'Round Merrily

The current carousel, one of six in parks around the city, is actually the fourth in Central Park. The very first, built in 1871, sat about 50 yards from the current site, midpark at 64th Street. The first Central Park carousel was "powered" by a blind mule and a horse, which pulled the carousel from a recessed circular track. Around 1912 a new carousel was electrified, but in 1924 it burned down and was replaced. The third carousel, too, burned down in 1950, leaving city officials to search mightily for a replacement. A 1908 carousel that had done long duty in Coney Island until the early 1940s was retrieved from storage, and expert carousel woodcarvers fabricated new horses. This renovation was financed with funds donated by the Michael Friedsam Foundation.

Today the Friedsam Memorial Carousel is open to the public year-round. A three-and-a-half-minute ride costs three dollars, but what a ride it is. This carousel's 57 horses are unusually large: three-quarters the size of real horses, and the carousel itself spins around at 12 miles an hour—twice as fast as your ordinary merry-go-round. What, you were expecting a *slow* merry-go-round? Hey, this is New York.

Great Moments in Kitsch History

Kitsch is a term used to describe objects of bad taste and poor quality. But despite its bad rap, plenty of people go to great lengths to collect kitsch and keep its "charms" alive.

* * * *

Pink Flamingos

IN 1957, UNION Products of Leominster, Massachusetts, introduced the ultimate in tacky lawn ornaments: the plastic pink flamingo. Designed by artist Don Featherstone, they were sold in the Sears mail-order catalog for $2.76 a pair with the instructions, "Place in garden, lawn, to beautify landscape." Authentic pink flamingos—which are sold only in pairs and bear Featherstone's signature under their tails—are no longer on the market (Union Products shuttered its factory in 2006), but knockoffs ensure the bird's survival.

Troll Dolls

Danish sculptor Thomas Damm created the popular troll doll as a handmade wooden gift for his daughter. After it caught the eye of the owner of a local toy shop, Dammit Dolls were born, and plastic versions with trademark oversize hairdos hit the mass market. The dolls swept the United States in the early 1960s and were lugged around as good-luck charms by people of all ages and walks of life, including Lady Bird Johnson.

Lava Lites

In the early 1960s, Englishman Edward Craven-Walker invented the Lava Lite, and Chicago entrepreneur Adolf Wertheimer bought the American distribution rights after seeing the psychedelic display at a trade show. Within five years, two million Lava Lites had been sold in the United States.

Cuckoo Collections

Tired of the usual snobby museum fare? Try these two oddball collections!

❊ ❊ ❊ ❊

Mr. Ed's Elephant Museum

If you still retain a pinch of youthful wonder, a love of kitsch, and don't have a peanut allergy, Mr. Ed's Elephant Museum in Orrtanna, Pennsylvania, is the place for you.

Ed Gotwalt's obsession with elephants began in 1967 when his sister-in-law gave him an elephant statuette as a wedding present. He opened his first elephant museum in 1975, and the present location started in 1984.

The museum's 6,000 items include an elephant potty chair, an elephant pulling a 24-karat gold circus wagon, and Miss Ellie, a 9 1/2-foot-tall fiberglass elephant that welcomes visitors from the side of the road.

International Banana Club Museum

If you tell Ken Bannister he's totally bananas, he'll probably take it as a compliment. The International Banana Club Museum in Hesperia, California, started when Bannister handed out banana stickers purely for fun and received a bunch of banana-related items in return. Since 1976, this storehouse of all things banana has grown vast enough to be listed in *Guinness World Records.*

The museum showcases banana trees, pipes, pins, belts, charms, glasses, rings, clocks, software, cups, and lights. Wacky (well, *wackier*) items include a Michael Jackson banana, and a "petrified" banana, which is hard as rock and dates back to 1975.

Unusual Art Projects

Artists are often known to "push the envelope." Whether they use paint, clay, or toy blocks, the art world wouldn't be the same if all those wacky artists didn't make some really wild art.

⁕ ⁕ ⁕ ⁕

Ronnie Nicolino's Breasts

FOR A GUY who claims he's not obsessed with female breasts, Ronnie Nicolino sure spent a lot of artistic energy on them. In March 1994, this California-based artist created a two-mile long sand sculpture of 21,000 size 34C breasts at Stinson Beach, near San Francisco. He also planned to string enough donated brassieres together to reach across the Grand Canyon, but despite collecting more than enough bras, Nicolino simply lost interest in the project.

Konzentrationslager

Polish artist Zbigniew Libera was known in Europe for his controversial video art in the 1980s, but in 1996, he gained notoriety on a larger scale when he created a concentration camp out of LEGO blocks. Libera often uses children's toys in his work to illustrate the ways kids learn about human suffering: Libera's LEGO set includes pieces such as gallows, inmates, and corpses.

The Wee World of Willard Wigan

You can't see the artwork of British artist Willard Wigan unless you happen to have a microscope handy. Wigan carves sculptures out of grains of sand, rice, or sugar, and paints them with hairs plucked from a housefly's back. Wigan has trained himself to work "between heartbeats," since working at that scale makes a trembling hand a disaster waiting to happen. Most of Wigan's sculptures fit on the head of a pin, but that doesn't mean they carry a small price tag. Much of his work is valued around $300,000 or more.

Marcel Duchamp's Fountain

No report on strange art projects would be complete without a nod to Marcel Duchamp's famous *Fountain*. In 1917, this famous French Dadaist placed a porcelain urinal on a pedestal, signed it, and called it art. Most people thought it was a total disgrace and denounced Duchamp's work. But the artist had a method to his madness: It was one of the first times "conceptual art" had come on the radar. He took a regular, nonart object and gave it artistic status. "I don't believe in art. I believe in artists," Duchamp said. This notion was outrageous, sure, but it caused a wave of change in the way people view art and the objects around them that still resonates.

Tracey Emin's Bed

In 1998, British artist Tracey Emin let the world in on a very private part of her life when she exhibited *My Bed*, an exact replica of, you guessed it, her bed. Emin took great pains to show her unmade bed exactly as it usually was, replete with rumpled, stained sheets, dirty clothes at the foot, prophylactic wrappers and liquor bottles strewn about, and dingy slippers nearby. The artwork offended many, especially one lady who reportedly came to the gallery with a bucket of cleaning materials, hoping to tidy up the work of art.

Damien Hirst's Shark(s)

In 1991, British art collector and gallery owner Charles Saatchi offered to fund whatever project Damien Hirst wanted to make. So Hirst had a 14-foot-long tiger shark caught and killed off the coast of Australia. He then had it shipped to London, where it sat floating in a giant tank filled with formaldehyde as the conceptual art piece, *The Physical Impossibility of Death in the Mind of Someone Living*.

But the shark wasn't preserved properly, and it eventually began to rot from the inside and the liquid grew murky. Attempts to fix the problem failed (including stretching the shark's skin over a fiberglass model), until in 2006, a billionaire funded a rehab that was rumored to have cost more than $100,000. This entailed catching, killing, and shipping another shark and injecting it with formaldehyde. But is it Art? And is it Art if the original art rotted and had to be replaced? Critics continue to debate the point.

The Louvre: Gift of the Revolution

Built by French kings over six centuries, the famed Louvre Museum found its true calling with the bloody end of the dynasty that built it.

⁂ ⁂ ⁂ ⁂

THE ART COLLECTION known throughout the world as the Musée de Louvre began as a moated, medieval arsenal erected to protect the city's inhabitants against the Anglo-Norman threat. Built by King Philippe Auguste in the late 12th century, the fortress lost much of its military value as the city expanded far beyond the castle walls over the next 150 years. In 1364, King Charles V had the Louvre redesigned as a royal residence.

During the next three centuries, French kings and queens remodeled and redesigned portions of the Louvre, connecting it with the nearby Tuileries palace. Apartments and galleries were added, and remnants of the medieval fortress were demolished.

During the reign of King Louis XIV, the famed "Sun King," the palace came into its own, as classical paintings and sculptures by artists of the day graced the palace's walls and ceilings. Work halted briefly in 1672, when Louis moved the French court to his fantastic palace at Versailles, but in 1692 Louis sent a set of antique sculptures back to the Louvre's Salle de Caryatides, inaugurating the first of the Louvre's many antique accessions. Academies of arts and sciences took up residence at the palace, and in 1699, the Académie Royale de Peinture et de Sculpture held its first exhibition in the Louvre's Grand Galerie.

The artistic treasures housed by the French monarchs were, of course, the property of the king and off limits to the masses. But in 1791, in the wake of the French Revolution, the French Assemblée Nationale declared all Bourbon property to be held by the state for the people of France. The government established a public art and science museum at the Louvre and Tuileries, and in 1793, the year King Louis XVI and his queen, Marie Antoinette, were sent to the guillotine, the Museum Central des Arts opened its doors to the public.

As Napoleon's armies marched through Italy, Austria, and Egypt in the late 1790s, the museum's collections grew with the spoils of war. Napoleon and Empress Josephine inaugurated an antiquities gallery at the Louvre in 1800, and three years later, the museum was briefly renamed the Musée Napoléon. With the emperor's fall in 1815, however, the Louvre's status diminished as many of its artifacts were returned to their rightful owners across Europe.

In the mid-19th century, the Louvre opened additional galleries to showcase Spanish, Algerian, Egyptian, Mexican, and other ethnographic artworks. As the Louvre soldiered on into the 20th century, its more exotic holdings—particularly Islamic and Middle Eastern art—expanded, and the museum was progressively remodeled to accommodate its growing collection.

At the outbreak of World War II, French officials worried that the Louvre's magnificent holdings would become an irresistible target for Nazi pillagers, so they removed most of the museum's treasures and dispersed them among chateaus in and around the Loire valley. Although the Nazis had the museum reopened in September 1940, there was little left for the subdued Parisians to view until the country was liberated four years later.

In 1945, the restored French government reorganized its national art collections, and in 1983 the government announced a sweeping reorganization and remodeling plan under the direction of the famed Chinese-American architect I. M. Pei. Impressionist and other late 19th-century works were moved to the Musée d'Orsay, while Pei's famous glass pyramid, which towers over the Cour Napoleon, signaled a new stage in the Louvre's life. Further renovations from 1993 to the present have given the Louvre its distinctive look, as well as its status as one of the world's premier museums.

Jackpot in the Desert

While rounding up a stray animal near Qumran, Israel, in early 1947, Bedouin shepherd Mohammed el-Hamed stumbled across several pottery jars containing scrolls written in Hebrew. It turned out to be the find of a lifetime.

* * * *

NEWS OF THE exciting discovery of ancient artifacts spurred archaeologists to scour the area of the original find for additional material. Over a period of nine years, the remains of approximately 900 documents were recovered from 11 caves near the ruins of Qumran, a plateau community on the northwest shore of the Dead Sea. The documents have come to be known as the Dead Sea Scrolls.

Tests indicate that all but one of the documents were created between the middle of the 2nd century B.C. and the 1st century A.D. Nearly all were written in one of three Hebrew dialects. The majority of the documents were written on animal hide.

The scrolls represent the earliest surviving copies of Biblical documents. Approximately 30 percent of the material is from the Hebrew Bible. Every book of the Old Testament is represented with the exception of the Book of Esther and the Book of Nehemiah. Another 30 percent of the scrolls contain essays on subjects including blessings, war, community rule, and the membership requirements of a Jewish sect. About 25 percent of the material refers to Israelite religious texts not contained in the Hebrew Bible, while 15 percent of the data has yet to be identified.

Since their discovery, debate about the meaning of the scrolls has been intense. One widely held theory subscribes to the belief that the scrolls were created at the village of Qumran and then hidden by the inhabitants. According to this theory, a Jewish sect known as the Essenes wrote the scrolls. Those subscribing to this theory have concluded that the Essenes hid the scrolls in nearby caves during the Jewish Revolt in A.D. 66, shortly before they were massacred by Roman troops.

A second major theory, put forward by Norman Golb, Professor of Jewish History at the University of Chicago, speculates that the scrolls were originally housed in various Jerusalem-area libraries and were spirited out of the city when the Romans besieged the capital in A.D. 68–70. Golb believes that the treasures documented on the so-called Copper Scroll could only have been held in Jerusalem. Golb also alleges that the variety of conflicting ideas found in the scrolls indicates that the documents are facsimiles of literary texts.

The documents were catalogued according to which cave they were found in and categorized into Biblical and non-Biblical works. Of the eleven caves, numbers 1 and 11 yielded the most intact documents, while number 4 held the most material-—an astounding 15,000 fragments representing 40 percent of the total material found. Multiple copies of the Hebrew Bible have been identified, including 19 copies of the Book of Isaiah, 30 copies of Psalms, and 25 copies of Deuteronomy. Also found were previously unknown psalms attributed to King David, and stories about Abraham and Noah.

Most of the fragments appeared in print between 1950 and 1965, with the exception of the material from Cave 4. Publication of the manuscripts was entrusted to an international group led by Father Roland de Vaux of the Dominican Order in Jerusalem.

Access to the material was governed by a "secrecy rule"—only members of the international team were allowed to see them. In late 1971, 17 documents were published, followed by the release of a complete set of images of all the Cave 4 material. The secrecy rule was eventually lifted, and copies of all documents were in print by 1995.

Many of the documents are now housed in the Shrine of the Book, a wing of the Israel Museum located in Western Jerusalem. The scrolls on display are rotated every three to six months.

The Mütter Museum's Cadaverous Collection

Sometimes mislabeled as a gallery of the gruesome and a mausoleum of the macabre, the Mütter Museum is much more than just skin and bones.

❋ ❋ ❋ ❋

IT IS, BEYOND any doubt, the oddest accumulation of artifacts on this planet, and that's including the National Museum of Scotland, which features more than a few anatomical anomalies preserved in jars of formaldehyde. The Mütter Museum, located at the College of Physicians in Philadelphia, is home to the world's most extensive collection of medical marvels, anatomical oddities, and biological enigmas.

Among the items one may view at the Mütter are the conjoined livers of the world-famous Siamese twins, Chang and Eng; a selection of 139 Central and Eastern European skulls from the collection of world-renowned anatomist Joseph Hyrtl; and the preserved body of the "Soap Lady." Also on display are an assortment of objects that have been extracted from people's throats; a bevy of brains, bones, and gallstones; and a cancerous growth that was removed from President Grover Cleveland. There are also pathological models molded in plaster, wax, papier-mâché, and plastic; memorabilia contributed by famous scientists and physicians, plus a medley of medical illustrations, photographs, prints, and portraits.

Dr. Thomas Dent Mütter, a flamboyant and slightly fanatical physician obsessed with devising new and improved surgical techniques to cure diseases and deformities, donated the collection to the College in 1858. The professor of surgery at Jefferson Medical College in Philadelphia until his retirement, Mütter was ostracized by the medical community for both his methods and his predilection to preserve the remains of some of his more peculiar patients. The Mütter Museum officially opened its doors in 1863, and it has intrigued curious visitors ever since.

Western Costume

Where does a filmmaker go when a movie that's set in Victorian England, the antebellum South, or the battlefields of World War I requires authentic clothing for 500 extras? Western Costume is one of the few companies in the world that can provide it all.

* * * *

ONE OF THE oldest film-related businesses in Hollywood, Western Costume was founded by Lou Burns in 1912. A former Indian trader, Burns had amassed a large collection of Native American clothing and artifacts before settling in Los Angeles. He approached William S. Hart, a pioneering actor and director of cowboy films, to point out how inaccurate the Native American clothing was in most films. Hart hired Burns as an advisor and supplier on his productions, and Western Costume was born.

The company quickly expanded to offer authentic outfits from any era or culture. It was involved in such important silent films as *The Birth of a Nation* (1915), *The Sheik* (1921), and *The Gold Rush* (1925). When sound was introduced to the movies,

Western Costume provided outfits for the first film with spoken dialogue, *The Jazz Singer* (1927). The advent of color films brought a new emphasis on opulent wardrobes, and Western Costume stepped up to provide memorable period clothing for *Gone with the Wind* (1939) and *The King and I* (1956); the company also created what may be the most famous article of clothing ever to appear in a film—the ruby slippers Dorothy wore in *The Wizard of Oz* (1939). The list of classics that the company has outfitted seems endless—*West Side Story* (1961), *The Godfather* (1972), *Young Frankenstein* (1974), *Jurassic Park* (1993), *L.A. Confidential* (1997), and *Gangs of New York* (2002), to name a few.

Western Costumes is not open to the public most of they year, but if you happen to be in Hollywood during the month of October, the company does open its doors for Halloween rentals, allowing you to choose from actual costumes worn by stars in films such as *Ben-Hur* (1959), *Grease* (1978), and *The Untouchables* (1987).

Art by the Numbers: The Vogel Collection

So you like art, but you're operating on a shoestring budget, huh? Take a few pointers from this New York couple, who managed to amass an outstanding collection in an unlikely way.

* * * *

YOU DON'T HAVE to be a mega-millionaire to collect great works of art. Take it from Herb and Dorothy Vogel, two New Yorkers who, back in the mid-1960s, began collecting art that they liked, could afford, and could fit into their modest apartment. And they liked a lot, could afford more than they thought, and managed to fit more than 2,000 works into their modest one-bedroom apartment.

In the Beginning, There Was Art

In 1962, Herb, a postal clerk, married Dorothy, a librarian. It was their mutual love of art that initially brought them together. Dorothy was once quoted as saying that when it came to their relationship, "It was art or nothing"—clearly a serious commitment to both one another and their shared interest. However, the Vogels were about as well off as the starving artists they patronized. But the couple made do: By living off Herb's salary and buying art with Dorothy's, they embarked on their art odyssey.

Although the art the Vogels could afford wasn't the sort sold at Sotheby's, that didn't stop them. This was the second half of the 20th century, and New York was bursting at the seams with an influx of minimalists, conceptualists, painters, sculptors, and artists of every kind, many at the beginning of their soon-to-be-lauded careers. Herb and Dorothy, aside from just loving art in general, happened to have a keen eye for the truly talented.

Around 1967,the two met conceptual artist Sol LeWitt and were the first people to buy his work. This kicked off their collection, and they subsequently bought many more pieces from LeWitt, who eventually made quite a mark on the American art world.

Their collection grew. Among the artists represented by the Vogels's collection were early works by the likes of Christo, Andy Warhol, Chuck Close, Carl Andre, and hundreds of others. They bought often and cheaply, and never, ever sold anything.

Not So User Friendly

Sure, the couple had an exceptional eye for artwork, but where to store it all? The Vogels were unfazed by their typically cramped New York headquarters and refused to move or rent additional space to house their collection. While almost every inch of their walls was covered in the pieces they had bought, most of the art Herb and Dorothy collected was hoarded away in closets or piled on top of shelves and boxes. As art lovers, the Vogels painstakingly maintained the upkeep of their collection, but some accidents were inevitable: Once, water from a fish tank splashed onto a Warhol canvas, and it had to be restored.

Bursting at the Seams

The Vogels might have been a bit preoccupied with art purchases, but they weren't hermits or recluses. In fact, they were often seen out and about with the artists they collected. For Herb and Dorothy, forging a connection with the artists who produced the art was a natural part of the art-buying process.

Finally, aware that they were running out of space, in 1992 the Vogels pledged their collection, comprised of a whopping 2,000 pieces, in installments to the National Gallery of Art in Washington, D.C.—where they spent part of their honeymoon decades ago. No longer stored in a coat closet, these incredible works of art could be enjoyed by all.

The public and art elite alike lauded the Vogels, embracing the couple and their collection for its scope, intensity, and foresight. In 2008, a documentary called *Herb and Dorothy* debuted about the couple and their lifelong ambition. The director, Megumi Sasaki, remarked, "I thought I was going to make a small film about beautiful, small people. But I learned...they are giants of the art world."

It's Still in the Cards

After more than a century, card collecting continues to be a simple pleasure.

* * * *

LONG BEFORE ESPN and MLB.com provided the opportunity to keep up-to-the-inning tabs on ballplayers, baseball cards allowed generations of fans to feel connected with their heroes. And while salaries may have now reached outrageous levels, for many devotees of the game, cardboard is still as good as gold.

Trade Cards, Tobacco, and the T-206

Today's cards are aimed primarily at the Little League crowd, but the first commercial versions (then called "trade cards") in the 1870s were marketed to adults, featuring pictures of top teams on one side and advertisements on the other. By the 1880s, the game's rising popularity prompted several tobacco companies to insert cards into packs of their products, and stars such as King Kelly and Cap Anson began appearing on their own cards.

The cards considered by many collectors to be the most beautiful ever produced appeared during the 1900–1915 era, featuring color drawings of players posed or in action. Usually smaller than today's standard 2 1/2 x 3 1/2 cards, they picked up a younger audience when candy companies also began producing them. The "*Mona Lisa* of baseball cards"—the T-206 White Borders Honus Wagner—was introduced during this time. Only about 50 of these cards were made before a dispute between the tobacco firm and the Pirates shortstop halted their production. In 2007, one of the cards reportedly sold to a private collector for $2.35 million.

Kids Get On Board

Back then, of course, nobody was worrying about a card's monetary value. By the 1930s, when Goudey and other gum companies began producing hundreds of individual player cards each year, kids began trading the cards—either to complete numbered sets or to pick up a favorite player. Many kids stuck cards into the spokes of their bikes to get just the right machine-gun sound.

Baseball's post–World War II boom extended to baseball cards, which reached new heights in popularity during the 1950s. The king now was the Topps Company, which had a virtual monopoly on collecting from 1952 through 1980 and produced as many as 800 new cards each year. Almost every major-leaguer signed an annual deal with Topps; each plastic-wrapped pack available at the corner drugstore contained several cards with players' vital statistics and complete playing records on the back, along with a rectangle of rock-hard pink bubblegum. Cards featured everything from league leaders and All-Stars to fathers and sons, boyhood photos, and even (briefly) umpires.

The Card Collecting Explosion

The Topps monopoly ended in 1981, when two companies (Fleer and Donruss) won the legal right to produce their own sets. Card collecting as a hobby exploded. Kids, faced with numerous choices, bought more cards than ever, and adults who had grown up with Topps paid $10, $20, even $100 to replace the ones they had bought for nickels as youngsters—and that their parents had long since thrown out. In addition to baseball card conventions, fans at one point had some 10,000 card shops in which to seek out treasures. Young collectors now put their top cards in protective cases rather than in their bike spokes, and gum (which damaged cards as well as teeth) faded from the scene.

The sheer glut of cards produced in the late 1980s and early '90s eventually led to a rapid decline in their monetary value, and this, plus fallout from the 1994 players' strike, forced many shops to close their doors. Recently, however, there has been another revival: Topps and numerous rivals now produce a seemingly endless array of specialty cards, including "relic" versions with shavings from actual game-used bats attached and "heritage" versions designed to look like those from the 1950s. Gum has even made a comeback, and the Internet—and especially eBay—has made it easier than ever to find an elusive card. The cards of such Golden Age players as Willie Mays and Mickey Mantle—at least those that have survived in decent condition—are valued in the thousands of dollars. And don't forget that T-206 Wagner!

A New Look at Old Icons

Archaeologists have uncovered the earliest known icons of the apostles Peter and Paul.

* * * *

TIMELESS TREASURES OFTEN turn up where you least expect them. Such was the case in 2009 and 2010 when Vatican archaeologists announced the discovery and restoration of the earliest known icons of the apostles Paul and Peter on a ceiling in a catacomb beneath a modern office building in Rome.

A Significant Find

The remarkable images are believed to date from the second half of the 4th century A.D. and were found on the ceiling of a tomb belonging to a Roman noblewoman in the famous St. Thekla catacomb, located near the Basilica of St. Paul. The square ceiling painting also included the earliest known images of the apostles John and Andrew, as well as a painting of Jesus as the Good Shepherd.

It took two years and cost an estimated 60,000 euros to restore the frescoes and paintings in the catacomb. A new laser technique was used to meticulously burn away several centimeters of white calcium carbonate deposits without harming the original artwork beneath it. The restoration is considered a triumph because the damp conditions within most ancient catacombs typically make the restoration and preservation of paintings within them extremely difficult.

St. Paul Restored

"The result was exceptional because from underneath all the dirt and grime we saw for the first time in 1,600 years the face of Saint Paul in a very good condition," Barbara Mazzei, who directed the work at the catacomb, told a British newspaper.

More than 40 catacombs—underground Christian burial places—are known to exist throughout Rome. They fall under the jurisdiction of Pontifical Commission of Sacred Archaeology because of their unique religious significance.

Treasures in Odd Places

Sometimes artifacts are hiding in places where you'd least expect them to be.

⁕ ⁕ ⁕ ⁕

⁕ A stolen painting by Mexican artist Tamayo was retrieved from an unlikely place: the trash. Out on a stroll, Elizabeth Gibson spotted it nestled among some garbage. After finding the painting posted on a Web site four years later, Gibson received a share of the one million dollars it fetched at Sotheby's.

⁕ John Wilkes Booth's body is buried in a Baltimore cemetery, but his third, fourth, and fifth vertebrae can be found on display in the National Health and Medicine Museum in Washington, D.C. They were removed for investigation after he was shot and killed while on the run following Lincoln's assassination.

⁕ Normally on display at the Elvis After Dark Museum, a handgun once owned by the King was stolen during a ceremony on the 30th anniversary of his death. Soon thereafter, the missing gun was found stashed inside a mucky portable toilet only yards from where the gun's exhibit case sat.

⁕ Egypt, which receives less rainfall than most places on Earth, doesn't seem like the kind of place where you'd find sea creatures. Yet, it was there that geologists discovered the nearly complete skeleton of a Basilosaurus, a prehistoric whale. During the whale's lifetime, 40 million years ago, Egypt's Wadi Hitan desert was underwater.

⁕ Something missing from Neil Armstrong's famous moon-stepping speech was found years later by a computer programmer. Using software that detects nerve impulses, the missing word a was found, which changed Armstrong's statement, "one small step for man" to his intended, different meaning: "one small step for a man."

* *Femme nue couchée*, a racy female nude painted by Gustave Courbet in the 19th century, turned up on a Slovakian doctor's wall in 2000, nearly six decades after it disappeared during World War II. It was believed to have been stolen by members of the Russian Red Army, but the country doctor claimed that he received the art in payment for services rendered.

* The pathologist who performed Albert Einstein's autopsy, Dr. Thomas Harvey, kept a pretty major souvenir: Einstein's brain. Thirty years postmortem, the brain segments were found in the doctor's Kansas home, stored in a pair of mason jars.

7 Notorious Art Thefts

Some people just can't keep their hands off other people's things—even the world's greatest art. Art thieves take their loot from museums, places of worship, and private residences. Because they would have trouble selling the fruits of their labor on the open market—auction houses and galleries tend to avoid stolen works—art burglars often either keep the art for themselves or try to ransom the hot property back to the original owner. Among the major robberies in the past hundred years are these daring thefts of very expensive art (values estimated at the time of the theft).

* * * *

1. **Boston, March 1990: $300 million:** Two men dressed as police officers visited the Isabella Stewart Gardner Museum in the wee hours of the morning. After overpowering two guards and grabbing the security system's

surveillance tape, they collected Rembrandt's only seascape, *Storm on the Sea of Galilee*, as well as Vermeer's *The Concert*, Manet's *Chez Tortoni*, and several other works. Authorities have yet to find the criminals despite investigating everyone from the Irish Republican Army to a Boston mob boss.

2. **Oslo, August 2004: $120 million:** Two armed and masked thieves threatened workers at the Munch Museum during a daring daylight theft. They stole a pair of Edvard Munch paintings, *The Scream* and *The Madonna*, estimated at a combined value of 100 million euros. In May 2006, authorities convicted three men who received between four and eight years in jail. The paintings were recovered three months later.

3. **Paris, August 1911: $100 million:** In the world's most notorious art theft to date, Vincenzo Peruggia, an employee of the Louvre, stole Leonardo da Vinci's *Mona Lisa* from the storied museum in the heart of Paris. Peruggia simply hid in a closet, grabbed the painting once alone in the room, hid it under his long smock, and walked out of the famed museum after it had closed. The theft turned the moderately popular *Mona Lisa* into the best-known painting in the world. Police questioned Pablo Picasso and French poet Guillaume Apollinaire about the crime, but they found the real thief—and the Mona Lisa—two years later when Peruggia tried to sell it to an art dealer in Florence.

4. **Oslo, February 1994: $60–75 million:** *The Scream* has been a popular target for thieves in Norway. On the day the 1994 Winter Olympics began in Lillehammer, a different version of Munch's famous work—he painted four—was taken from Oslo's National Art Museum. In less than one minute, the crooks came in through a window, cut the wires holding up the painting, and left through the same window. They attempted to ransom the painting to the Norwegian government, but they had left a piece of the frame at a bus

stop—a clue that helped authorities recover the painting within a few months. Four men were convicted of the crime in January 1996.

5. **Scotland, August 2003: $65 million:** Blending in apparently has its advantages for art thieves. Two men joined a tour of Scotland's Drumlanrig Castle, subdued a guard, and made off with Leonardo da Vinci's *Madonna with the Yarnwinder*. Alarms around the art were not set during the day, and the thieves dissuaded tourists from intervening, reportedly telling them: "Don't worry... we're the police. This is just practice." Escaping in a white Volkswagen Golf, the perpetrators have never been identified—and the painting remains missing.

6. **Stockholm, December 2000: $30 million:** Caught! Eight criminals each got up to six and half years behind bars for conspiring to take a Rembrandt and two Renoirs—all of them eventually recovered—from Stockholm's National Museum. You have to give the three masked men who actually grabbed the paintings credit for a dramatic exit. In a scene reminiscent of an action movie, they fled the scene by motorboat. Police unraveled the plot after recovering one of the paintings during an unrelated drug investigation four months after the theft.

7. **Amsterdam, December 2002: $30 million:** Robbers climbed onto the roof of the Van Gogh Museum, then broke in and stole two of the master's paintings, *View of the Sea at Scheveningen* and *Congregation Leaving the Reformed Church in Nuenen*, together worth $30 million. Police told the press that the thieves worked so quickly that, despite setting off the museum's alarms, they had disappeared before police arrived. Authorities in the Netherlands arrested two men in 2003, based on DNA from hair inside two hats left at the scene, but they have been unable to recover the paintings, which the men deny taking.

John Myatt, Master Forger

When people hear the word forgery, they usually think of money. But legal currency isn't the only thing that can be faked.

❋ ❋ ❋ ❋

MONET, MONET, MONET. Sometimes I get truly fed up doing Monet. Bloody haystacks." John Myatt's humorous lament sounds curiously Monty Pythonesque, until you realize that he can do Monet—and Chagall, Klee, Le Corbusier, Ben Nicholson, and almost any other painter you can name, great or obscure. Myatt, an artist of some ability, was probably the world's greatest art forger. He took part in an eight-year forgery scam in the 1980s and '90s that shook the foundations of the art world.

Despite what one might expect, art forgery is not a victimless crime. Many of Myatt's paintings—bought in good faith as the work of renowned masters—went for extremely high sums. One "Giacometti" sold at auction in New York for $300,000, and as many as 120 of his counterfeits are still out there, confusing and distressing the art world. But Myatt never set out to break the law.

Initially, Myatt would paint an unknown work in the style of one of the cubist, surrealist, or impressionist masters, and he seriously duplicated both style and subject. For a time, he gave them to friends or sold them as acknowledged fakes. Then he ran afoul of John Drewe.

The Scheme Begins

Drewe was a London-based collector who had bought a dozen of Myatt's fakes over two years. Personable and charming, he ingratiated himself with Myatt by posing as a rich aristocrat. But one day he called and told Myatt that a cubist work the artist had done in the style of Albert Gleizes had just sold at Christies for £25,000 ($40,000)—as a genuine Gleizes. Drewe offered half of that money to Myatt.

The struggling artist was poor and taking care of his two children. The lure of the money was irresistible. So the scheme developed that he would paint a "newly discovered" work by a famous painter and pass it to Drewe, who would sell it and then pay Myatt his cut—usually about 10 percent. It would take Myatt two or three months to turn out a fake, and he was only making about £13,000 a year (roughly $21,000)—hardly worthy of a master criminal.

One of the amazing things about this scam was Myatt's materials. Most art forgers take great pains to duplicate the exact pigments used by the original artists, but Myatt mixed cheap emulsion house paint with a lubricating gel to get the colors he needed. One benefit is that his mix dried faster than oil paints.

The Inside Man

But Drewe was just as much of a master forger, himself. The consummate con man, he inveigled his way into the art world through donations, talking his way into the archives of the Tate Gallery and learning every trick of *provenance*, the authentication of artwork. He faked letters from experts and, on one occasion, even inserted a phony catalog into the archives with pictures of Myatt's latest fakes as genuine.

But as the years went by, Myatt became increasingly worried about getting caught and going to prison, so at last he told Drewe he wanted out. Drewe refused to let him leave, and Myatt realized that his partner wasn't just in it for the money. He loved conning people.

The Jig Is Up

The scam was not to last, of course. Drewe's ex-wife went to the police with incriminating documents, and when the trail led to Myatt's cottage in Staffordshire, he confessed.

Myatt served four months of a yearlong sentence, and when he came out of prison, Detective Superintendent Jonathan

Searle of the Metropolitan Police was waiting for him. Searle suggested that since Myatt was now infamous, many people would love to own a real John Myatt fake. As a result, Myatt and his second wife Rosemary set up a tidy business out of their cottage. His paintings regularly sell for as much as £45,000 ($72,000), and Hollywood has shown interest in a movie—about the real John Myatt.

A Pretty Penny: Very Rare U.S. Coins

Why are certain coins so valuable? Some simply have very low mintages, and some are error coins. In some cases (with gold, in particular), most of the pieces were confiscated and melted. Better condition always adds value.

❋ ❋ ❋ ❋

THE CURRENT MINTS and marks are Philadelphia (P, or no mark), Denver (D), and San Francisco (S). Mints in Carson City, Nevada (CC); Dahlonega, Georgia (D); and New Orleans (O) shut down long ago, which adds appeal to their surviving coinage. Here are the most prized and/or interesting U.S. coins, along with an idea of what they're worth:

1787 Brasher gold doubloon: It was privately minted by goldsmith Ephraim Brasher before the U.S. Mint's founding in 1793. The coin was slightly lighter than a $5 gold piece, and at one point in the 1970s it was the most expensive U.S. coin ever sold. Seven known; last sold for $625,000.

1792 half-disme (5¢ piece): Disme was the old terminology for "dime," so half a disme was five cents. George Washington supposedly provided the silver for this mintage. Was Martha the model for Liberty's image? If so, her hairdo suggests she'd been helping Ben Franklin with electricity experiments. Perhaps 1,500 minted; sells for up to $1.3 million.

1804 silver dollar: Though actually minted in 1834 and later, the official mint delivery figure of 19,570 refers to the 1804 issue. Watch out—counterfeits abound. Only 15 known; worth up to $4.1 million.

1849 Coronet $20 gold piece: How do you assess a unique coin's value? The Smithsonian owns the only authenticated example, the very first gold "double eagle." Why mint only one? It was a trial strike of the new series. Rumors persist of a second trial strike that ended up in private hands; if true, it hasn't surfaced in more than 150 years. Never sold; literally priceless.

1870-S $3 gold piece: Apparently, only one (currently in private hands) was struck, though there are tales of a second one placed in the cornerstone of the then-new San Francisco Mint building (now being renovated as a museum). If the building is ever demolished, don't expect to see it imploded. One known; estimated at $1.2 million.

1876-CC 20-cent piece: Remember when everyone confused the new Susan B. Anthony dollars with quarters? That's what comes of ignoring history. A century before, this 20-cent coin's resemblance to the quarter caused similar frustration. Some 18 known; up to $175,000.

1894-S Barber dime: The Barber designs tended to wear quickly, so any Barber coin in great condition is scarce enough. According to his daughter Hallie, San Francisco Mint director John Daggett struck two dozen 1894-S coins, mostly as gifts for his rich banker pals. Dad gave little Hallie three of the dimes, and she used one to buy herself the costliest ice cream in history. Twenty-four minted, ten known; as high as $1.3 million.

1907 MCMVII St. Gaudens $20 gold piece: This is often considered the loveliest U.S. coin series ever. Its debut featured the year in Roman numerals, unique in U.S. coinage. The first, ultra-high-relief version was stunning in its clarity and beauty, but it proved too time-consuming to mint, so a less striking (but still impressive) version became the standard. About 11,000 minted, but very few in ultra-high relief; those have sold for $1.5 million.

1909-S VDB Lincoln cent: It's a collectors' favorite, though not vanishingly rare. Only about a fourth of Lincoln pennies from the series' kickoff year featured designer Victor D. Brenner's initials on the reverse; even now, an occasional "SVDB" will show up in change. There were 484,000 minted; worth up to $7,500.

1913 Liberty Head nickel: This coin wasn't supposed to be minted. The Mint manufactured the dies as a contingency before the Buffalo design was selected for 1913. Apparently, Mint employee Samuel W. Brown may have known that the Liberty dies were slated for destruction and therefore minted five of these for his personal gain. One of the most prized U.S. coins—and priced accordingly at $1.8 million.

1913-S Barber quarter: Forty thousand of these were made—the lowest regular-issue mintage of the 20th century. Some Barbers wore so flat that the head on the obverse was reduced to a simple outline. Quite rare in good condition; can bring up to $24,000.

1915 Panama-Pacific $50 gold piece: This large commemorative piece was offered in both octagonal and round designs. Approximately 1,100 were minted; prices range from $40,000 to $155,000.

1916 Liberty Standing quarter: This coin depicts a wardrobe malfunction . . . except by design! Many were shocked when the new coin displayed Lady Liberty's bared breast. By mid-1917, she was donning chain mail. Like the Barber quarter before it, the Liberty Standing wore out rapidly. With only 52,000 minted in 1916, the series' inaugural year, a nice specimen will set you back nearly $40,000.

1933 St. Gaudens $20 gold piece: This coin is an outlaw. All of the Saint's final mintage were to be melted down—and most were. Only one specific example is legal to own; other surviving

1933 Saints remain hidden from the threat of Treasury confiscation. The legal one sold in 2002 for an incredible $7.6 million.

1937-D "three-legged" Buffalo nickel: A new employee at the Denver Mint tried polishing some damage off a die with an emery stick. He accidentally ground the bison's foreleg off, leaving a disembodied hoof. No telling exactly how many were struck, but they sure look funny. Up to $30,000.

1943 bronze Lincoln cent: This was the exciting year of the steel penny—except someone flubbed and minted a few on standard bronze planchets (coin blanks) left over from 1942. Surely the dozen known examples can't be all that exist—you might find this one in your pocket! A bronze Lincoln cent sold for $112,500 in 2000.

The Taj Mahal

Known as one of the Wonders of the World, the Taj Mahal was a shrine to love and one man's obsession. Today an average of three million tourists a year travel to see what the United Nations has declared a World Heritage site.

* * * *

Taj Mahal: Foundations

THE MUGHAL (OR "Mogul") Empire occupied India from the mid-1500s to the early 1800s. At the height of its success, this imperial power controlled most of the Indian subcontinent and much of what is now Afghanistan, containing a population of around 150 million people.

During this era, a young prince named Khurram took the throne in 1628, succeeding his father. Six years prior, after a military victory Khurram was given the title Shah Jahan by his emperor father. Now, with much of the subcontinent at his feet, the title was apt: *Shah Jahan* is Persian for "King of the World." (17th-century emperors were nothing if not modest.)

When Khurram Met Arjumand

Being shah had a lot of fringe benefits—banquets, and multiple wives, among other things. Shah Jahan did have several wives, but one woman stood out from the rest. When he was age 15, he was betrothed to 14-year-old Arjumand Banu Begam. Her beauty and compassion knocked the emperor-to-be off his feet; five years later, they were married. The bride took the title of *Mumtaz Mahal,* which means, according to various translations, "Chosen One of the Palace," "Exalted One of the Palace," or "Beloved Ornament of the Palace." You get the point.

Court historians have recorded the couple's close friendship, companionship, and intimate relationship. The couple traveled extensively together, Mumtaz often accompanying her husband on his military jaunts. But tragedy struck in 1631, when on one of these trips, Mumtaz died giving birth to what would have been their 14th child.

Breaking Ground

Devastated, Shah Jahan began work that year on what would become the Taj Mahal, a palatial monument to his dead wife and their everlasting love. While there were surely many hands on deck for the planning of the Taj, the architect who is most often credited is Ustad Ahmad Lahori. The project took until 1648 to complete and enlisted the labor of 20,000 workers and 1,000 elephants. This structure and its surrounding grounds covers 42 acres. The following are the basic parts of Mumtaz's giant mausoleum.

The Gardens: To get to the structural parts of the Taj Mahal, one must cross the enormous gardens surrounding it. Following classic Persian garden design, the grounds to the south of the buildings are made up of four sections divided by marble canals (reflecting pools) with adjacent pathways. The gardens stretch from the main gateway to the foot of the Taj.

The Main Gateway: Made of red sandstone and standing approximately 100 feet high and 150 feet wide, the main gateway is composed of a central arch with towers attached to each of its corners. The walls are richly adorned with calligraphy and floral arabesques inlaid with gemstones.

The Tomb: Unlike most Mughal mausoleums, Mumtaz's tomb is placed at the north end of the Taj Mahal, above the river and in between the mosque and the guesthouse. The tomb is entirely sheathed in white marble with an exterior dome that is almost 250 feet above ground level. The effect is impressive: Depending on the light at various times of the day, the tomb can appear pink, white, or brilliant gold.

The Mosque and the Jawab: On either side of the great tomb lie two smaller buildings. One is a mosque, and the other is called the *jawab,* or "answer." The mosque was used as a place of worship; the jawab was often used as a guesthouse. Both buildings are made of red sandstone so as not to take away too much from the grandeur of the tomb. The shah's monument to the love of his life still stands, and still awes, more than 360 years later.

✳ Chapter 14

Outré Space

Gimme a Star!

We've all been there: Your spouse's birthday is fast approaching and you haven't a clue what to buy. You wander the malls hopelessly, avoiding the roving packs of wild teenagers as your palms sweat and your heart pounds. Time is slipping away, and you can't find anything. You leave the mall and raise your fists to the heavens, cursing the stars... wait a minute! The stars! Can't you buy those things?

✳ ✳ ✳ ✳

NO, YOU CAN'T. Sorry. Despite what dozens of fly-by-night companies may tell you, the names of stars are not for sale. In fact, there is only one organization that has the right to give stars their names: the International Astronomical Union (IAU). And sadly, it probably won't accept your suggestion of naming a star after your wife or husband. The IAU has very strict and serious guidelines regarding how it names newly identified stars. Indeed, everything about the IAU is serious—founded in 1919 and headquartered in Paris, the IAU is made up of nearly ten thousand highly educated astronomers in eighty-seven countries. In other words, it ain't hawking star names on late-night TV.

The official star-naming process begins when a star is discovered. This happens more often than one might think—with an estimated one sextillion (1,000,000,000,000,000,000,000) stars in the universe, there is no shortage of work for the IAU. Of course, most of those stars are in distant galaxies, far from the reach of our most powerful telescopes.

Still, there are billions of stars from which to choose within our own galaxy. Coming up with a witty name for each one, as you might imagine, would be a chore. Thus, the IAU came up with a decidedly boring but nevertheless effective system—or we should say systems, as there are at least a dozen different catalogs that the IAU uses to assign names, depending on a star's distance from Earth and its brightness.

These catalogs often assign names to stars based on their coordinates in the sky, which works well because it makes them easy to find (sort of like when the government looks you up by your social security number). It also means that a typical star will be called something like BD +75 deg 752. Not quite as catchy as Huggy Bear, is it?

Take the Elevator Instead!

Rockets are for suckers. Looking for a cheaper way to build space stations and launch satellites, NASA scientists have an idea that sounds ludicrous but that they swear is feasible: an elevator that reaches from Earth's surface to outer space.

* * * *

"HOW COULD AN elevator to outer space be possible?" you ask. The answer is nanotubes. Discovered in 1991, nanotubes are cylindrical carbon molecules that make steel look like a 98-pound weakling. A space elevator's main component would be a 60,000-odd-mile nanotube ribbon, measuring about as thin as a sheet of paper and about three feet wide.

It gets weirder. That ribbon would require a counterweight up at the top to keep it in place. The counterweight, hooked to the nanotube ribbon, would be an asteroid pulled into Earth's orbit or a satellite. Once secured, the ribbon would have moving platforms attached to it. Each platform would be powered by solar-energy-reflecting lasers and could carry several thousand tons of cargo up to the top. The trip would take about a week. Transporting materials to outer space in this fashion would supposedly reduce the cost of, say, putting a satellite into orbit from about $10,000 a pound to about $100 a pound.

The base of the elevator would be a platform situated in the eastern Pacific Ocean, near the equator, safe from hurricanes and many miles clear of commercial airline routes. The base would be mobile so that it could be moved out of the path of potentially damaging debris orbiting Earth. Although there are a lot of theoretical kinks to work out, the more optimistic of the scientists who have hatched this scheme believe the whole thing could be a reality within a couple of decades.

The Area That Doesn't Exist

Who killed JFK? Did Americans really land on the moon? Conspiracy theorists have been debating these questions for years. But they all agree on one thing—these conspiracies pale in comparison to the mother of all conspiracies: Area 51.

* * * *

ALIEN AUTOPSIES. COVERT military operations. Tests on bizarre aircraft. These are all things rumored to be going on inside Area 51—a top secret location inside the Nevada Test and Training Range (NTTR) about an hour northwest of Las Vegas. Though shrouded in secrecy, some of the history of Area 51 is known. For instance, this desert area was used as a bombing test site during World War II, but no facility existed on the site until 1955. At that time, the area was chosen as the perfect location to develop and test the U-2 spy plane. Originally known as Watertown, it came to be called Area 51 in 1958 when 38,000 acres were designated for military use. The entire area was simply marked "Area 51" on military maps. Today, the facility is rumored to contain approximately

575 square miles. But you won't find it on a map because, officially, it doesn't exist.

An Impenetrable Fortress

Getting a clear idea of the size of Area 51, or even a glimpse of the place, is next to impossible. Years ago, curiosity seekers could get a good view of the facility by hiking to the top of two nearby mountain peaks known as White Sides and Freedom Ridge. But government officials soon grew weary of people climbing up there and snapping pictures, so in 1995, they seized control of both. Currently, the only way to legally catch a glimpse of the base is to scale 7,913-foot-tall Tikaboo Peak. Even if you make it that far, you're still not guaranteed to see anything because the facility is more than 25 miles away and is only visible on clear days with no haze.

The main entrance to Area 51 is along Groom Lake Road. Those brave (or foolhardy) souls who have ventured down the road to investigate quickly realize they are being watched. Video cameras and motion sensors are hidden along the road, and signs alert the curious that if they continue any further, they will be entering a military installation, which is illegal "without the written permission of the installation commander." If that's not enough to get unwanted guests to turn around, one sign clearly states: "Use of deadly force authorized." Simply put, take one step over that imaginary line in the dirt, and they will get you.

Camo Dudes

Just exactly who are "they"? They are the "Camo Dudes," mysterious figures watching trespassers from nearby hillsides. If they spot something suspicious, they might call for backup—Blackhawk helicopters that will come in for a closer look. All things considered, it would probably be best to just go back home. And lest you think about flying over Area 51, the entire area is considered restricted air space, meaning that unauthorized aircraft are not permitted to fly over, or near, the facility.

Who Works There?

Most employees are general contractors who work for companies in the area. But rather than allow these workers to commute individually, the facility has them ushered in secretly and en masse in one of two ways. The first is a mysterious white bus with tinted windows that picks up employees at several unmarked stops before whisking them through the front gates of the facility. Every evening, the bus leaves the facility and drops the employees off.

The second mode of commuter transport, an even more secretive way, is JANET, the code name given to the secret planes that carry workers back and forth from Area 51 and Las Vegas McCarran Airport. JANET has its own terminal, which is located at the far end of the airport behind fences with special security gates. It even has its own private parking lot. Several times a day, planes from the JANET fleet take off and land at the airport.

Bob Lazar

The most famous Area 51 employee is someone who may or may not have actually worked there. In the late 1980s, Bob Lazar claimed that he'd worked at the secret facility he referred to as S-4. In addition, Lazar said that he was assigned the task of reverse engineering a recovered spaceship in order to determine how it worked. Lazar had only been at the facility for a short time, but he and his team had progressed to the point where they were test flying the alien spaceship. That's when Lazar made a big mistake. He decided to bring some friends out to Groom Lake Road when he knew the alien craft was being flown. He was caught and subsequently fired.

During his initial interviews with a local TV station, Lazar seemed credible and quite knowledgeable as to the inner workings of Area 51. But when people started trying to verify the information Lazar was giving, not only was it next to impossible to confirm most of his story, his education and employment history could not be verified either. Skeptics immediately proclaimed that Lazar was a fraud. To this day, Lazar contends that everything he said was factual and that the government deleted all his records in order to set him up and make him look like a fake. Whether or not he's telling the truth, Lazar will be remembered as the man who first brought up the idea that alien spaceships were being experimented on at Area 51.

What's Really Going On?

So what really goes on inside Area 51? One thing we do know is that they work on and test aircraft. Whether they are alien spacecraft or not is still open to debate. Some of the planes worked on and tested at Area 51 include the SR-71 Blackbird and the F-117 Nighthawk stealth fighter.

If you want to try and catch a glimpse of some of these strange craft being tested, you'll need to hang out at the "Black Mailbox" along Highway 375, also known as the Extraterrestrial Highway. It's really nothing more than a mailbox along the side of the road. But as with most things associated with Area 51, nothing is as it sounds, so it should come as no surprise that the "Black Mailbox" is actually white. It belongs to a rancher, who owns the property nearby. Still, this is the spot where people have been known to camp out all night just for a chance to see something strange floating in the night sky.

The Lawsuit

In 1994, a landmark lawsuit was filed against the U.S. Air Force by five unnamed contractors and the widows of two others. The suit claimed that the contractors had been present at Area 51 when large quantities of "unknown chemicals" were burned in trenches and pits. As a result of coming into contact with the fumes of the chemicals, the suit alleged that two of the contractors died, and the five survivors suffered respiratory problems and skin sores. Reporters worldwide jumped on the story, not only because it proved that Area 51 existed but also because the suit was asking for many classified documents to be entered as evidence. Would some of those documents refer to alien beings or spacecraft? The world would never know because in September 1995, while petitions for the case were still going on, President Bill Clinton signed Presidential Determination No. 95–45, which basically stated that Area 51 was exempt from federal, state, local, and interstate hazardous and solid waste laws. Shortly thereafter, the lawsuit was dismissed due to a lack of evidence, and all attempts at appeals were rejected. In 2002, President George W. Bush renewed Area 51's exemptions, ensuring once and for all that what goes on inside Area 51 stays inside Area 51.

So at the end of the day, we're still left scratching our heads about Area 51. We know it exists and we have some idea of what goes on there, but there is still so much more we don't know. More than likely, we never will know everything, but then again, what fun is a mystery if you know all the answers?

The Moon Man

Through an unlikely turn of events, the scientist in charge of developing the Nazi's V-2 rocket would help the Americans reach the moon.

* * * *

BORN IN 1912, Wernher Magnus Maximilian Freiherr von Braun had a pedigree of greatness. His father Baron Magnus von Braun was the minister of agriculture under the Weimar Republic, and his mother was descended from Swedish and German aristocrats. At age 12, inspired by Fritz von Opel's land speed records, he took a "rocket-propelled" trip after lighting six large skyrockets fitted onto a wagon. "The wagon careened crazily about," Braun recalled, "trailing a tail of fire like a comet." The propellants made a thunderous noise; alarmed police briefly put Braun under custody.

From Skyrockets to the Vengeance Weapon

Inspired by scientist Hermann Oberth's landmark 1923 work The Rocket Into Interplanetary Space, in 1930 Braun attended Berlin's Charlottenburg Institute, where he worked with Oberth on liquid-fueled rockets. As Nazi Germany rearmed in the 1930s and banned research on civilian rockets, Braun began work conducting missile tests for the Wehrmacht's Ordnance Corps. He formed a long-term friendship with an artillery captain, Walter Dornberger, who arranged funding for Braun's doctorate. Braun and Dornberger went on to work at Peenemünde, as technical and military directors. Along the way Braun joined the Nazi Party. He later stated, "My refusal [to join] . . . would have meant abandon[ing] the work of my life."

Braun's team designed the A-4 ballistic missile, renamed by Josef Goebbels as the Vergeltungswaffe 2, the "Vengeance Weapon 2," or V-2. (The shorter-range V-1 "buzz bomb" was designed by engineer Robert Lusser under Luftwaffe supervision.) The designer of the V-2 engine was Walter Thiel, who was killed during a 1943 British air raid on Peenemünde.

The 46-foot-long, ethanol-and-water-fueled missile could hurl its 2,800-pound warhead some 200 miles. Its highest recorded altitude was 117 miles, making it the first craft to reach outer space. But its guidance system was inaccurate: Chances were about even that it would come within 10 miles of a target. Ironically, Braun regarded the V-2 primarily as a device for space travel, and was briefly imprisoned by the Gestapo in 1944 for his presumed disinterest in weaponry.

Death from the Skies

Hurtling down from an altitude of 60 miles at supersonic speeds no fighter could catch, the V-2 struck without warning. "The V-2 was a truly remarkable machine for its time," recalled Braun. From September 1944 to March 1945, 3,172 V-2s were successfully launched: 1,610 hit the vital port of Antwerp; 1,358 hit London. In one incident, 567 people died after a

V-2 landed in a Belgian movie house. The attacks on the British capital killed 2,754 and wounded 6,523. The Nazis relied on the weapons even when impractical: 11 V-2s were fired toward the Remagen bridge while Patton's troops poured over the Rhine into Germany.

The British incessantly bombed fixed sites for launching the missiles, so the V-2s were fired instead from mobile launchers, the equipment and fuel borne by a caravan of 30 trucks. It took two hours to set up the rocket, launch the V-2, and repack the equipment and crew.

The German missile program also successfully tested a "rocket U-boat," which fired V-2s from a submarine-towed platform. Had the method been developed sooner, Germany may have used the subs to launch rockets from off the U.S. coastline. German scientists worked on chemical weapons agents for the V-2. Braun's technicians also adapted the V-2 into the Wasserfall, perhaps the first antiaircraft missile.

Created in Deplorable Conditions

Much of this ordnance was built with slave labor. The V-2's chief assembly plants, called Mittelwerk, were in the Harz Mountains near the town of Nordhausen. There, Russian, Polish, and French inmates from the nearby Dora concentration camp dug out an underground factory from an old mine. The V-2's top engineer, Arthur Rudolph, helped arrange for the transfer of the inmates, after SS General Hans Kammler, an engineer who had built Auschwitz, came up with the idea of using slave labor for rockets.

Braun visited the Mittelwerk plant often enough to know, he later admitted, that many laborers died from brutal treatment and wretched conditions. Perhaps 15,000 perished. One eyewitness noted, "You could see piles of prisoners every day who had not survived the workload and had been tortured to death by the vindictive guards . . . But Professor Wernher von Braun just walked past them, so close that he almost touched the bodies." At the time Braun remarked, "It is hellish. My spontaneous reaction was to talk to one of the SS guards, only to be told with unmistakable harshness that I should mind my own business, or find myself in the same striped fatigues! . . . I realized that any attempt of reasoning on humane grounds would be utterly futile."

A New Life for the Nazi Scientist

As the war ground to a close, Braun again faced the SS. In spring 1945, Soviet troops closed to 100 miles of Peenemünde. Most of the V-2 staff decided to surrender to the Western Allies, but in the meantime the SS was ordered to liquidate the rocket engineers and burn their records. With forged documents, Braun and 500 of his staff put together dozens of train cars, as well as about 1,000 automobiles and trucks, and headed toward advancing American troops. At the end of the journey, Braun's brother Magnus buttonholed a GI: "My name is Magnus von Braun. My brother invented the V-2. We want to surrender." The Americans took Braun and his staff into custody, recovered their hidden records, and seized hundreds of freightloads of V-2 components.

Under Operation Paperclip, Braun, his team, their families, caches of scientific records, and enough V-2 components for 100 missiles, were brought to America. Since their Nazi associations would have barred many from visas, the scheme was hush-hush. Stationed at Fort Bliss, Texas, they called themselves "Prisoners of Peace." At the White Sands Proving Grounds in New Mexico, they continued their missile work; progress was rapid. By October 24, 1945, one of their reconstituted V-2s snapped photos from space.

Thereafter, Braun was transformed into an honored, naturalized American citizen, and fulfilled his boyhood dreams. He married his German sweetheart and had three children. In 1950 his group moved to Huntsville, Alabama, and designed the army's Jupiter ballistic missile, which later launched the first

U.S. satellite. He made television programs with Walt Disney that argued for manned space flight. In 1960, in the wake of *Sputnik,* he was made head of the NASA team that built the Saturn V rocket, which ferried Americans to the moon. He retired in 1972 when NASA opted for the earthbound space shuttle instead of a piloted mission to Mars.

How Was the Moon Born?

Here's one thing we know for certain about the moon: It isn't made of cheese. Most everything else, including its origins, is a matter of scientific reasoning and speculation.

✻ ✻ ✻ ✻

Mooning Over the Moon

OUR PLANET'S MOON, our only true natural satellite, has stimulated romance, mystery, and scientific curiosity. And no wonder—besides the sun, the moon is the most noticeable member of our solar system, measuring about one-quarter the size of Earth. Only one side faces our planet, and every month, because of its orbit around us, we watch the moon change phases, from full to quarter to gibbous to new and back again. The moon is also the subject of various origin theories, which alternately laud it as a deity or discount it as a flying chunk of rock, depending on the culture.

Blinded by Science

The list of scientific theories concerning the moon's origin is a bit smaller. One theory suggests that the moon was "captured" by Earth's gravity as it traveled by our planet; another theory posits that our planet and its satellite formed side by side as the solar system developed some 4.56 billion years ago. The moon has simply tagged along with us ever since.

The most recently accepted theory has its origins in the 19th century. In 1879, the son of British astronomer George Darwin (son of Charles Darwin) suggested that a rapidly spinning

Earth threw off material from the Pacific Ocean, creating the moon. The idea drew criticism on and off for decades. But thanks to the advent of modern computers, scientists have created a similar theoretical scenario that makes parts of Darwin's suggestion more reasonable. The data suggests that while Earth was still in a semimolten state, it was hit by a space body—a protoplanet, or planetesimal—almost the size of Mars, or about half the size of Earth. The massive collision would have sent a huge chunk of broken material into orbit around Earth; over time, those larger pieces could have gathered together—thanks to gravity—creating our moon.

It's All Relative?

Why do scientists now agree with the "Moon, daughter of Earth" theory? One of the main reasons is the U.S. moon missions. Astronauts gathered and delivered more than 800 pounds of lunar material back to Earth. The dates of those rocks—ranging from 3.2 to 4.2 billion years old for material gathered from the flat, dark maria (lava seas) and 4.3 to 4.5 billion years old for rocks from the highlands—along with their composition, have led scientists to believe that the moon is definitely related to Earth.

The evidence is in the fact that the rocks are similar to Earth's mantle material—the moving, molten layer of our planet just under the crust. If a huge planetary body struck our planet, it would make sense that the resulting material would be similar to rock deep below Earth's surface. In addition, moon rocks have exactly the same oxygen isotope composition as Earth's rocks. Materials from other parts of the solar system have different oxygen isotope compositions, which means that the moon probably formed around Earth's neighborhood.

Is the moon our only satellite? Scientists know there are other space bodies circling our planet, but none of the objects can be considered a moon. They are more likely asteroids caught in the Earth's and moon's gravitation. For example, the asteroid 3753 Cruithne looks like it's following Earth in the orbit around the sun; the asteroid 2002 AA29 follows a horseshoe path near Earth. Neither is a moon, and so far neither rock has been in danger of striking our planet. Another object once caught scientists' eyes: Nearby J002E3 was considered a possible new moon of Earth until it was determined to be the third stage of a Saturn V rocket.

The Avrocar: Not a Bird, Not a Plane, and Not an Alien Either

Not all UFOs are alien spaceships. One top-secret program was contracted out by the U.S. military to an aircraft company in Canada.

* * * *

OH, THE 1950S—A time of sock hops, drive-in movies, and the Cold War between America and the Soviet Union, when each superpower waged war against the other in the arenas of scientific technology, astronomy, and politics. It was also a time when discussion of life on other planets was rampant, fueled by the alleged crash of an alien spaceship near Roswell, New Mexico, in 1947.

Watch the Skies

Speculation abounded about the unidentified flying objects (UFOs) spotted nearly every week by everyone from farmers to airplane pilots. As time passed, government authorities began to wonder if the flying saucers were, in fact, part of a secret Russian program to create a new type of air force. Fearful that such a craft would upset the existing balance of power, the U.S. Air Force decided to produce its own saucer-shape ship.

In 1953, the military contacted Avro Aircraft Limited of Canada, an aircraft manufacturing company that operated in Malton, Ontario, between 1945 and 1962. Project Silverbug was initially proposed simply because the government wanted to find out if UFOs could be manufactured by humans. But before long, both the military and the scientific community were speculating about its potential. Intrigued by the idea, designers at Avro—led by British aeronautical engineer John Frost—began working on the VZ-9-AV Avrocar. The round craft would have been right at home in a scene from the classic science fiction film The Day the Earth Stood Still. Security for the project was so tight that it probably generated rumors that America was actually testing a captured alien spacecraft—speculation that remains alive and well even today.

Of This Earth

By 1958, the company had produced two prototypes, which were 18 feet in diameter and 3.5 feet tall. Constructed around a large triangle, the Avrocar was shaped like a disk, with a curved upper surface. It included an enclosed 124-blade turbo-rotor at the center of the triangle, which provided lifting power through an opening in the bottom of the craft. The turbo also powered the craft's controls. Although conceived as being able to carry two passengers, in reality a single pilot could barely fit inside the cramped space. The Avrocar was operated with a single control stick, which activated different panels around the ship. Airflow issued from a large center ring, which was controlled by the pilot to guide the craft either vertically or horizontally.

The military envisioned using the craft as "flying Jeeps" that would hover close to the ground and move at a maximum speed of 40 mph. But that, apparently, was only going to be the beginning. Avro had its own plans, which included not just commercial Avrocars, but also a family-size Avrowagon, an Avrotruck for larger loads, Avroangel to rush people to the hospital, and a military Avropelican, which, like a pelican hunting for fish, would conduct surveillance for submarines.

But Does It Fly?

The prototypes impressed the U.S. Army enough to award Avro a $2 million contract. Unfortunately, the Avrocar project was canceled when an economic downturn forced the company to temporarily close and restructure. When Avro Aircraft reopened, the original team of designers had dispersed. Further efforts to revive the project were unsuccessful, and repeated testing proved that the craft was inherently unstable. It soon became apparent that whatever UFOs were spotted overhead, it was unlikely that they came from this planet. Project Silverbug was abandoned when funding ran out in March 1961, but one of the two Avrocar prototypes is housed at the U.S. Army Transportation Museum in Fort Eustis, Virginia.

Sir Arthur C. Clarke

The three people often considered the Big Three of science fiction are writer Robert Heinlein, biochemist-turned-novelist Isaac Asimov, and Sir Arthur C. Clarke. The last living of these was Clarke, who passed away at age 90 in March 2008 at his home in Sri Lanka.

* * * *

CLARKE WAS BORN in Minehead, England, in 1917. He developed a love for science during his grade school years and would regularly devour books by such fantasy writers as H. G. Wells and Jules Verne. In 1945, Clarke authored a paper that explained how communication satellites could work.

In 1947, Clarke published his first science fiction novel, *Prelude to Space*, and would produce scores more throughout his career. Developing an interest in undersea exploration during the 1950s, Clarke moved to Sri Lanka and concentrated his writing efforts around the depths of the Indian Ocean.

A 1962 polio attack left Clarke's body paralyzed, but his fertile imagination remained unimpaired. Much of the author's best work was yet to come.

A 1964 novel about space travel led to the epic 1968 film *2001: A Space Odyssey*, directed by Stanley Kubrick. The 1980s saw Clarke fronting two television shows: *Arthur C. Clarke's Mysterious World* (1980) and *World of Strange Powers* (1985). These introduced the master of imagination to an entirely new generation and furthered his fame.

Clarke is noted for his predictions concerning technology. Satellite television and the ability to land space probes on asteroids are among his direct hits. The world will have to wait until 2023 to see if dinosaurs will indeed be cloned from computer-generated DNA, and 2050 to see if time travel will become possible through cryonic suspension.

The Speed of Dark

Most of us believe that nothing is faster than the speed of light. In high school physics, we learned that something traveling faster could theoretically go back in time. This would allow for the possibility that you could go back in time and kill your grandfather and, thus, negate your existence—a scenario known as the Grandfather Paradox. Or, more horrifyingly, you could go back in time in order to set up your future parents as you skateboard around to the musical stylings of Huey Lewis and the News.

* * *

YET THERE IS something that may be faster than the speed of light: the speed of dark. Unless the speed of dark doesn't exist. When you're talking about astrophysics and quantum mechanics, nothing is certain (indeed, uncertainty might be said to be the defining principle of modern physics).

Observations and experiments in recent years have helped astrophysicists shape a more comprehensive understanding of how the universe operates, but even the most brilliant scientists are operating largely on guesswork. To understand how the speed of dark theoretically might—or might not—exceed the speed of light, we'll have to get into some pretty wild concepts.

As with much of astronomy, our explanation is rooted in the Big Bang. For those of you who slept through science class or were raised in the Bible Belt, the Big Bang is the prevailing scientific explanation for the creation of the universe. According to the Big Bang theory, the universe started as a pinpoint of dense, hot matter. About fourteen billion years ago, this infinitely dense point exploded, sending the foundations of the universe into the outer reaches of space.

The momentum from this initial explosion caused the universe to expand its boundaries outward. For most of the twentieth century, the prevailing thought was that the rate of expansion was slowing down and would eventually grind to a halt. Seemed logical enough, right?

In 1998, however, astronomers who were participating in two top-secret-sounding projects—the Supernova Cosmology Project and the High-Z Supernova Search—made a surprising discovery while observing supernovae events (exploding stars) in the distant reaches of space. Supernovae are handy for astronomers because just prior to exploding, these stars reach a uniform brightness. Why is this important? The stars provide a standard variable, allowing scientists to infer other statistics, such as how far the stars are from Earth. Once scientists know a star's distance from Earth, they can use another phenomenon known as a redshift (a visual analogue to the Doppler effect in which light appears differently to the observer because an object is moving away from him or her) to determine how much the universe has expanded since the explosion.

Still with us? Now, based on what scientists had previously believed, certain supernovae should have appeared brighter than what the redshift indicated. But to the scientists' amazement, the super-novae appeared dimmer, indicating that the expansion of the universe is speeding up, not slowing down. How could this be? And if the expansion is quickening, what is it that's driving it forward and filling up that empty space?

Initially, nobody had any real idea. But after much discussion, theorists came up with the idea of dark energy. What is dark energy? Ultimately, it's a made-up term for the inexplicable and incomprehensible emptiness of deep space. For the purposes of our question, however, dark energy is theoretically far faster than the speed of light—it's so fast, in fact, that it is moving too quickly for new stars to form in the empty space. No, it doesn't make a whole heck of a lot of sense to us either, but rest assured, a lot of very nerdy people have spent a long time studying it.

Of course, there may be a far simpler answer, one posited by science-fiction writer Terry Pratchett: The speed of dark must be faster than the speed of light—otherwise, how would dark be able to get out of the way?

John Lennon's UFO Sighting

Lucy in the sky with warp drive.

* * * *

IN MAY 1974, former Beatle John Lennon and his assistant/mistress May Pang returned to New York City after almost a year's stay in Los Angeles, a period to which Lennon would later refer as his "Lost Weekend." The pair moved into Penthouse Tower B at 434 East 52nd Street. As Lennon watched television on a hot summer night, he noticed flashing lights reflected in the glass of an open door that led onto a patio. At first dismissing it as a neon sign, Lennon suddenly realized that since the apartment was on the roof, the glass *couldn't* be reflecting light from the street. So—sans clothing—he ventured onto the terrace to investigate. What he witnessed has never been satisfactorily explained.

Speechless

As Pang recollected, Lennon excitedly called for her to come outside. Pang did so. "I looked up and stopped mid-sentence," she said later. "I couldn't even speak because I saw this thing up there . . . it was silvery, and it was flying very slowly. There was a white light shining around the rim and a red light on the top . . . [it] was silent. We started to watch it drift down, tilt slightly, and it was flying below rooftops. It was the most amazing sight." She quickly ran back into the apartment, grabbed her camera, and returned to the patio, clicking away.

Lennon friend and rock photography legend Bob Gruen picked up the story: "In those days, you didn't have answering machines, but a service [staffed by people], and I had received a call from 'Dr. Winston.'" (Lennon's original middle name was Winston, and he often used the alias "Dr. Winston O'Boogie.") When Gruen returned the call, Lennon explained his incredible sighting and insisted that the photographer pick up and develop the film. "He was serious," Gruen said. "He wouldn't call me in the middle of the night to joke around." Gruen noted that although Lennon had been known to partake in mind-altering substances in the past, during this period he was totally straight. So was Pang, a nondrinker who never took drugs and whom Gruen characterized as "a clear-headed young woman."

The film in Pang's camera was a unique type supplied by Gruen, "four times as fast as the highest speed then [commercially] available." Gruen had been using this specialty film, usually employed for military reconnaissance, in low-light situations such as recording studios. The same roll already had photos of Lennon and former bandmate Ringo Starr, taken by Pang in Las Vegas during a recording session.

Gruen asked Lennon if he'd reported his sighting. "Yeah, like I'm going to call the police and say I'm John Lennon and I've seen a flying saucer," he scoffed. Gruen picked up the couple's phone and contacted the police, *The Daily News,* and the *New York Times.* The photographer claims that the cops and the *News* admitted that they'd heard similar reports, while the *Times* just hung up on him.

It Would Have Been the Ultimate Trip

Gruen's most amusing recollection of Lennon, who had been hollering "UFO!" and "Take me with you!" was that none of his neighbors either saw or heard the naked, ex-Beatle screaming from his penthouse terrace. And disappointingly, no one who might have piloted the craft responded to Lennon's pleas.

Gruen took the film home to process, "sandwiching" it between two rolls of his own. Gruen's negatives came out perfectly, but the film Pang shot was "like a clear plastic strip," Gruen says. "We were all baffled . . . that it was completely blank."

Lennon remained convinced of what he'd seen. In several shots from a subsequent photo session with Gruen that produced the iconic shot of the musician wearing a New York City T-shirt (a gift from the photographer), John points to where he'd spotted the craft. And on his *Walls and Bridges* album, Lennon wrote in the liner notes: "On the 23rd Aug. 1974 at 9 o'clock I saw a U.F.O.—J.L."

Who's to say he and May Pang didn't? Certainly not Gruen, who still declares—more than 35 years after the fact—"I believed them."

Black Holes

How do we know something is there if we can't see it? With apologies to the often discussed and dubiously sighted Loch Ness Monster and Sasquatch, the mystical black holes that populate the great beyond do exist—but they're invisible.

❋ ❋ ❋ ❋

THE SCIENTIFIC JARGON needed to properly explain the principles behind the concept of black holes would put the average person to sleep faster than a heavy meal on a warm afternoon. Let's describe them this way: Black holes are so dense that not even light can escape their gravity. If an object does not emit any kind of light, it cannot be seen. Because nothing can travel faster than light, nothing can escape from inside a black hole. What we do know is that black holes absorb and suck up all sorts of matter, gases, and outer-space junk into their mysterious confines. That activity allows scientists to determine exactly where the black holes are. What you are seeing is not the black hole itself but the elements that define its shape.

If you're seeking a black hole that is actually visible, you should visit the library at Warner Bros. Cartoons, Inc., where you'll find one of the most clever and comedic caricatures in the animators' vaults. "The Hole Idea" tells the story of meek and mild inventor Calvin Q. Calculus, who designs a "portable hole," presumably to revolutionize the storage of dog bones and other debris and to bring immense and instant gratification to putt-weary golfers. In a classic case of life-meets-art, Mr. Calculus actually conceived the idea of the hole as a hatch to escape the hostilities perpetrated by his loud and loquacious wife, who feels that all of his inventions are worthless. Calamity ensues with a comedic conclusion: Calvin's wife falls into the hole, only to be thrown back out. It seems she's not wanted there, either.

Tracking Tektite Truths

The origin of strangely shaped bits of glass called tektites has been debated for decades—do they come from the moon? From somewhere else in outer space? It seems the answer is more down-to-earth.

* * * *

THE FIRST TEKTITES were found in 1787 in the Moldau River in the Czech Republic, giving them their original name, "Moldavites." They come in many shapes (button, teardrop, dumbbell, and blobs), have little or no water content, and range from dark green to black to colorless.

Originally, many geologists believed tektites were extraterrestrial in origin, specifically from the moon. They theorized that impacts from comets and asteroids—or even volcanic eruptions—on the moon ejected huge amounts of material. As the moon circled in its orbit around our planet, the material eventually worked its way to Earth, through the atmosphere, and onto the surface.

One of the first scientists to debate the tektite-lunar origin idea was Texas geologist Virgil E. Barnes, who contended that tektites were actually created from Earth-bound soil and rock. Many scientists now agree with Barnes, theorizing that when a comet or asteroid collided with the earth, it sent massive amounts of material high into the atmosphere at hypervelocities. The energy from such a strike easily melted the terrestrial rock and burned off much of the material's water. And because of the earth's gravitational pull, what goes up must come down—causing the melted material to rain down on the planet in specific locations. Most of the resulting tektites have been exposed to the elements for millions of years, causing many to be etched and/or eroded over time.

Unlike most extraterrestrial rocks—such as meteorites and micrometeorites, which are found everywhere on Earth—tektites are generally found in four major regions of the world called *strewn* (or splash) fields. The almost 15-million-year-old Moldavites are mainly found in the Czech Republic, but the strewn field extends into Austria; these tektites are derived from the Nordlinger Ries impact crater in southern Germany. The *Australites, Indochinites,* and *Chinites* of the huge Australasian strewn field extend around Australia, Indochina, and the Philippines; so far, no one has agreed on its source crater. The *Georgiaites* (Georgia) and *Bediasites* (Texas) are North American tektites formed by the asteroid impact that created the Chesapeake Crater around 35 million years ago. And finally, the 1.3-million-year-old *Ivorites* of the Ivory Coast strewn field originate from the Bosumtwi crater in neighboring

Ghana. Other tektites have been discovered in various places around the world but in very limited quantities compared to the major strewn fields.

Mystery Orb

If Texas were a dartboard, the city of Brownwood would be at the center of the bull's-eye. Maybe that's how aliens saw it, too.

* * * *

Brownwood is a peaceful little city with about 20,000 residents and a popular train museum. A frontier town at one time, it became the trade center of Texas when the railroad arrived in 1885. Since then, the city has maintained a peaceful lifestyle. Even the massive tornado that struck Brownwood in 1976 left no fatalities.

An Invader from the Sky

In July 2002, however, the city's peace was broken. Brownwood made international headlines when a strange metal orb fell from space, landed in the Colorado River, and washed up just south of town. The orb looked like a battered metal soccer ball—it was about a foot across, and it weighed just under ten pounds. Experts described it as a titanium sphere. When it was x-rayed, it revealed a second, inner sphere with tubes and wires wrapped inside.

That's all that anybody knows. No one is sure what the object is, and no one has claimed responsibility for it. The leading theory is that it's a cryogenic tank from some kind of spacecraft from Earth, used to store a small amount of liquid hydrogen or helium for cooling purposes. Others have speculated that it's a bomb, a spying device, or even a weapon used to combat UFOs.

It's Not Alone

The Brownwood sphere isn't unique. A similar object landed in Kingsbury, Texas, in 1997, and was taken by the Air Force for testing. So far, no further announcements have been made.

Of course, the Air Force probably has a lot to keep it busy. About 200 UFOs are reported each month, and Texas is among the top three states where UFOs are seen. But until anything is known for sure, those in Texas at night should keep an eye on the skies.

Sci-Fi Settings

While sci-fi movies are often set in regions of space entirely alien to our own, a simple fact is consistent with them all—no matter the setting, every one is filmed right here on Earth. The stories and locations may be out of this world, but here are some real places that stood in for a galaxy far, far away.

Ape World

ALTHOUGH CHARLTON HESTON'S character returned home at the end of *Planet of the Apes* (1968), he didn't actually make it to the East Coast of the United States. The crash scene at the beginning of the movie was filmed in Glen Canyon, Utah, while the Statue of Liberty scenes, along with much of the rest of the film, were shot in Malibu, California.

More Monkeys

In Tim Burton's 2001 remake of *Planet of the Apes*, astronaut Mark Wahlberg crash-lands in an unknown time and place that looks remarkably like Hawaii. Actually, it was Hawaii... some of it anyhow. Additional footage was shot at California's Trona Pinnacles and at Lake Powell, which straddles the Utah–Arizona border.

One for the Conspiracy Theorists

Capricorn One (1977) took its story from the conspiracy theory that the 1969 *Apollo* mission to the moon had been faked by NASA and the U.S. government. In Hollywood's version, three astronauts become pawns of the space program when their mission to Mars is canceled due to faulty equipment and lack of funds. They are ordered to fake it in the desert, which was actually Red Rock Canyon State Park in California. This is a clever twist on the use of locations because *Capricorn One*'s fictional American public is fooled into believing they are seeing Mars, just as real-life moviegoers suspend their disbelief regarding locations when they watch sci-fi movies.

Tatooine, Home Planet of Luke Skywalker

Luke Skywalker may have been a poor moisture farmer from a truly backwater planet, but it was actually the upscale Sidi Driss Hotel in Tunisia that served as the backdrop for his boyhood home on the planet Tatooine in the original *Star Wars* (1977), as well as the later prequels. Other North African locations, including Chott el Djerid, were also used, and Death Valley National Park in California was a stand in for the planet as well.

The Ewoks' Forests of Endor

Whether you love 'em or hate 'em, the Ewoks of Endor did save the day for the Rebel Alliance at the end of *Return of the Jedi* (1983), and the tall trees of the Redwood National and State Parks in northern California served as stand-ins for those of the forest moon.

Chill Out on Ice Planet Hoth

Luke Skywalker and the Rebel Alliance cooled off on the frozen world of Hoth before fighting off an invasion by the Empire's giant AT-AT walkers. The real Hoth locations—Finse and the nearby Hardangerjøkulen, the fifth largest glacier in mainland Norway—were actually part of the Nazi occupation of the Scandinavian nation during World War II.

Dune's Planet Arrakis

The desert world known as Dune in the 1984 film was actually the Samalayuca Dunes in the Mexican state of Chihuahua. Located near the Texas border, they are among the largest and deepest sand dunes in North America, but don't expect to find any giant sand worms there.

Mars Invasion

In one of a few recent films about the colonization of Mars, Val Kilmer led a mission to the *Red Planet* (2000), but instead of training for interplanetary travel, he merely had to travel to Coober Pedy in South Australia, while Gary Sinise's *Mission to Mars* (2000) took its cast on a journey to Jordan.

A World of Aliens

James Cameron's 1986 blockbuster *Aliens* was set on a planet known as LV-426, but most of the film was actually shot at Pinewood Studios in Buckinghamshire, England. The climactic scenes at the atmosphere-processing station were filmed in London at the Acton Lane Power Station. No aliens were harmed during the production of the film.

Total Arnold

A favorite among Arnold Schwarzenegger fans, *Total Recall* (1990) sends Arnold to Mars, but California's future "Governator" didn't have to venture too far from home—most of the film was shot in Mexico.

Strange Lights in Marfa

According to a 2007 poll, approximately 14 percent of Americans believe they've seen a UFO. How many of them have been in Marfa?

* * * *

IF ANYONE IS near Marfa at night, they should watch for odd, vivid lights over nearby Mitchell Flat. Some believe that the lights from UFOs or even alien entities can be seen. The famous Marfa Lights are about the size of basketballs and are usually white, orange, red, or yellow. These unexplained lights only appear at night and usually hover above the ground at about shoulder height. Some of the lights—alone or in pairs—drift and fly around the landscape. From cowboys to truck drivers, people traveling in Texas near the intersection of U.S.

Route 90 and U.S. Route 67 in southwest Texas have reported the Marfa Lights. And these baffling lights don't just appear on the ground. Pilots and airline passengers claim to have seen the Marfa Lights from the skies. So far, no one has proved a natural explanation for the floating orbs.

Eyewitness Information

Two 1988 reports were especially graphic. Pilot R. Weidig was about 8,000 feet above Marfa when he saw the lights and estimated them rising several hundred feet above the ground. Passenger E. Halsell described the lights as larger than the plane and noted that they were pulsating. In 2002, pilot B. Eubanks provided a similar report.

In addition to what can be seen, the Marfa Lights may also trigger low-frequency electromagnetic (radio) waves—which can be heard on special receivers—similar to the "whistlers" caused by lightning. However, unlike such waves from power lines and electrical storms, the Marfa whistlers are extremely loud. They can be heard as the orbs appear, and then they fade when the lights do.

A Little Bit About Marfa

Marfa is about 60 miles north of the Mexican border and about 190 miles southeast of El Paso. This small, friendly Texas town is 4,800 feet above sea level and covers 1.6 square miles.

In 1883, Marfa was a railroad water stop. It received its name from the wife of the president of the Texas and New Orleans Railroad, who chose the name from a Russian novel that she was reading. A strong argument can be made that this was Dostoyevsky's *The Brothers Karamazov.* The town grew slowly, reaching its peak during World War II when the U.S. government located a prisoner of war camp, the Marfa Army Airfield, and a chemical warfare brigade nearby. (Some skeptics suggest that discarded chemicals may be causing the Marfa Lights, but searchers have found no evidence of such.)

Today, Marfa is home to about 2,500 people. The small town is an emerging arts center with more than a dozen artists' studios and art galleries. However, Marfa remains most famous for its light display. The annual Marfa Lights Festival is one of the town's biggest events, but the mysterious lights attract visitors year-round.

The Marfa Lights are seen almost every clear night, but they never manifest during the daytime. The lights appear between Marfa and nearby Paisano Pass, with the Chinati Mountains as a backdrop.

Widespread Sightings

The first documented sighting was by 16-year-old cowhand Robert Reed Ellison during an 1883 cattle drive. Seeing an odd light in the area, Ellison thought he'd seen an Apache campfire. When he told his story in town, however, settlers told him that they'd seen lights in the area, too, and they'd never found evidence of campfires.

Two years later, 38-year-old Joe Humphreys and his wife, Sally, also reported unexplained lights at Marfa. In 1919, cowboys on a cattle drive paused to search the area for the origin of the lights. Like the others, they found no explanation for what they had seen.

In 1943, the Marfa Lights came to national attention when Fritz Kahl, an airman at the Marfa Army Base, reported that airmen were seeing lights that they couldn't explain. Four years later, he attempted to fly after them in a plane but came up empty again.

Explanations?

Some skeptics claim that the lights are headlights from U.S. 67, dismissing the many reports from before cars—or U.S. 67—were in the Marfa area. Others insist that the lights are swamp gas, ball lightning, reflections off mica deposits, or a nightly mirage.

At the other extreme, a contingent of people believe that the floating orbs are friendly observers of life on Earth. For example, Mrs. W. T. Giddings described her father's early 20th-century encounter with the Marfa Lights. He'd become lost during a blizzard, and according to his daughter, the lights "spoke" to him and led him to a cave where he found shelter.

Most studies of the phenomenon, however, conclude that the lights are indeed real but cannot be explained. The 1989 TV show Unsolved Mysteries set up equipment to find an explanation. Scientists on the scene could only comment that the lights were not made by people.

Share the Wealth

Marfa is the most famous location for "ghost lights" and "mystery lights," but it's not the only place to see them. Here are just a few of the legendary unexplained lights that attract visitors to dark roads in Texas on murky nights.

* In southeast Texas, a single orb appears regularly near Saratoga on Bragg Road.
* The Anson Light appears near Mt. Hope Cemetery in Anson, by U.S. Highway 180.
* Since 1850, "Brit Bailey's Light" glows five miles west of Angleton near Highway 35 in Brazoria County.
* In January 2008, Stephenville attracted international attention when unexplained lights—and perhaps a metallic spaceship—flew fast and low over the town.

The Marfa Lights appear over Mitchell Flat, which is entirely private property. However, the curious can view the lights from a Texas Highway Department roadside parking area about nine miles east of Marfa on U.S. Highway 90. Seekers should arrive before dusk for the best location, especially during bluebonnet season (mid-April through late May), because this is a popular tourist stop.

The Men on the Moon

On July 20, 1969, millions of people worldwide watched in awe as U.S. astronauts became the first humans to step on the moon. However, a considerable number of conspiracy theorists contend that the men were just actors performing on a soundstage.

* * * *

THE NATIONAL AERONAUTICS and Space Administration (NASA) has been dealing with this myth for nearly 40 years. In fact, it has a page on its official Web site that scientifically explains the pieces of "proof" that supposedly expose the fraud. These are the most common questions raised.

If the astronauts really did take photographs on the moon, why aren't the stars visible in them? The stars are there but are too faint to be seen in the photos. The reason for this has to do with the fact that the lunar surface is so brightly lit by the sun. The astronauts had to adjust their camera settings to accommodate the brightness, which then rendered the stars in the background difficult to see.

Why was there no blast crater under the lunar module? The astronauts had slowed their descent, bringing the rocket on the lander from a maximum of 10,000 pounds of thrust to just 3,000 pounds. In addition, the lack of atmosphere on the moon spread the exhaust fairly wide, lowering the pressure and diminishing the scope of a blast crater.

If there is no air on the moon, why does the flag planted by the astronauts appear to be waving? The flag appears to wave because the astronauts were rotating the pole on which it was mounted as they tried to get it to stand upright.

When the lunar module took off from the moon back into orbit, why was there no visible flame from the rocket? The composition of the fuel used for the takeoff from the surface of the moon was different in that it produced no flame.

Conspiracy theorists present dozens of "examples" that supposedly prove that the moon landing never happened, and all of them are easily explained. But that hasn't kept naysayers from perpetuating the myth.

Twenty-three years after the moon landing, on February 15, 2001, Fox TV stirred the pot yet again with a program titled *Conspiracy Theory: Did We Land on the Moon?* The show trotted out the usual array of conspiracy theorists, who in turn dusted off the usual spurious "proof." And once again, NASA found itself having to answer to a skeptical but persistent few.

Many people theorize that the landing was faked because the United States didn't have the technology to safely send a crew to the moon. Instead, it pretended it did as a way to win the final leg of the space race against the Soviet Union. But consider the situation: Thousands of men and women worked for almost a decade (and three astronauts died) to make the success of *Apollo 11* a reality. With so many people involved, a hoax of that magnitude would be virtually impossible to contain, especially after almost four decades.

For additional proof that the moon landing really happened, consider the hundreds of pounds of moon rocks brought back by the six *Apollo* missions that were able to retrieve them. Moon rocks are unique and aren't easily manufactured, so if they didn't come from the moon, what is their source? Finally, there's no denying the fact that the *Apollo* astronauts left behind a two-foot reflecting panel equipped with dozens of tiny mirrors. Scientists are able to bounce laser pulses off the mirrors to pinpoint the moon's distance from Earth.

The myth of the faked moon landing will probably never go away. But the proof of its reality is irrefutable. In the words of astronaut Charles Duke, who walked on the moon in 1972 as part of the *Apollo 16* mission: "We've been to the moon nine times. Why would we fake it nine times, if we faked it?"

Teleportation: Not Just the Stuff of Science Fiction

Scientists say that it's only a matter of time before we're teleporting just like everyone does on Star Trek.

✻ ✻ ✻ ✻

Beam Us Up

WE'RE CLOSER THAN you might think to being able to teleport, but don't squander those frequent-flyer miles just yet. There's a reason why Captain Kirk is on TV late at night shilling for a cheap-airfare Web site and not hawking BeamMeToHawaiiScotty.com. For the foreseeable future, jet travel is still the way to go.

If, however, you're a photon and need to travel a few feet in a big hurry, teleportation is a viable option. Photons are subatomic particles that make up beams of light. In 2002, physicists at the Australian National University were able to disassemble a beam of laser light at the subatomic level and make it reappear about three feet away. There have been advances since, including an experiment in which Austrian researchers teleported a beam of light across the Danube River in Vienna via a fiber-optic cable—the first instance of teleportation taking place outside of a laboratory.

These experiments are a far cry from dematerializing on your spaceship and materializing on the surface of a planet to make out with an alien who, despite her blue skin, is still pretty hot. But this research demonstrates that it is possible to transport matter in a way that bypasses space—just don't expect teleportation of significant amounts of matter to happen until scientists clear a long list of hurdles, which will take many years.

Here, Gone, There

Teleportation essentially scans and dematerializes an object, turning its subatomic particles into data. The data is transferred to another location and used to recreate the object. This is not unlike the way your computer downloads a file from another computer miles away. But your body consists of trillions upon trillions of atoms, and no computer today could be relied on to crunch numbers powerfully enough to transport and precisely recreate you elsewhere.

As is the case with many technological advances, the most vexing and long-lasting obstacle probably won't involve creation of the technology, but rather the moral and ethical issues surrounding its use. Teleportation destroys an object and recreates a facsimile somewhere else. If that object is a person, does the destruction constitute murder? And if you believe that a person has a soul, is teleportation capable of recreating a person's soul within the physical body it recreates? These are questions with no easy answers.

Space Jam

No one gives much thought to all the stuff we launch into space and don't bring back, but it creates a major hazard.

* * * *

Steer Clear of that Satellite

IF YOU THINK it's nerve-wracking when you have to swerve around a pothole as you cruise down the highway, just imagine how it would feel if you were miles above the Earth, where the stakes couldn't be higher. That's what the crew of the International Space Station (ISS) faced in 2008, when it had to perform maneuvers to avoid debris from a Russian satellite.

And that was just one piece of orbital trash—all in all, there are tens of millions of junky objects that are larger than a millimeter and are in orbit. If you don't find this worrisome, imagine the little buggers zipping along at up to 17,000 miles per hour. Worse, these bits of flotsam and jetsam constantly crash into each other and shatter into even more pieces.

The junk largely comes from satellites that explode or disintegrate; it also includes the upper stages of launch vehicles, burnt-out rocket casings, old payloads and experiments, bolts, wire clusters, slag and dust from solid rocket motors, batteries, droplets of leftover fuel and high-pressure fluids, and even a space suit. (No, there wasn't an astronaut who came home naked—the suit was packed with batteries and sensors and was set adrift in 2006 so that scientists could find out how quickly a spacesuit deteriorates in the intense conditions of space.)

The U.S. and Russia: Space's Big Polluters

So who's responsible for all this orbiting garbage? The two biggest offenders are Russia—including the former Soviet Union—and the United States. Other litterers include China, France, Japan, India, Portugal, Egypt, and Chile. Each of the last three countries has launched one satellite during the past twenty years.

Most of the junk orbits Earth at between 525 and 930 miles from the surface. The ISS operate a little closer to Earth. It maintains an altitude of about 250 miles, and therefore rarely sees the worst of it. Still, the ISS's emergency maneuver in 2008 was a sign that the situation is getting worse.

NASA and other agencies use radar to track the junk and are studying ways to get rid of it for good. Ideas such as shooting at objects with lasers or attaching tethers to some pieces to force them back to Earth have been discarded because of cost considerations and the potential danger to people on the ground. Until an answer is found, NASA practices constant vigilance, monitoring the junk and watching for collisions with working satellites and vehicles as they careen through outer space.

Hazardous driving conditions, it seems, extend well beyond Earth's atmosphere.

The Tunguska Event

What created an explosion 1,000 times greater than the atomic bomb at Hiroshima, destroyed 80 million trees, but left no hole in the ground?

✻ ✻ ✻ ✻

The Event

On the morning of June 30, 1908, a powerful explosion ripped through the remote Siberian wilderness near the Tunguska River. Witnesses, from nomadic herdsmen and passengers on a train to a group of people at the nearest trading post, reported seeing a bright object streak through the sky and explode into an enormous fireball. The resulting shockwave flattened approximately 830 square miles of forest. Seismographs in England recorded the event twice, once as the initial shockwave passed and then again after it had circled the planet. A huge cloud of ash reflected sunlight from over the horizon across Asia and Europe. People reported there being enough light in the night sky to facilitate reading.

A Wrathful God

Incredibly, nearly 20 years passed before anyone visited the site. Everyone had a theory of what happened, and none of it good. Outside Russia, however, the event itself was largely unknown. The English scientists who recorded the tremor, for instance, thought that it was simply an earthquake. Inside Russia, the unstable political climate of the time was not conducive to mounting an expedition. Subsequently, the economic and social upheaval created by World War I and the Russian Revolution made scientific expeditions impossible.

Looking for a Hole in the Ground

In 1921, mineralogist Leonid A. Kulik was charged by the MineralogicalMuseum of St. Petersburg with locating meteorites that had fallen inside the Soviet Union. Having read old newspapers and eyewitness testimony from the Tunguska region, Kulik convinced the Academy of Sciences in 1927 to fund an expedition to locate the crater and meteorite he was certain existed.

The expedition was not going to be easy, as spring thaws turned the region into a morass. And when the team finally reached the area of destruction, their superstitious guides refused to go any further. Kulik, however, was encouraged by the sight of millions of trees splayed to the ground in a radial pattern pointing outward from an apparent impact point. Returning again, the team finally reached the epicenter where, to their surprise, they found neither a meteor nor a crater. Instead, they found a forest of what looked like telephone poles—trees stripped of their branches and reduced to vertical shafts. Scientists would not witness a similar sight until 1945 in the area below the Hiroshima blast.

Theories Abound

Here are a few of the theories of what happened at Tunguska.

Stony Asteroid: Traveling at a speed of about 33,500 miles per hour, a large space rock heated the air around it to 44,500 degrees Fahrenheit and exploded at an altitude of about 28,000 feet. This produced a fireball that utterly annihilated the asteroid.

Kimberlite Eruption: Formed nearly 2,000 miles below the Earth's surface, a shaft of heavy kimberlite rock carried a huge quantity of methane gas to the Earth's surface where it exploded with great force.

Black Holes & Antimatter: As early as 1941, some scientists believed that a small antimatter asteroid exploded when it encountered the upper atmosphere. In 1973, several theorists proposed that the Tunguska event was the result of a tiny black hole passing through the Earth's surface.

Alien Shipwreck: Noting the similarities between the Hiroshima atomic bomb blast and the Tunguska event, Russian novelist Alexander Kazantsev was the first to suggest that an atomic-powered UFO exploded over Siberia in 1908.

Tesla's Death Ray: Scientist Nikola Tesla is rumored to have test-fired a "death ray" on June 30, 1908, but he believed the experiment to be unsuccessful—until he learned of the Tunguska Event.

Okay, but What Really Happened?

In June 2008, scientists from around the world marked the 100-year anniversary of the Tunguska event with conferences in Moscow. Yet scientists still cannot reach a consensus as to what caused the event. In fact, the anniversary gathering was split into two opposing factions—extraterrestrial versus terrestrial—who met at different sites in the city.

You Say Uranus, I Say George

Its name has been the butt of countless bad jokes, but was the planet Uranus—the dimmest bulb in our solar system and nothing more than a celestial conglomeration of hydrogen, helium, and ice—first known as George?

* * * *

THERE'S ACTUALLY MORE truth than rumor in this story, but the lines of historical fact and fiction are blurred just enough to make the discovery and naming of the seventh planet fascinating. The heavenly globe that eventually was saddled with the name Uranus had been seen for years before it was given its just rewards. For decades, it was thought to be simply another star and was even cataloged as such under the name 34 Tauri (it was initially detected in the constellation Taurus). Astronomer William Herschel first determined that the circulating specimen was actually a planet. On the evening of March 13, 1791, while scanning the sky for the odd and unusual, Herschel spotted what he first assumed was a comet.

After months of scrutiny, Herschel announced his discovery to a higher power, in this case the Royal Society of London for the Improvement of Natural Knowledge, which agreed that the scientist had indeed plucked a planet out of the night sky. King George III was duly impressed and rewarded Herschel with a tidy bursary to continue his research. To honor his monarch, Herschel named his discovery Georgium Sidus, or George's Star, referred to simply as George. This caused some consternation among Herschel's contemporaries, who felt the planet should be given a more appropriate—and scientific—appellation. It was therefore decided to name the new planet for Uranus, the Greek god of the sky.

How Are Stars and Planets Different?

Even astronomers quibble over this one. In the most general terms, stars and planets can be differentiated by two characteristics: what they're made of and whether they produce their own light. According to the Space Telescope Science Institute, a star is "a huge ball of gas held together by gravity." At its core, this huge ball of gas is super-hot. It's so hot that a star produces enough energy to twinkle and glow from light-years away. You know, "like a diamond in the sky."

* * * *

In case you didn't know, our own sun is a star. The light and energy it produces are enough to sustain life on Earth. But compared to other stars, the sun is only average in terms of temperature and size. Talk about star power! It's no wonder that crazed teenage girls and planets revolve around stars. In fact, the word "planet" is derived from the Greek *plan te* ("wanderer"). By definition, planets are objects that orbit around stars. As for composition, planets are made up mostly of rock (Earth, Mercury, Venus, and Mars) or gas (Jupiter, Saturn, Neptune, and Uranus).

Now hold your horoscopes! If planets can be gaseous, then just what makes Uranus different from the stars that form Ursa Major? Well, unlike stars, planets are built around solid cores. They're cooler in temperature, and some are even home to water and ice. Remember what the planet Krypton looked like in the *Superman* movies? All right, so glacial Krypton is not a real planet, but you get the point: Gaseous planets aren't hot enough to produce their own light. They may appear to be shining, but they're actually only reflecting starlight.

So back to the astronomers: Just what are they quibbling about? Well, it's tough agreeing on exact definitions for stars and planets when there are a few celestial objects that fall somewhere in between the two. Case in point: brown dwarfs.

Brown dwarfs are too small and cool to produce their own light, so they can't be considered stars. Yet they seem to form in the same way stars do, and since they have gaseous cores, they can't be considered planets either. So what do we call them? Some say "failed stars," "substars," or even "planetars." In our vast universe, there seems to be plenty of room for ambiguity.

Looking for the Great Wall

An old Ripley's Believe It or Not cartoon claims that the Great Wall of China is the only man-made object visible from the moon. Well, you can't see it, so there's no reason to believe it.

* * * *

THIS MYTH ACTUALLY originated with Richard Halliburton's *Second Book of Marvels* (1938). But as any astronaut can tell you, there are many artificial creations that can be spied from space—if by "space" you mean low-Earth orbit, approximately 100 miles up. These include the lines of major highways and railroads; the sprawl of large cities; and giant, individual constructions such as the Pentagon and the manufactured islands of Dubai. During the flight of *Gemini V* in 1965, space jockeys Charles Conrad and Gordon Cooper detected the network of roads around the Nile River, as well as the aircraft carrier that was scheduled to pick up their returning capsule.

But from orbit, you can barely see the Great Wall. Often, you can't make it out at all. China's first astronaut, Colonel Yang Liwei, confirmed this on returning to Earth in 2003. "Earth looked very beautiful in space, but I did not see our Great Wall," Liwei said. United States astronaut William Pogue thought he saw it from *Skylab*, until he realized he was looking at the ancient Grand Canal outside Beijing. And a space station scientist remarked that the Great Wall is less visible than many other objects; you really have to know where to look.

The wall is difficult to see because large parts of it have crumbled away or have been buried by dirt and sand. It also is rather narrow, less than 20 feet across in many places—a mere microdot from space. In addition, the wall is hard to spot because it is made from materials that are the same color and texture as its surroundings.

Space Travel Demystified

Like nature, humans abhor a vacuum, and we've been filling the void of scientific knowledge with near-truths and outright falsehoods ever since we broke the grip of Earth's gravity. Here are a few.

* * * *

There is no gravity in space. There is a difference between "weightlessness" and "zero-g" force. Astronauts may effortlessly float inside a space shuttle, but they are still under the grasp of approximately 10 percent of Earth's gravity. Essentially, gravity will decrease as the distance from its source increases—but it never just vanishes.

Gravitational forces are powerful enough to distort a person's features. This popular notion can be traced to the fertile minds of Hollywood filmmakers, who quickly learned the value of "artistic license" when dealing with the subject of outer space. In 1955's *Conquest of Space,* director Byron Haskin portrayed space travelers stunned and frozen by the forces of liftoff, pressed deep into their seats with their faces grotesquely distorted. When humankind actually reached space in 1961, the truth became known: Although gravitational forces press against the astronauts, they are perfectly capable of performing routine tasks, and their faces do not resemble Halloween masks.

An ill-suited astronaut will explode. Filmmakers would have you believe that an astronaut who is exposed to the vacuum of space without the protection of a spacesuit would expand like a parade float. With eyes bulging and the body swelling like a big balloon, the poor soul would soon blow up. It would be a gruesome sight, indeed, but that's not the way it would happen. The human body is too tough to distort in a vacuum. The astronaut would double over in pain and eventually suffocate, but that unfortunate occurrence would likely not make the highlight reel.

Stranded space travelers will be asphyxiated. The film world's take on space dangers has occasionally spilled into reality. In movies such as *Marooned,* astronauts are stuck in space as their oxygen supply runs out. Although the danger of being stranded in space is very real (*Apollo 13* comes to mind), astronauts in such a situation would not die from lack of oxygen. Carbon dioxide in a disabled spacecraft could build up to life-threatening levels long before the oxygen ran out.

The world watched as the Challenger "exploded." Myth even lies in one of the most tragic spaceflights in U.S. history—the *Challenger* disaster of January 1986. Stories tell of the millions of horrified viewers who watched as the spacecraft and its solid-rocket boosters broke apart on live television. Except for cable network CNN, however, the major networks had ceased their coverage of the launch. Because crew member Christa McAuliffe was to be the first teacher in space, NASA had arranged for public schools to show the launch on live TV. Consequently, many of those who actually saw it happen were schoolchildren. It was only when videotaped replays filled the breaking newscasts that "millions" of people were able to view the catastrophe. Another misconception about the *Challenger* is that it actually "exploded." It didn't, at least not in the way most people assume. The shuttle's fuel tank ripped apart, but there was no blast.

A City for the Space Age

Anyone who's ever watched a space launch on television is probably familiar with the Kennedy Space Center at Cape Canaveral, Florida, but any Texan knows that Houston is where the hardest work is being done.

❋ ❋ ❋ ❋

THE LYNDON B. Johnson Space Center (JSC) is the home of NASA's Mission Control Center, which coordinates and monitors all space flights. The center was the hub for all of the Gemini and Apollo missions, and that building is now designated a National Historic Landmark. Today, all activities aboard the International Space Station are monitored from the Johnson Space Center.

The center consists of 100 buildings scattered across 1,620 acres in southeast Houston. Originally known as the Manned Spacecraft Center, the facilities opened in 1963, and in 1973, the center was renamed for the late president.

Setting It Up

In 1962, President John F. Kennedy made it a goal to put an American on the moon by the end of the decade. The administrator of NASA, James E. Webb, headed a selection team to find a site where test facilities and research laboratories could be built to mount the space program. Requirements included the availability of water transport and an all-weather airport, proximity to a major telecommunications network, availability of established industrial workers and contractor support, an available supply of water, a mild climate permitting year-round outdoor work, and a culturally attractive community. Another factor Houston had over the competition—and one of the reasons it was initially considered—was its proximity to the U.S. Army San Jacinto Ordnance Depot, the Houston Shipping Channel, and regional universities.

Today, roughly 3,000 civilians and 110 astronauts are employed at JSC. The bulk of the workforce is the 15,000 contract workers representing about 50 contracting firms.

What It Takes to Be an Astronaut

In addition to being the Mission Control Center, the JSC is the home of the astronaut corps and is responsible for training astronauts from both the United States and its international partners. Astronauts receive training on the shuttle system and in basic sciences, which include mathematics, guidance and navigation, oceanography, astronomy, and physics. Candidates are put through military water survival training, SCUBA certification, and flying instruction, and they learn to handle emergencies associated with atmospheric pressure and space flight.

Astronauts begin their formal training by reading manuals and taking computer-based training programs. In the recent past, they would then move on to the orbiter systems trainer to practice orbiter landings and prepare for malfunctions and corrective actions, and then to shuttle mission simulators, which provided training on shuttle operations. The neutral buoyancy laboratory, a large pool containing 6.2 million gallons of water, allows astronauts to practice tasks in an environment that simulates zero gravity conditions. It also prepares them for space walks.

Research and Development

But the center is much more than an astronaut training facility—the Johnson Space Center leads NASA's flight-related scientific and medical research programs. The technologies that support space flight are now in use in civilian medicine, the energy industry, transportation, agriculture, communications, and electronics. Current research studies include the prebreathe reduction program, which is intended to help walks in space from the International Space Station become safer and more efficient.

A Tourist Site

The visitor center and grounds contain historical and archival information that chronicles the history of the astronaut program and its contributions to NASA. One can see the lunar receiving laboratory where the first astronauts were quarantined after returning from space. The center's landing and recovery division was responsible for retrieving astronauts after splashdown during the Gemini and Apollo missions and is housed at JSC. The majority of moon rocks and lunar samples are also stored at the complex.

One of the artifacts displayed at JSC is a Saturn V rocket made of actual surplus flight-ready materials. An incomplete Apollo Capsule Service Module (CSM), intended to fly on the cancelled Apollo 19 mission, is also displayed on the grounds. An educational center provides student internships, day camps, and materials for educators, and it trains volunteers, as well.

Security Is Vital

The Johnson Space Center has its own security headquarters and maintains a high level of monitoring due to the sensitive nature of its business and the equipment housed at its facilities. Only one reported security incident has occurred in its several decades of history when, in 2007, a hostage situation occurred in the communication and tracking development laboratory. A gunman killed one employee and injured another before taking his own life.

The center is also vulnerable to the effects of nature. In 2008, Hurricane Ike hit as a Category 3, destroying several airplane hangars and damaging a number of buildings.

The Johnson Space Center will continue to provide services to NASA and the International Space Station as the space program moves forward. Those missions may take humans back to the moon, to Mars, or into the deep reaches of space. One thing is certain: The crew at JSC Houston will be there to make sure everything runs smoothly.

The Kecksburg Incident

Did visitors from outer space once land in a rural western Pennsylvania thicket?

✻ ✻ ✻ ✻

Dropping in for a Visit

On December 9, 1965, an unidentified flying object (UFO) streaked through the late-afternoon sky and landed in Kecksburg—a rural Pennsylvania community about 40 miles southeast of Pittsburgh. This much is not disputed. However, specific accounts vary widely from person to person. Even after closely examining the facts, many people remain undecided about exactly what happened. "Roswell" type incidents—ultra-mysterious in nature and reeking of a governmental cover-up—have an uncanny way of causing confusion.

Trajectory-Interruptus

A meteor on a collision course with Earth will generally "bounce" as it enters the atmosphere. This occurs due to friction, which forcefully slows the average space rock from 6 to 45 miles per second to a few hundred miles per hour, the speed at which it strikes Earth and officially becomes a meteorite. According to the official explanation offered by the U.S. Air Force, it was a meteorite that landed in Kecksburg. However, witnesses reported that the object completed back and forth maneuvers before landing at a very low speed—moves that an un-powered chunk of earthbound rock simply cannot perform. Strike one against the meteor theory.

An Acorn-Shape Meteorite?

When a meteor manages to pierce Earth's atmosphere and make its way to the surface, it has the physical properties of exactly what it is: a space rock. That is to say, it will generally be unevenly shaped, rough, and darkish in color, much like rocks found on Earth.

But at Kecksburg, eyewitnesses reported seeing something far, far different. The unusual object they described was bronze to golden in color, acorn-shape, and as large as a Volkswagen Beetle automobile. Unless the universe has started to produce uniformly shaped and colored meteorites, the official explanation seems highly unlikely. Strike two for the meteor theory.

Markedly Different

Then there's the baffling issue of markings. A meteorite can be chock-full of holes, cracks, and other such surface imperfections. It can also vary somewhat in color. But it should never, ever have markings that seem intelligently designed. Witnesses at Kecksburg describe intricate writings similar to Egyptian hieroglyphics located near the base of the object. A cursory examination of space rocks at any natural history museum reveals that such a thing doesn't occur naturally. Strike three for the meteor theory.

Logically following such a trail, could an unnatural force have been responsible for the item witnessed at Kecksburg? At least one man thought so.

Reportis Rigor Mortis

Just after the Kecksburg UFO landed, reporter John Murphy arrived at the scene. Like any seasoned pro, the newsman immediately snapped photos and gathered eyewitness accounts of the event. Strangely, FBI agents arrived, cordoned off the area, and confiscated all but one roll of his film. Undaunted, Murphy assembled a radio documentary entitled *Object in the Woods* to describe his experience. Just before the special was to air, the reporter received an unexpected visit by two men.

According to a fellow employee, a dark-suited pair identified themselves as government agents and subsequently confiscated a portion of Murphy's audiotapes.

A week later, a clearly perturbed Murphy aired a watered-down version of his documentary. In it, he claimed that certain interviewees requested their accounts be removed for fear of retribution at the hands of police, military, and government officials. In 1969, John Murphy was struck dead by an unidentified car while crossing the street.

Resurrected by Robert Stack

In all likelihood the Kecksburg incident would have remained dormant and under-explored had it not been for the television show *Unsolved Mysteries*. In a 1990 segment, narrator Robert Stack took an in-depth look at what occurred in Kecksburg, feeding a firestorm of interest that eventually brought forth two new witnesses. The first, a U.S. Air Force officer stationed at Lockbourne AFB (near Columbus, Ohio), claimed to have seen a flatbed truck carrying a mysterious object as it arrived on base on December 10, 1965. The military man told of a tarpaulin-covered conical object that he couldn't identify and a "shoot to kill" order given to him for anyone who ventured too close. He was told that the truck was bound for Wright–Patterson AFB in Dayton, Ohio, an installation that's alleged to contain downed flying saucers.

The other witness was a building contractor who claimed to have delivered 6,500 special bricks to a hanger inside Wright–Patterson AFB on December 12, 1965. Curious, he peeked inside the hanger and saw a "bell-shaped" device, 12-feet high, surrounded by several men wearing anti-radiation style suits. Upon leaving, he was told that he had just witnessed an object that would become "common knowledge" in the next 20 years.

Will We Ever Know the Truth?

Like Roswell before it, we will probably never know for certain what occurred in western Pennsylvania back in 1965. The more that's learned about the case, the more confusing and contradictory it becomes. For instance, the official 1965 meteorite explanation contains more holes than Bonnie and Clyde's death car, and other explanations, such as orbiting space debris (from past U.S. and Russian missions) reentering Earth's atmosphere, seem equally preposterous.

In 2005, as the result of a new investigation launched by the Sci-Fi Television Network, NASA asserted that the object was a Russian satellite. According to a NASA spokesperson, documents of this investigation were somehow misplaced in the 1990s. Mysteriously, this finding directly contradicts the official air force version that nothing at all was found at the Kecksburg site. It also runs counter to a 2003 report made by NASA's own Nicholas L. Johnson, Chief Scientist for Orbital Debris. That document shows no missing satellites at the time of the incident. This includes a missing Russian Venus Probe (since accounted for)—the very item that was once considered a prime crash candidate.

Brave New World

These days, visitors to Kecksburg will be hard-pressed to find any trace of the encounter—perhaps that's how it should be. Since speculation comes to an abrupt halt whenever a concrete answer is provided, Kecksburg's reputation as "Roswell of the East" looks secure, at least for the foreseeable future.

But if one longs for proof that something mysterious occurred there, they need look no further than the backyard of the Kecksburg Volunteer Fire Department. There, in all of its acorn-shape glory, stands an full-scale mock-up of the spacecraft reportedly found in this peaceful town on December 9, 1965. There too rests the mystery, intrigue, and romance that have accompanied this alleged space traveler for more than 40 years.

5 Memorable Meteor Crashes

Every day hundreds of meteors, commonly known as shooting stars, can be seen flying across the night sky. Upon entering Earth's atmosphere, friction heats up cosmic debris, causing streaks of light that are visible to the human eye. Most burn up before they ever reach the ground. But if one actually survives the long fall and strikes Earth, it is called a meteorite. Here are some of the more memorable meteor falls in history.

✻ ✻ ✻ ✻

1. The Ensisheim Meteorite, the oldest recorded meteorite, struck Earth on November 7, 1492, in the small town of Ensisheim, France. A loud explosion shook the area before a 330-pound stone dropped from the sky into a wheat field, witnessed only by a young boy. As news of the event spread, townspeople gathered around and began breaking off pieces of the stone for souvenirs. German King Maximilian even stopped by Ensisheim to see the stone on his way to battle the French army. Maximilian decided it was a gift from heaven and considered it a sign that he would emerge victorious in his upcoming battle, which he did. Today bits of the stone are located in museums around the world, but the largest portion stands on display in Ensisheim's Regency Palace.

2. Many scientists believe that a meteorite was responsible for the extinction of the dinosaurs. The theory holds that approximately 65 million years ago, a six-mile-wide asteroid crashed into Earth, causing a crater about 110 miles across and blowing tons of debris and dust into the atmosphere. Scientists believe the impact caused several giant tsunamis, global fires, acid rain, and dust that blocked sunlight for weeks or months, disrupting the food chain and eventually wiping out the dinosaurs. The theory is controversial, but believers point to the Chicxulub Crater

in Yucatan, Mexico, as the striking point of the asteroid. Skeptics say the crater predates the extinction of dinosaurs by 300,000 or so years. Others believe dinosaurs may have been wiped out by several distinct asteroid strikes, rather than just the widely credited Chicxulub impact. Scientists will likely be debating this one for centuries.

3. As Colby Navarro sat at his computer on March 26, 2003, he had no idea that a meteorite was about to come crashing through the roof of his Park Forest, Illinois, home, strike his printer, bounce off the wall, and land near a filing cabinet. The rock, about four inches wide, was part of a meteorite shower that sprinkled the area, damaging at least six houses and three cars. Scientists said that before the rock broke apart, it was probably the size of a car.

4. The Hoba Meteorite, found on a farm in Namibia in 1920, is the heaviest meteorite ever found. Weighing in at about 66 tons, the rock is thought to have landed more than 80,000 years ago. Despite its gargantuan size, the meteorite left no crater, which scientists credit to the fact that it entered Earth's atmosphere at a long, shallow angle. It lay undiscovered until 1920 when a farmer reportedly hit it with his plow. Over the years, erosion, vandalism, and scientific sampling have shrunk the rock to about 60 tons, but in 1955 the Namibian government designated it a national monument, and it is now a popular tourist attraction.

5. Santa had to compete for airspace on Christmas Eve 1965, when Britain's largest meteorite sent thousands of fragments showering down on Barwell, Leicestershire. Museums immediately started offering money for fragments of the rock, causing the previously sleepy town to be inundated with meteorite hunters and other adventurers from around the world. Decades later, the phenomenon continues to captivate meteorite enthusiasts, and fragments can often be found for sale online.